Scientific Communication

Yu and Northcut have blazed a new, important, timely, and practicable trail in the field of science communication.

—*Paul Dombrowski, University of Central Florida*

For faculty (and grad students) who want to initiate courses in science writing, or for those who want to enrich their approaches, Yu and Northcut's new work has much to offer. The volume offers the best current thinking to support the teaching of science writing.

—*Stephen A. Bernhardt, University of Delaware, Emeritus*

This book addresses the roles and challenges of people who communicate science, who work with scientists, and who teach STEM majors how to write. In the Practice and Theory section, chapters address themes encountered by scientists and communicators, including ethical challenges, visual displays, communication with publics, as well as changed and changing contexts and genres. The Pedagogy and Curriculum section covers topics important to instructors' everyday teaching as well as longer-term curricular development. Chapters address delivery of rhetorically informed instruction, communication from experts to the publics, writing assessment, online teaching, and communication-intensive pedagogies and curricula.

Han Yu is a Professor of technical communication in the English Department, Kansas State University, USA. She is co-editor (with Gerald Savage) of *Negotiating Cultural Encounters: Narrating Intercultural Engineering and Technical Communication*, author of *The Other Kind of Funnies: Comics in Technical Communication*, and author of *Communicating Genetics: Visualizations and Representations.*

Kathryn Northcut is a Professor of technical communication in the Department of English and Technical Communication at Missouri S&T, USA. She teaches courses in technical communication at the undergraduate and graduate levels. She co-edited (with Eva Brumberger) *Designing Texts: Teaching Visual Communication.*

Routledge Studies in Technical Communication, Rhetoric, and Culture
Series Editors: Miles A. Kimball and Charles H. Sides

For a full list of titles in this series, please visit www.routledge.com.

This series promotes innovative, interdisciplinary research in the theory and practice of technical communication, broadly conceived as including business, scientific, and health communication. Technical communication has an extensive impact on our world and our lives, yet the venues for long-format research in the field are few. This series serves as an outlet for scholars engaged with the theoretical, practical, rhetorical, and cultural implications of this burgeoning field. The editors welcome proposals for book-length studies and edited collections involving qualitative and quantitative research and theoretical inquiry into technical communication and associated fields and topics, including user-centered design; information design; intercultural communication; risk communication; new media; social media; visual communication and rhetoric; disability/accessibility issues; communication ethics; health communication; applied rhetoric; and the history and current practice of technical, business, and scientific communication.

4 **Topic-Driven Environmental Rhetoric**
Edited by Derek G. Ross

5 **Rhetoric and Communication Perspectives on Domestic Violence and Sexual Assault**
Policy and Protocol Through Discourse
Amy D. Propen and Mary Lay Schuster

6 **The Neuroscience of Multimodal Persuasive Messages**
Persuading the Brain
Dirk Remley

7 **Writing Postindustrial Places**
Technoculture amid the Cornfields
Michael J. Salvo

8 **Scientific Communication**
Practices, Theories, and Pedagogies
Edited by Han Yu and Kathryn Northcut

Scientific Communication

Practices, Theories, and Pedagogies

Edited by Han Yu and
Kathryn Northcut

Routledge
Taylor & Francis Group
NEW YORK AND LONDON

First published 2018
by Routledge
605 Third Avenue, New York, NY 10017

and by Routledge
2 Park Square, Milton Park, Abingdon, Oxon OX14 4RN

Routledge is an imprint of the Taylor & Francis Group, an informa business

Library of Congress Cataloging-in-Publication Data
CIP data has been applied for.

ISBN: 978-1-138-06478-2 (hbk)
ISBN: 978-1-315-16019-1 (ebk)

DOI: 10.4324/9781315160191

Typeset in Sabon
by codeMantra

Contents

Acknowledgements

The editors would like to thank Miles Kimball for his assistance in his role as series editor. We would also like to thank the blind reviewers who helped extensively with thoughtful recommendations for stylistic improvements. Staff at Routledge have been helpful and enthusiastic; we appreciate their professionalism and support for the project. Dr. Melanie Mormile of Missouri S&T provided a scientist's perspective as we framed the introductory chapter. Of course, the authors of the chapters cannot be thanked enough, as their contributions convey the message that constitutes the book. They have been a pleasure to work with, and we hope their work has found a fitting home in this volume.

Han Yu and Kathryn Northcut

High stakes and great responsibility

An introduction to scientific communication

Han Yu and Kathryn Northcut

Before his death in 2002, Stephen Jay Gould, one of the best-selling scientist authors of our time, reflected upon his first two books, which coincidentally appeared in the same year, 1977. The Harvard professor of zoology and geology and museum curator noted that his first "technical" book (*Ontogeny and Phylogeny*) and his first "popular" book (*Ever Since Darwin*) were featured by *The New York Times* because the idea of a scientist publishing successfully in two separate genres seemed quite surprising (Gould, 2002, pp. 1–2). However, in looking back at his own publication history, Gould apologized for having perpetuated a false distinction between scientific and non-scientific prose:

> I no longer view this conjunction of technical and "popular" as anomalous, or even as interesting or unusual.... For, beyond some obvious requirements of stylistic tuning to expected audiences—avoidance of technical jargon in popular essays as the most obvious example—I have come to believe... that the conceptual depth of technical and general writing should not differ, lest we disrespect the interest and intelligence of millions of potential readers who lack advanced technical training in science, but who remain just as fascinated as any professional, and just as well aware of the importance of science to our human and earthly existence.
>
> (Gould, 2002, p. 2)

In the eight-year tenure of Barack Obama, despite cataclysmic problems in U.S. banking and industry, recognition of the importance of science was borne out in strongly positive language and modest fiscal boosts across scientific enterprises (Hourihan, 2017). At the 2016 Democratic National Convention, Hillary Clinton's declaration, "I believe in science!" received cheers from the crowd, but not enough votes in key districts to be elected. On securing the presidency, Donald Trump immediately made decisions that reinforced his avowed dismissal of certain initiatives, including climate science, renewable energy, environmental sustainability, and public health, among others. For example, the Trump budget aims to cut $250 million in research activity sponsored

DOI: 10.4324/9781315160191

by the National Oceanic and Atmospheric Administration (NOAA) and $900 million from the Office of Science in the Department of Energy (Achenbach, 2017). Science, it seems, fell out of political favor with the pendulum swing of partisan politics.

For scientists and citizens whom Gould identified as "well aware," 2017 thus dawns with serious concern for "our human and earthly existence." As the world looks on, the institutions Americans took for granted, like the Environmental Protection Agency (EPA) and the National Institutes of Health (NIH), face debilitating budget cuts. Federally affiliated websites are monitored and purged of climate change related content, social media posts, and blog entries (Davenport, 2017). And appointments to positions of national science leadership seem antithetical to the purposes of their organizations—like the appointment of Scott Pruitt, a fossil fuel industry advocate, as the head of EPA.

Our colleagues in science are deeply concerned, and some of their life-long research pursuits are at stake. For the time being, those of us in the interdisciplinary (and relatively unknown) field of technical communication might seem spared. However, that's really not the case: even though the fallout does not appear to directly target us, we are all connected, and anything that affects science inevitably affects us. A brief look at our investment and involvement in science should make that evident.

Our Current Involvement in Science

Although technical communication lacks a precise definition, it is concerned with transactions where someone needs to obtain information in order to make decisions or take action. Contexts for such transactions vary, but in research, practice, and teaching, we engage most extensively with discourses of technical fields (such as engineering and computing), scientific fields (such as biology and physics), and general business contexts (like finance or marketing).

Recent studies by our scholars, for example, examined image construction in science (Buehl, 2014; Northcut, 2011; Welhausen, 2015), the argumentation and representation of science (Kitalong, Moody, Middlebrook, & Ancheta, 2009; Whithaus, 2012), and the communication of various topics from climate change to genetics to drug therapy (Cagle & Tillery, 2015; Mogull & Balzhiser, 2015; Turner, 2005; Yu, 2017). Our journals also devoted special issues to engage with the topic (Johnson-Sheehan & Stewart, 2003a,b; Northcut, 2007).

Although few of our programs may explicitly use "science" or "scientific" in their program titles, our experience and conversations with colleagues tell us that many programs, especially those at land-grant universities, regularly offer classes with science-related titles that serve students from life sciences, agricultural sciences, and medical sciences. The general goal of these classes is to prepare students for the kinds of communication tasks expected of them in their academic programs and/or future careers.

Such classes are prevalent because nationwide, our colleagues in science recognize that communication competency determines whether their graduates are able to publish, obtain grant funding, and secure professional or academic positions. Similarly, for science students to succeed in industry, government, and nonprofits where they work in positions ranging from manufacturing, R&D, to policy making, they must communicate with management, marketing, and legislative bodies. Academic and industry scientists also need the ability to work with popular media and communicate with a broader audience about the relevance and worth of their work. Indeed, in today's political climate, the ability to invite, engage, and persuade public audiences may be the best way to preserve science.

Opportunities We Are Missing

Despite our deep involvements and investments, our field has produced surprisingly little focused and systematic research in scientific communication the way it has in other flavors of technical and professional communication (although, for a welcome exception, see Gross & Buehl, 2016).

In important research areas such as intercultural communication (e.g., Thatcher & St. Amant, 2011; Yu & Savage, 2013), genre studies (e.g., Spinuzzi, 2003; Winsor, 2003), and new media research (e.g., Brewer, 2015; DeWinter & Moeller, 2014; Lamberti & Richards, 2011), our books focus exclusively or primarily on technical and professional discourses and contexts. Journal publications are no more encouraging. Database searches using the key words "scientific communication" and "science communication" yield enormous literature. But major publication venues are outside of our discipline and often situated in the sciences (e.g., *Nature* and *Science*), communication studies (e.g., *Journal of Science Communication* and *Public Understanding of Science*), and science education (e.g., *International Journal of Science Education* and *Journal of Research in Science Teaching*). Our scholars' research on science, for all intents and purposes, disappears in this enormous literature.

Certainly, we do benefit by drawing upon other fields and subfields. Charles Darwin's *Origin of Species* provides the classical example of how scientists persuade audiences using rhetorical strategies. Edward O. Wilson's *Consilience* (1998) helps us understand science as an evolved enterprise. Bruno Latour and Steve Woolgar's *Laboratory Life* (1979, 1986) long served as a staple for introducing our students to the culture of science. Works in the veins of discourse analysis, cultural studies, and ethnography that refer extensively to science and scientists (e.g., Baake, 2003; Fahnestock, 1999; Gross, 1990) also inform much of our research.

However, these works, valuable as they are, frequently have different concerns and focuses than what our scholars, and many in our expected readership, value. To start, most published works are interested in scientific communication as "historical phenomena—created, recognized, mobilized, and given force within the mind of each writer and reader at specific social-historical moments" (Bazerman, 2000, p. 318). As such, they are frequently focused on historical cases and classical figures such as Francis Bacon, Isaac Newton, and Charles Darwin (see, e.g., Bazerman, 2000; Fahnestock, 1999; Gross, 1990; Gross, Harmon, & Reidy, 2009). Such studies are therefore of limited use to our scholars who are interested in contemporary scientific discourses and practices.

In addition, these works are rooted in rhetorical, social, cultural, and philosophical discussions but less interested in how to enhance (in a more or less normative sense) scientific discourses. Because of this difference, they are not immediately useful to scholars who are interested in pedagogical challenges: namely, how can we teach science students to produce more effective and ethical communication. Also, how can we teach students in communication and rhetoric to be critical readers, writers, editors, and critics of scientific discourses?

Filling the Gap

Such missed opportunities inspired this volume. When we proposed the book, we sought to address any number of challenges that we imagine will emerge when the institution of science meets the context of writing. We profoundly empathize with science communication instructors because we are two of them. We recognize that resources on this subject that we attempt to teach and theorize are very limited, while much more information is readily available concerning generic "technical writing" or "technical communication." We wanted to help prospective scientists who need to write and writers who are immersed in the world of science. We wanted to know what and how these students should be taught. And we wanted to know how writing instructors and scholars broach science. In short, we began with a strong commitment to compiling a useful book, but only a vague sense of what the volume would look like when it was done.

Between 2014 and 2017, this volume developed. Every time the authors of the 13 chapters submitted a new version, our confidence grew. Arguments evolved. Methodologies unfolded. Connections among chapters emerged. After years of refinement, each chapter has a story to tell, an argument (or several) to make, and a body of knowledge that is likely to inspire or educate.

Now, our collection can claim to effectively achieve the following objectives, winnowed down from the 24 originally listed in the call for chapter proposals:

1 Define and describe science communication and scientific communication, explicating and complicating those terms.
2 Demonstrate how the writing practices of scientists change over time, as do the constraints (both internal and external) under which they work.
3 Examine assumptions and expectations about how scientists *should* communicate and how ethical practice is (and is not) enforced in broader social contexts.
4 Provide specific, extended examples of genres in which science is heavily invested.
5 Explicate the roles and social responsibilities of scientists, science communicators, and other agents as they invent, repurpose, and deploy scientific information for different purposes to different audiences.
6 Describe courses in science communication, including descriptions of and observations about the instructors, the students, and what they do.
7 Interrogate the role of rhetoric in science communication pedagogy.

Not all chapters demonstrate every objective, but each makes a substantial contribution.

Reid explains in Chapter 1 that, despite scientists cordoning themselves off through exclusionary professionalization, non-scientists are involved in scientific endeavors for various reasons. The changing role of the citizen-scientist perhaps tells us even more about a science-infused society than it does about particular science projects. In such a society, the boundary between insiders and outsiders, experts and non-experts, is blurred. This blurred boundary is made clear in Reid's description of a biology lab and echoed in Chapter 10, as Maddalena and Reilly describe science communication service courses.

Two studies examine how science as an institution operates in the larger regulatory context of the U.S. In Chapter 2, Katz and Linvill interrogate how the federal government oversees (or demurs from regulating) the ethical activities of science. Mogull, in Chapter 3, reveals how the financial motivations of pharmaceutical companies compete with a strictly correct interpretation of research data. These bird's-eye accounts help us see scientists out of the romanticized setting of the isolated lab, but rather as actors in a heavily regulated industry operating in a litigious social environment.

Rich examples of science communication genres are highlighted in most of the chapters in the Practice and Theory section. These genres frequently reflect the changing context of science and implicate both insiders and outsiders, experts and publics. Maps of the Zika virus give us insight into how visualizations communicate health risks to publics and how that communication is bound up with people's perception of risk (Welhausen, Chapter 4). The visual trope of evolutionary biology, the tree of life, is interrogated not only for scientific accuracy, but for anthropocentric implications (Yu, Chapter 5). Trending Twitter hashtags enable us to observe public perceptions of and responsiveness to our deteriorating atmosphere (Cagle & Tillery, Chapter 6). And museum displays are evolving to engage visitors as active co-creators of scientific knowledge (Schneider-Bateman, Chapter 7).

However, knowing a bit about what scientists do, should do, or don't do is still a long way from knowing how to effectively teach students to appreciate ethical, effective practice as scientists, writers, and communicators. The second section of the book, Pedagogy and Curriculum, begins with a strong argument for taxonomizing the world of science communication. In Chapter 8, Buehl and FitzGerald take a scientific approach practiced in biology and anthropology to explain the role and function of scientific communication species, niches, and terrains across the university.

Chapters that follow then give us a peek into actual species, niches, and terrains. In Chapter 9, Gigante discusses the kind of knowledge to teach students and debunks "quick-fix" approaches to teaching scientific communication. In Maddalena and Reilly's service classes (Chapter 10), we learn how to produce unexpected research questions, dialogic literature reviews, and functional research posters. In Davis and Frost (Chapter 11), we see how a rhetorically informed scientific communication pedagogy is enacted in face-to-face and online classrooms, the latter becoming an increasingly popular fixture in our programs. Harding and Studer (Chapter 12) then guide us through a series of workshops that train unexperienced graduate lab assistants to teach science writing to undergraduate students. Finally, in Chapter 13, Carmichael and Klock give guidelines on how to integrate Wikipedia into scientific communication classes to provide an authentic authoring experience to students.

How Does This Book Help Me to Teach Science Communication?

What helps an instructor to teach scientists to write, or to teach writers to write about science? The short answer is "a lot" and "it depends." This volume engages with that question and, however imperfectly, provides a basis for learning much of what needs to be known. No single book is sufficient, but we hope that this volume offers a necessary component in professional development sought by motivated and competent instructors.

In order to teach science, we need to understand scientists. A qualified science communication instructor needs at least some interaction with scientists or scientists-in-training. Scientists in different disciplines form different hypotheses and research questions, use different tools and equipment, and work with different materials and samples. However, the training of scientists does have some common themes, which defy complete explication in this essay but are part of a current national conversation: recognition of the importance of reproducible data, systematic observation, and logical reasoning upon which to build theories; assumptions that science exists for the purpose of examining and explaining the natural world as well as the built environment; and concern for ethical practice and exploration.

A qualified science communication instructor is one aware of and fascinated by these practices and themes, as Gould (2002) would recognize. She has probably taken a university science course, and visited or worked in a laboratory. He probably watches TED talks about scientific topics, or reads *Scientific American* and watches NOVA for fun.

Such fascinations will teach us some key lessons about communication, though probably very little indispensable knowledge about science. That doesn't mean that science *per se* isn't important, because it is. But no single discipline can be isolated as the one that we need to understand in order to have the academic equivalent of a science communication credential. Science is socially performed, not mechanically formulated, and scientists are people, not data. Therefore, like communicating in other contexts, we are communicating for, with, and about human activities. False attempts at objectivity (see Chapter 9) and universalism may be where we start as we try to broach science, but understanding situated science communication contexts and their unique stakeholders, audiences, purposes, conventions, and expectations is, as always, key.

Because of this, the ideal science communication instructor has undertaken research where she studied individual scientists; their motivations and concerns; their subject matters; their approaches and methods; and/or their data, findings, and communication products. This research will show us that data collection and analysis are central to the work of most scientists (Chapter 1), but at the same time, science is *not* merely about finding robust data to better answer research questions (Chapters 2 and 3). Scientists work within social, cultural, political, and economic contexts, and their work is subject to the values of those contexts. To identify "scientist" as a professional title *only* is reductive.

Instead, scientists are citizens, experts, agents, and rhetors. Many who identify as "scientist" also identify with any number of other labels: artist, poet, historian, activist, congregant, and student, to name a few. Labels aside, all scientists must continually share their research both within their communities and across broader audiences. The publication/dispersal of their work is not an optional last step; it's inherently part of doing the science and being a scientist. Scientists sometimes work alone in their communication work but frequently with a mixed group of experts: other scientists, technical communicators, science writers, journalists, graphic designers, medical illustrators, marketing/sales specialists, and publishers. Depending on the skillsets of the scientists and their colleagues, their works run the spectrum from masterful and transcendent to clumsy and misguided. Our authors, Welhausen, Yu, Cagle and Tillery, and Schneider-Bateman, show us these complex possibilities in their respective engagements with scientists' communication products.

The more time we spend around scientists, the less likely we are to generalize—as is true with most intercultural experiences. Science is not homogeneous, and its practitioners are complicated individuals with overlapping and sometimes contradictory values and beliefs. Numerous subcultures exist, even within disciplines. Like cultures, science also changes over time. The paradigms, methods, ethics, and norms of science evolve. The views toward science shared by broader social and cultural groups also change, and are also inherently complex.

To teach science communication, then, is to present how the heterogeneous group of scientists, in their respective ways, establish credibility to explain and promote their version of science. Doing so, we can then help students look beyond jargon, research methods, quantitative data, proper English, writing conventions, and move toward legitimate participation in the interdisciplinary communities where science and communication—and concomitant social change—happen.

At some points in the development of these ideas and of this book, we wondered: given the challenges, the complexity, and the difficulty in defining even the basic terms of what we do, who would *want* to teach science communication? Then we read draft after draft from our pedagogy and curriculum authors, Buehl and FitzGerald, Gigante, Maddalena and Reilly, Davis and Frost, Harding and Studer, and Carmichael and Klock, and we stopped worrying. Their attention to everything about teaching science and communication demonstrates the persistence and engagement that will propel this enterprise in productive directions without further comment by us.

What's in a Name: Technical, Science, and Scientific?

Despite our authors doing the heavy lifting, two terms deserve some initial unpacking as they encompass, in several ways, the premise of this volume.

Technical Communication vs. Scientific Communication

At this point, some readers may question *if* we need to be concerned by our field's lack of explicit research on scientific discourse. Wouldn't our studies of engineering, manufacturing, and other professional discourses transfer to the science? Although the content of these discourses differs, surely the essence of our work applies if all of them are, for the lack of an alternative descriptor, "technical" in nature. In short, how is "technical communication" and "scientific communication" so very different, if at all?

These questions surfaced, although were by no means settled, in our early attempts to define the field. Britton (1965), for one, treated the two terms synonymously and spoke of them conjunctively as "technical and scientific writing," a practice that is still common today. For him, the two concepts are one and the same because their primary characteristic "lies in the effort of the author to convey one meaning and only one meaning" (p. 114). Kelley (1976) had a similar view. Subsuming technology under science as "applied science," he defined technical writing as "Writing about subjects in the sciences in which the writer informs the reader through an objective presentation of facts" (p. 3).

Of course, there were positions to the contrary. For Zappen (1983), both scientists and technologists address a range of contexts, some "basic" and some "applied." Technical writing, then, is not simply applied scientific writing. Dobrin (1983), more pointedly, argued that technical writing and scientific writing are distinct: technical writing makes accurate, individual statements (e.g., nut A fits bolt B); scientific writing makes a universal truth claim that is provisional given certain terms (e.g., given a particular experiment setup, certain findings emerge).

Such debates and attempts to taxonomize are relics of a bygone era, in which disciplines could lay claim to territory and distinguish themselves from others by what they did and how they did it. In the current era of flattened hierarchies, interdisciplinary academic pursuits, and rapidly changing priorities, the work itself often takes precedence over how the work is defined, getting us back to the question of how people—whoever they are—write science for various audiences.

For these reasons, and the fact that our field has become more established and diversified and that no concise definition could easily describe it, debates over definitions waned. Rather, "technical communication" becomes the umbrella term, loosely defined to describe a range of technical, scientific, business, and professional communication products and processes.

As editors of this collection, we have no intention to (re)define the field or (re)define terms. Instead, we highlight some distinct characteristics of technical and scientific communication—not so much to offer dichotomous definitions but to demonstrate that the latter deserves focused attention. Without such attention, it is wishful for us to think that our understandings of, say, the financial report genre would automatically apply to understanding research reports, or that our pedagogy with engineering students would automatically apply in a class enrolled with pre-med students.

What are the differences? First and probably most obvious is the ultra-specialized nature of contemporary scientific investigation. Each discipline, sub-discipline, and research area involves extensive insider knowledge, implicit convention, and nomenclature. Together, these factors make it difficult for scientists, and communicators who work with them, to convey their work to non-scientists or even scientists from other fields.

Second, compared with technical information, scientific information is more resistant to adaptation for varying audiences, or it takes more layers of complexity in working through that adaptation (see Chapters 1, 3, 4, 5, and 7). To borrow from Dobrin (1983), this is because scientific discourses must connect with the norms and premises of a research paradigm to make truth claims. Scientific claims are contingent upon multiple elements, including previous findings, experimental design (which is in turn built upon previous work), and analytical assumptions. It is therefore less easy to determine what are essential findings, what is non-essential background, or how to disentangle the why, the what, and the how. For example, a patient trying to read literature on a treatment for heart disease will want to know more than just how to obtain or take a drug, but how we know what we know about the drug's activity, including side effects.

With technical information, it is relatively easier to separate background of why a piece of technology works from how it works, or what users can do with the technology from how the technology does it. Again, to borrow from Dobrin (1983), this is because technical writing is more situation-specific and relatively free from theoretical or methodological baggage. Leveraging this feature, a writer can choose and combine specific statements to suit a target audience's need. For example, instructional manuals usually omit the "why" and focus on "how," whereas an engineering testing report will forgo the laws of physics and focus on quantitative data of product performance.

Last, technical and scientific discourses also elicit different affective and social responses. In the realm of technology, audiences are frequently technologists who engage in technical work or consumers who purchase technologies. While there can be profound safety and ethical issues in communicating technologies to these stakeholders (see Dombrowski, 2000; Katz, 1992), in today's techno-centric world, audiences generally assume the "good" of the technology or see it as a neutral means to an end. In terms of Feenberg's critical theory of technology (1991), we are likely to either take an instrumental attitude toward technology or to valorize it.

In contract, skepticism of science is rampant. Feenberg's determinist attitude (1991) may prevail as the publics remain wary of uncomfortable and contradictory scientific claims. For any number of science disciplines, stakeholders—including members of the publics, activists, legislative bodies, and companies/laboratories with competing interests—may not agree with the inherent "good" of a piece of science, seeing it more like a runaway train where the discovery and creation process is one step ahead of policies that protect the public interest. The progression of science, whether for the betterment or detriment of humans and the planet, seems inevitable and frightening. Stem cell research, genetically modified food, climate change, and even natural evolution are a few examples where factors such as education, emotion, religion, economics, values, ethics, and politics complicate the understanding and communication of science.

In short, we repeat: the communication of science deserves our focused attention.

Scientific Communication vs. Science Communication

Another pair of terms deserving explanation here is "scientific" vs. "science," as in "scientific communication" vs. "science communication." Conventionally and generally, "scientific communication" means communication that transpires between fellow scientists and specialists (as in peer-to-peer communication). It results in such products as research proposals, research articles, and conference presentations. By contrast, "science communication" means communication that transpires between scientists and non-scientists—or more precisely, accommodations and popularizations that transmit *from* scientists *to* non-scientists. Its results include newspaper and magazine reports or TV programs of science.

This conventional distinction is by no means universal or consistent. Knowingly or otherwise, researchers and teachers use them interchangeably or differently (see Chapter 8). As for our position, we think that their difference was born out of outdated rhetorical contexts and academic traditions. We maintain that in contemporary use, a demarcation between the two is not only unnecessary but indeed problematic.

First, the supposed difference between the two terms is semantically unattainable. When we use the term "scientific" to prefix "communication," we could mean that the communicated content is scientific in nature or that the act of communication is somehow scientific—whatever that means. But if we accept these premises, nothing changes when we shift from "scientific" the adjective to "science" the noun. Semantically, the "science" prefix would yield the same connotations.

Certainly, one may argue that in "science communication," the communicated content is *about* science but *not* actually scientific. This argument may be at the bottom of the conventional differentiation between "scientific communication" and "science communication." Communication between fellow scientists and experts, as the idea goes, is the *real* science, pure and precise. What is trickled downstream to the publics, however, is diluted, simplified, distorted, and ultimately *not* scientific.

But this dichotomy is difficult to maintain as it is impossible to pinpoint the place where content becomes simplified, language becomes imprecise, and knowledge becomes derived (Hilgartner, 1990; also see Chapters 3, 6 and 10). Canonically, the communication of science starts with scientists doing experiments and reporting their findings to fellow experts in peer-reviewed publications; once validated through peer reviews, those findings may then be disseminated by mainstream media to the publics. This canonical model, however, has started to crumble when economic incentives, institutional pressures, and the intention to engage the publics are driving scientists to bypass peer review and work directly with the popular press (Chubin & Hackett, 1990; Russo, 2000). In addition, through such formats as citizen science projects, social movements, and policy debates, today's publics are increasingly involved in the discussion and performance of science (Chapter 1).

More fundamentally, we will do well to recognize that the supposed demarcation between "experts" and "non-experts" is situation-dependent, politically charged, and potentially problematic. Brian Wynne's (2004) famous study of the Chernobyl Nuclear Power Plant explosion demonstrates how sheep farmers' knowledge of the local environment and farming management could have informed scientists' attempt to assess and reduce contamination. In this and other cases (see Irwin & Michael, 2003), the boundary between insiders and outsiders is blurred as science becomes a social enterprise, a cultural phenomenon, and an economic necessity.

This is not to say that formal knowledge of science is *not* relevant and important. It is, especially in today's political climate. But also important is publics' local knowledge and value stances (Holliman, Whitelegg, Scanlon, Smidt, & Thomas, 2009; Irwin & Wynne, 2004)—or, in Gould's (2002) words, their interest, intelligence, awareness, and fascination.

For these reasons, in this collection, we treat "scientific communication" and "science communication" interchangeably. We did not, however, prescribe that our authors use one term over the other because both have historical currency and may be preferred by individual scholars, instructors, and programs. Ultimately, what is important is not semantic rigidity but a consensus in the way we understand science as social discourse: created and used by people in dynamic and unstable contexts, for varying and ever-changing reasons.

Concluding the Beginning

We who spend much of our time at universities are privileged, surrounded by researchers in a wide range of academic disciplines, working with motivated, ambitious, and capable students. If we choose to, we can place ourselves on the front lines of the fight to reduce oceanic pollution, improve efficiency of vehicles, and solve the other problems of a small, hot, crowded planet. Those of us who—however awkwardly—straddle various disciplines of humanities and science are in the enviable position to encourage students to pursue critical questions of their disciplines. In one of the ironies of the modern university, those of us credentialed in liberal arts and humanities (self-labeled "non-scientists") are extremely likely to teach students majoring in science, technology, engineering, and math (STEM).

While 2017 may not seem like an ideal historical moment to valorize science and argue for the merits of science communication, perhaps this is an ideal time for this volume to be published. We hope it will bolster those who continually strive to improve science education, technological competence, and greater access to information and education. As a species, we remain ignorant of the promises and practices of science at our peril. Or perhaps, as another icon of science has said, "[T]here is no shame in not knowing. The problem arises when irrational thought and attendant behavior fill the vacuum left by ignorance" (Tyson, 2004, p. 38). By way of closing, we resort to riffing on Neil deGrasse Tyson's infamous words: "Science communication is important whether we want it to be or not."

References

Achenbach, J. (2017, March 16). Trump's budget calls for seismic disruption in medical and science research. *The Washington Post*. Retrieved March 31, 2017, from www.washingtonpost.com/national/health-science/trumps-budget-would-slash-scientific-and-medical-research/2017/03/15/d3261f98-0998-11e7-a15f-a58d4a988474_story.html?utm_term=.6186ada6e60a.

Baake, K. (2003). *Metaphor and knowledge: The challenges of writing science*. Albany, NY: State University of New York Press.

Bazerman, C. (2000). *Shaping written knowledge: The genre and activity of the experimental article in science*. Colorado State University, CO: The WAC Clearinghouse.

Brewer, P. (2015). *International virtual teams: Engineering global success*. Hoboken, NJ: Wiley.

Britton, W. (1965). What is technical writing? *College Composition and Communication, 16*(2), 113–116.

Buehl, J. (2014). Toward an ethical rhetoric of the digital scientific image: Learning from the era when science met Photoshop. *Technical Communication Quarterly, 23*(3), 184–206.

Cagle, L., & Tillery, D. (2015). Climate change research across disciplines: The value and uses of multidisciplinary research reviews for technical communication. *Technical Communication Quarterly, 24*(2), 147–163.

Chubin, D., & Hackett, E. (1990). *Peerless science: Peer review and U.S. science policy*. Albany, NY: State University of New York Press.

Davenport, C. (2017, January 25). Federal agencies told to halt external communications. *The New York Times*. Retrieved March 26, 2017, from www.nytimes.

com/2017/01/25/us/politics/some-agencies-told-to-halt-communications-as-trump-administration-moves-in.html?_r=0.

DeWinter, J., & Moeller, R. (2014). *Computer games and technical communication: Critical methods & applications at the intersection.* Farnham, U.K.: Ashgate Publishing Company.

Dobrin, D. N. (1983). What's technical about technical writing? In P. Anderson, R. Brockmann, & C. Miller (Eds.), *New essays in technical and scientific communication: Research, theory, practice* (pp. 227–250). Farmingdale, NY: Baywood Publishing Company.

Dombrowski, P. (2000). Ethics and technical communication: The past quarter century. *Journal of Technical Writing and Communication, 30*(1), 3–29.

Fahnestock, J. (1999). *Rhetorical figures in science.* New York: Oxford University Press.

Gould, S. J. (2002). *I have landed: The end of a beginning in natural history.* New York: Harmony Books.

Gross, A. (1990). *The rhetoric of science.* Cambridge, MA: Harvard University Press.

Gross, A., & Buehl, J. (2016). Science and the Internet: Communicating knowledge in a digital age. Amityville, NY: Baywood Publishing Company.

Gross, A., Harmon, J., & Reidy, M. (2009). *Communicating science: The scientific article from the 17th century to the present.* Anderson, SC: Parlor Press.

Hilgartner, S. (1990). The dominant view of popularization: Conceptual problems, political uses. *Social Studies of Science, 20*(3), 519–539.

Holliman, R., Whitelegg, E., Scanlon, E., Smidt, S., & Thomas, J. (2009). *Investigating science communication in the information age: Implications for public engagement and popular media.* Oxford: Oxford University Press.

Hourihan, M. (2017, January 19). Science and technology funding under Obama: A look back. *AAAS News.* Retrieved March 26, 2017, from www.aaas.org/news/science-and-technology-funding-under-obama-look-back.

Irwin, A., & Michael, M. (2003). *Science, social theory and public knowledge.* Philadelphia, PA: Open University Press.

Irwin, A., & Wynne, B. (Eds.). (2004). *Misunderstanding science? The public reconstruction of science and technology.* Cambridge: Cambridge University Press.

Johnson-Sheehan, R., & Stewart, K. (2003a). Special issue of *Technical Communication Quarterly, 12*(3), 245–355.

Johnson-Sheehan, R., & Stewart, K. (2003b). Special issue of *Technical Communication Quarterly, 12*(4), 365–480

Katz, S. B. (1992). The ethic of expediency: Classical rhetoric, technology, and the Holocaust. *College English, 54*(3), 255–275.

Kelley, P. (1976). *A basic definition of technical writing.* Paper presented at the Annual Meeting of the Conference on College Composition and Communication, Philadelphia, PA. March 25–27, 1976. Retrieved June 2, 2016, from http://files.eric.ed.gov/fulltext/ED124969.pdf.

Kitalong, K., Moody, J., Middlebrook, R., & Ancheta, G. (2009). Beyond the screen: Narrative mapping as a tool for evaluating a mixed-reality science museum exhibit. *Technical Communication Quarterly, 18*(2), 142–165.

Lamberti, A., & Richards, A. (2011). *Complex worlds: Digital culture, rhetoric, and professional communication.* Amityville, NY: Baywood Publishing Company.

Latour, B., & Woolgar, S. (1979). *Laboratory life: The social construction of scientific facts*. Beverly Hills, CA: Sage Publications.

Latour, B., & Woolgar, S. (1986). *Laboratory life: The construction of scientific facts*. Princeton, NJ: Princeton University Press.

Mogull, S. A., & Balzhiser, D. (2015). Pharmaceutical companies are writing the script for health consumerism. *Communication Design Quarterly, 3*(4), 35–49.

Northcut, K. (2007). Visual communication in life sciences. Special issue of *Journal of Technical Writing and Communication, 37*(4), 375–478.

Northcut, K. (2011). Insights from illustrators: The rhetorical invention of paleontology representations. *Technical Communication Quarterly, 20*(3), 303–326.

Russo, E. (2000, March 6). Bypassing peer review. *The Scientist*. Retrieved June 3, 2016, from www.the-scientist.com/?articles.view/articleNo/12713/title/Bypassing-Peer-Review/.

Spinuzzi, C. (2003). *Tracing genres through organizations: A sociocultural approach to information design*. Cambridge, MA: MIT Press.

Thatcher, B., & St. Amant, K. (2011). *Teaching intercultural rhetoric and technical communication: Theories, curriculum, pedagogies, and practice*. Amityville, NY: Baywood Publishing Company.

Turner, S. S. (2005). Critical junctures in genetic medicine: The transformation of DNA lab science to commercial pharmacogenomics. *Journal of Business and Technical Communication, 19*(3), 328–352.

Tyson, N. d. (2004). *The sky is not the limit: Adventures of an urban astrophysicist*. Amherst, NY: Prometheus Books.

Welhausen, C. (2015). Visualizing a non-pandemic: Considerations for communicating public health risks in intercultural contexts. *Technical Communication, 62*(4), 244–257.

Whithaus, C. (2012). Claim-evidence structures in environmental science writing: Modifying Toulmin's model to account for multimodal arguments. *Technical Communication Quarterly, 21*(2), 105–128.

Wilson, E. O. (1998). *Consilience: The unity of knowledge*. New York, NY: Alfred A. Knopf.

Winsor, D. (2003). *Writing power: Communication in an engineering center*. Albany, NY: State University of New York Press.

Wynne, B. (2004). Misunderstood misunderstandings: Social identities and public uptake of science. In A. Irwin & B. Wynne (Eds.), *Misunderstanding science? The public reconstruction of science and technology* (pp. 19–46). Cambridge: Cambridge University Press.

Yu, H. (2017). *Communicating genetics: Visualizations and representations*. London: Palgrave Macmillan.

Yu, H., & Savage, G. (2013). *Negotiating cultural encounters: Narrating intercultural engineering and technical communication*. Hoboken, NJ: Wiley.

Zappen, J. (1983). A rhetoric for research in sciences and technologies. In P. Anderson, R. Brockmann, & C. Miller (Eds.), *New essays in technical and scientific communication: Research, theory, practice* (pp. 123–138). Farmingdale, NY: Baywood Publishing Company.

Part I

Practice and Theory

1 Shifting Networks of Science

Citizen Science and Scientific Genre Change

Gwendolynne Reid

This chapter reports on a study of communication related to the emerging scientific practice of citizen science, a practice with multiple definitions, but which, in simplest terms, denotes "participation by the public in a scientific project" (McKinley et al., 2015, p. 3). Many of the earliest and most widely known citizen science projects have focused on birds, such as the National Audubon Society's Christmas Bird Count and Cornell University's eBird project – a project that has produced dynamic visualizations of bird migration patterns across continents (Audubon and Cornell Lab of Ornithology, 2016). Recently, these projects have greatly diversified, with featured projects in SciStarter – a database of citizen science projects – ranging from those focused on genome and environment interactions, to those on flu symptoms, and to those on monarch butterfly counts.

The number of citizen science projects has also grown significantly in recent years. Jonathan Silvertown (2009), in his influential article on citizen science, attributes its current rise to widespread access to the Internet and mobile technologies, scientists' increasing realization of the public's interest in and availability for research (p. 467), and the fact that funding organizations routinely build in outreach as an outcome for funded research (p. 469). These factors have led to a radical climb in the number of projects labeled "citizen science," a trend documented in recent reports, such as ecologist Duncan McKinley's et al. (2015) report on citizen science contributions to environmental protection and natural resource management. Mapping the number of peer-reviewed publications per year that are indexed in the Web of Science database as relevant to "citizen science" for the last two decades, McKinley et al. show growth from near-zero results between 1995 and 2005 to an almost vertical climb past the 200-per-year mark in 2015 (p. 5). Silvertown (2009), moreover, points out that while not always labeled "citizen science," science conducted by citizens has been an integral part of its history (e.g., Benjamin Franklin and Charles Darwin).

But what does all this mean for scientific communication? Why pay attention to an emerging practice like citizen science when seeking to understand, practice, or teach scientific communication? This collection

DOI: 10.4324/9781315160191-1

is in part intended to address the question of how to define scientific communication, including whether "scientific communication" is distinct from "science communication." In this chapter, I argue that, while roughly distinguishable as communication between expert scientists and communication about science with non-experts, the two are more interrelated than commonly understood and that this interrelationship is a force for mutual change and influence. The canonical model of scientific communication between experts and the public has a particular sequence and direction, a particular rhythm: internal communication within the scientific community occurs first, and transmission and popularization of scientific findings for the public occurs second (Bucchi, 1998, p. 5; Hilgartner, 1990, p. 519). While this model has been challenged and complicated (Bucchi, 1998; Hilgartner, 1990; Lewenstein, 1995; Myers, 2003), citizen science, by including communication with the public *during* a study rather than after its publication, overtly disrupts and changes the rhythm inherent in the model, a model many engaged in science and scientific communication continue to operate under. This shift in timing, while seemingly simple, both responds to and creates pressure for genre change in scientific research articles and their genre networks, changes that are especially relevant to those seeking to participate in this context or preparing others to do so. In short, the emerging writing practices related to citizen science impact traditional scientific writing practices, including those related to the research article genre and communication with the public.

Taking a case study approach (Gomm, Hammersley, & Foster, 2000), this chapter draws on my ethnographic work with the Heartbeats Project (Hine, 2015; Marcus, 1998), a citizen science project run by a biology lab actively engaged in innovating with scientific communication and embracing the blurred boundaries between expert and non-expert that citizen science encourages. The project is conceived of as a response to the limited data behind the well-known "rule" that, on average, mammals' hearts beat one billion times per lifetime and seeks additional data to test whether the original rule holds as well as to extend the analysis to other biological classes, like birds and amphibians. To participate, citizen scientists submit species heartrate data found in the scientific literature, which is then vetted by members of the project team before inclusion in the data set. The findings reported here were developed from eighteen months of ethnographic engagement with the project, an engagement that included analysis of project-related writing, public speaking, interviews, observations, and digital artifacts (e.g., data spreadsheets, data submission tools, etc.). The study received IRB approval, and pseudonyms are used throughout to protect participants' confidentiality. In analyzing the relationship of the team's citizen science communication with their scientific communication and traditional forms of public communication, I found a rhetorical genre framework coupled with the concepts of *uptake* (Austin, 1962) and *recontextualization* (Linell, 1998) proved useful. This framework and these concepts allowed me to trace the relationships between genres as well as to model broader changes to the scientific genre chains and networks. While multidirectional, interconnected models of scientific communication have been theorized (Bucchi, 1998; Hilgartner, 1990; Lewenstein, 1995; Myers, 2003), examining the relationships between genres provides a specific mechanism for how this multidirectionality occurs and recurs, and demonstrates the potential for genre change that citizen science presents. Specifically, citizen science, by bookending the dominant genre of scientific discourse—the research article—with public-facing genres, responds to some of scientists' most pressing rhetorical exigences while simultaneously exerting pressure for professional scientific genres, to change. In addition, analysis of the Heartbeats Project's communication suggests that scientists engaging with such projects benefit from them in unanticipated ways, namely at the level of rhetorical invention, and that this inventional work provides an additional mechanism through which citizen science influences professional scientific writing. I conclude with a discussion of the implications for those engaged in or teaching scientific communication, including the rhetorical and ethical considerations presented by revised relationships with members of the public.

Theoretical Framework

The relationship between scientific discourse and public discourse about science has been theorized for many decades, with the most prevalent model showing scientific knowledge moving unidirectionally from science to the public through the mass media, generally losing precision along the way (Bucchi, 1998, p. 5; Hilgartner, 1990, p. 519). Several theorists, however, have challenged the accuracy of this model, calling for more bidirectional (or multidirectional) accounts of how ideas move between discourses. Based on his study of communication related to cold fusion, for example, Bruce Lewenstein (1995) troubled the idea of linear dissemination of information to the public, instead proposing a "web of science communication contexts" that includes everything from journals, grant proposals, and talks to mass media, textbooks, and policy reports, "with all forms of communication leading to each other" (p. 426). Around the same time, Massimiano Bucchi (1998), in *Science and the Media*, offered the "continuity model" of scientific communication as an alternative that, instead of theorizing science and the public as discrete spheres, theorizes a continuous and reciprocal movement of scientific information through four stages: the intraspecialistic, interspecialistic, pedagogical, and popular stages (p. 13). More recently, Bucchi (2004) presented the double helix as a metaphor that can model how scientific and public discourses develop in parallel, mutually acting on one another at "junctions," with influence moving both ways (p. 279). Focusing here on the specialist and public discourse on genes, Bucchi noted the long history of public discourse on heredity, "as documented, for example, by the famous claim by French novelist Emile Zola—thirty years before the rediscovery of Mendel's laws of heredity—that 'heredity has its laws, just like gravitation'" (p. 274). Rather than an "impoverished" version of specialist ideas, Bucchi points out that public ideas on heredity have evolved in parallel to scientific discourse, with the two intersecting at various points in a mutually reinforcing pattern. In this same vein, Greg Myers (2003) made a similar case against the dominant model of popularization, arguing that, instead, "scientific discourses are embedded in and intertwined with other discourses" (p. 271).

The study I describe here contributes to this troubling of the dominant account of the relationship between scientific discourse and the public. I use rhetorical genre theory as a framework that provides a view of the recurring types of communication within these two larger spheres, as well as the routine patterns of interaction that knit the two together. Rhetorical genre theory, rather than focusing on shared formal features as the basis for genre, focuses on the pragmatic *actions* genres perform for communities. To quote Carolyn Miller's (1984) influential definition, genres are "typified rhetorical actions based in recurrent situations" (p. 159). In this paradigm, genres are a complex of the "substantive, situational, and stylistic" (Campbell & Jamieson, 1978, p. 18). Mediating between individual action and culture, genres help rhetorical communities, "the relationships we carry around in our heads, to reproduce and reconstruct themselves, to continue their stories" (Miller, 1994, p. 75).

While studies of individual genres have yielded important insights into a range of rhetorical communities and events, a number of genre theorists have found it productive to expand their focus to *groups* of genres, such as genre sets, systems, and networks, for a fuller account of the processes at work in the production of events, texts, and communities (Bazerman, 1994; Berkenkotter, 2001; Devitt, 1991; Swales, 2004). In this same vein, the concept of "intertextual chains" has been useful (Fairclough, 1992; Linell, 1998). These chains, once routinized into fairly predictable, recurrent patterns of interaction, might best be thought of as "genre chains" (Swales, 2004). The relationships between texts in these chains can vary in nature. Some texts directly prompt or form the exigence for another, while others bear more implicit traces of each other in what Mikhail Bakhtin (1986) has called the "dialogic overtones" of all language—the "echoes and reverberations of other utterances" that fill our own utterances (p. 91).

For the first, more directly linked relationships, I have drawn on John Austin's (1962) concept of *uptake*, particularly as it has been put into conversation with genre theory by Anne Freadman (2002), who theorizes that "a text is contrived to secure a certain class of uptakes" and that "the uptake text confirms [the first text's] generic status by conforming itself to this contrivance" and responding in the expected way (p. 40). While warning against the impulse to systematize these utterance-uptake relationships into rigid sets of rules, Freadman offers a rationale for looking at pairs of texts (and ultimately genre chains) as a productive way to understand both the relationships between genres in a given context and, through this, the social actions of individual genres. For the second, more implicit intertextual relationship, I have drawn on Linell's (1998) recontextualization, defined roughly as "the extrication of some part or aspect from a text or discourse, or from a genre of texts or discourses, and fitting of this part or aspect into another context" (p. 145). This can involve direct quoting or reworking of material for another context, as well as "vague influences" between texts (p. 148).

While these are different types of relationships between texts, both can become routinized into recurring patterns that can be usefully mapped into genre chains. For my purposes, they have provided useful analytical tools for concretely mapping some of the paths professional scientists and citizen scientists travel in interacting with each other. They therefore also provided some concrete mechanisms for how scientific discourse and public discourse about science influence each other. As John Swales (2004) points out, however, we should be careful about overly systematizing these relationships with a term like *genre system*, which "suggests that we have a greater understanding of how everything fits together in a 'system' than is likely the case" (p 23). I have therefore opted for Swales's term, *genre network*.

Studying the Heartbeats Project

The Heartbeats Project is one project of many at a biology lab in a large Southeastern U.S. public university and includes data collected by citizen scientists as well as data collected by members of the lab. The project is conceived of as a response to the limited data behind the rule that on average mammals' hearts beat one billion times per lifetime. The project's researchers have sought to add substantially to this data, to test whether the original rule holds, and to look at the relationship between heart rate and lifespan for other biological classes, like birds and amphibians. In part in order to speed up the process of gathering heart rate data from the scientific literature, the lab created a webpage that solicits submission of relevant species heartrate data and research articles by citizen scientists, which are then vetted by members of the project team before inclusion or exclusion in the data set. This lab has a strong commitment to public science and experience running citizen science projects, with some of those projects reaching participation in the tens of thousands of citizen scientists. At the time of writing, however, the Heartbeats Project had only garnered a little over 100 citizen science contributions, a number high enough to help address the limited data of the original research in this area, but that calls attention to the project's difficulty in engaging citizen scientists, a fact also reinforced by the one-star rating given it in the SciStarter database. The team has hypothesized a number of reasons for this low engagement, with the dominant explanation that this is related to the project's focus on the scientific literature, which is often inaccessible to the public and which does not fit the schema of "doing science" many citizen scientists are seeking. The team running the project consisted of five core people, though other members of the lab and campus resources were tapped as needed:

- Clay – the postdoc leading the Heartbeats Project and lead author on the first research article manuscript.
- Jada – the undergraduate researcher assigned to the project.
- Summer – a lab manager and research associate assisting with project visualizations.
- Soren – the principal investigator (PI) of the lab who began the project.
- Rachel – a former member of the lab assisting with statistical analysis.

The findings I report on here were developed from eighteen months of ethnographic engagement with the Heartbeats Project's writing and focus on the team's inventional work, meaning the process of developing scientific findings, ideas, and material for their research article. This study employed qualitative research methods and proceeded inductively, with ongoing analysis driving further data collection. In order to situate the team's writing practices within a larger context, I took an ethnographic approach informed by the "connective" practices described by Christine Hine (2015) in *Ethnography for the Internet*, practices that integrate mediated forms of engagement with participants and the field into ethnographic inquiry in order to avoid increasingly problematic divisions between online/offline activity. In practical terms, this means that digital artifacts and interactions (e.g., emails, the project website, Twitter events, etc.) were included in my analysis alongside interviews and observations (see Table 1.1).

I analyzed data for this chapter through three approaches. First, I mapped the relationships between texts, both those with direct utterance-uptake relationships and those with signs of recontextualization. Second, I performed rhetorical analysis of the texts included in this mapping. Third, I coded collected data using MAXQDA, a qualitative data analysis tool. While my larger study of this team included several other codes, this chapter focuses on three main codes, along with their subcodes: (1) public communication, (2) scientific communication, and (3) genre talk.

Table 1.1 Types and Quantities of Data Collected

Within Team	*Beyond Team*
9 Interviews (9 hours)	SciStarter observations (4 projects)
5 Observations (6 hours)	AnAge database observations (3 entries)
Drafts & writing samples (9 texts)	Twitter #CitiSciChat events (3 events)
Project data spreadsheets (5 spreadsheets)	CitSci listserv observations (270 listserv posts)
Project website/lab blog posts (14 posts)	Citizen science participation (2 projects)
Emails (70 emails)	Local citizen science events (3 events)
Media coverage of the project/lab (12 articles)	National Science Foundation documents and databases (4 documents)
Lab social media account subscriptions (3 platforms)	Citizen Science Association website and blog (2 site visits)
Participation in the Heartbeats Project (1 submission)	

Overlapping Models of Communication

Before examining the Heartbeats Project team's innovative communication practices, it is important to note that the team engages in traditional scientific communication practices like writing research articles and communicating with the public after the publication of research articles alongside their citizen science writing. In fact, the dominant model of scientific communication between experts and the public was strongly present in their work, both as a standard to be pushed against and as a resource to be enlisted. Clay, the postdoc leading the project, referenced a typical chain of genres resembling the dominant model and the desired uptake between those genres many times over the eighteen months of this study (see Figure 1.1), an order that amounted to important procedural knowledge for acting effectively as a scientist.

Clay explained that the way to set this chain in motion in his institutional context was to contact the university public information officer assigned to them, Brendan Cross, as soon as a research article was accepted. The desired uptake from the press release Brendan produced was widespread coverage of their work through both traditional news media and through social media, though this uptake was by no means guaranteed. Having a press release ready to send out while the research article was still under embargo or on the day the article was published was one way to increase the chances of success, since, as Brendan explained to me, "Depending on the discipline, [the research article's] shelf life can

Figure 1.1 Traditional chain of genres communicating science to the public.

be as short as days." Yet while the genre chain to the news media and to social media was a fairly concrete path that participants could readily outline for me, the desired uptake by the public itself was more nebulous. This uptake was often represented by quantified "views," "shares," and references to "impact." That numeric representation, however, seemed to stand for a host of desired outcomes not easily reduced to the number of times a given article was viewed. Stated goals for communicating with the public, for example, ranged from specific goals like publicizing the research or addressing a particular problem to broader goals like educating the public about science, influencing public policy, and creating more favorable conditions for science. The team was also sensitive to the injustice of having publicly-funded research rendered inaccessible to the public by journal and database paywalls. Regardless of the ambiguity of the end of the genre chain, this well-traveled path underscores how the canonical model Bucchi (1998) and others have described serves as a blueprint for scientists in their work, something Clay confirmed for me. When presented with a visual of the model (Bucchi, 1998, p. 5), Clay admitted how the directionality of the model is "one every scientist acts under."

The team's strategies for communication, however, also demonstrate how multiple models of communication overlap and coexist, with the traditional genre chain outlined in Figure 1.1 existing alongside the team's direct communication with the public through websites and blogs. This traditional chain of genres reflects what Bucchi (2013) has more recently termed Science Communication 1.0, a model that includes mediators like the press release and news genres in a sequential pattern of communication (p. 906). At the same time that the Heartbeats Project team has made use of this traditional pattern, however, it has also made use of resources presented by what Bucchi (2013) calls Science Communication 2.0, in which scientists communicate directly with the public in a horizontal, simultaneous relationship (p. 906). Public-facing videos are examples of such communication. Explaining his resistance to some journals' requests for videos as supplementary materials for research articles, Clay described how in his experience those videos are "buried" and don't get watched or shared, a reason for his preference to post videos "on my own website" or "on a Vimeo account" where they are more likely to be found, viewed, and shared. While adept at enlisting the traditional chain of genres for communicating with the public, Clay also routinely circumvents the traditional mediators of science-related communication. The Heartbeats Project, as a less readily visual project, has not lent itself to videos, but the team's regular use of blogs, Facebook, Twitter, and public events to encourage the public to participate is evidence of a direct and simultaneous approach to communication overlapping with a traditional sequential approach. During the time I studied the Heartbeats Project, both the canonical model and the "direct" model of communication with the public were simultaneously present in the team's work, with the canonical model understood as the "standard" to be pushed and worked against, and with both models enlisted as resources. In Clay's words, the team has put much effort into considering "what scientists can do to influence the end game," with citizen science and direct communication with the public key to those efforts.

Recontextualizing Science for Science

While the traditional chain of genres used to communicate with the public after a research article's publication can be traced through explicit evidence of uptake, those involved in the Heartbeats Project's citizen science before its publication are better traced through intertextuality, specifically through evidence of recontextualization. In concrete terms, the press release in the genre chain (see Figure 1.1) directly "takes up" and responds to the exigence presented by the research article, and the news articles take up the invitation offered by the press release. Each of these explicitly signals the text it is taking up through textual features such as a direct reference to that text. The research article, however, does not have an overt utterance-uptake relationship with any publicly-oriented genre—this relationship, when it exists, is occluded. While press releases and news articles are meant to call attention to the research article as their exigence, the research article genre has embedded in it the logic that the exigence for scientific work lies in the scientific literature, not in publicly-oriented genres or even in the lab (Berkenkotter & Huckin, 1995). Observing the relationships between texts and genres related to the Heartbeats team's citizen science and its scientific writing requires looking at slightly different traces than for those participating in the traditional science-to-public genre chain.

Examining the Heartbeats Project's writing for recontextualized material reveals a striking aspect of this type of citizen science project, namely that, in direct contrast with the conventional wisdom about how scientific ideas are recontextualized from scientific conversations for public fora (e.g., Luzón, 2013), *citizen science recontextualizes science for science*. By this, I mean that citizen science practices encourage the recontextualization of scientific activity and science-related communication occurring in public contexts for scientific discourse (the research article). In the case of the Heartbeats Project, the project's origin story begins with publicly-oriented writing in the form of a popular science book Soren (the lab's principal investigator) was working on. This required recontextualizing scientific findings on the billion beats rule for a public audience (the conventional order of things), but then prompted the inception of the Heartbeats Project when the scientific literature was found lacking (a departure from the conventional). From that point forward in the team's writing, citizen science activity and writing developed in tandem with their scientific writing, with recontextualization often occurring from the publicly-oriented texts to the scientifically-oriented texts. For example, when the team compiled and analyzed data in part originating from citizen scientists, they first communicated those to citizen scientists and the broader public using the project website and the lab blog. This material then provided a starting point for working on research articles related to the project.

The project website from February 2015 and an early research article manuscript from this same time period provide a useful example of this recontextualization (see Table 1.2), with material originally developed for a public audience making its way into writing intended for an expert audience. For instance, the examples developed for citizen scientists of species with extremely fast and extremely slow heartrates—the shrew and the grey whale, respectively (sentence 2)—show up in the manuscript draft (sentence 5), as do the explanations of the billion beats rule (sentences 1 and 3, respectively). Interestingly, both documents also begin by gesturing at the scientific literature.

But while the two texts show a clear relationship, they also depart from each other in important ways. The research article manuscript makes extensive use of hedges, like the verbs "consider" and "suggest,"

Table 1.2 Introductions to the Heartbeats Project, Public and Scientific

Heartbeats Project Website, 2015	*Draft Article, 2015*
1 "Studies have concluded" – the rule	1 "A large literature considers" – the variables
2 Shrew and grey whale examples	2 Generally understood relationship between variables
3 Exceptions to the rule	3 Relationship, continued – the rule
4 Outliers with more than 1 billion beats	4 Empirical support for the rule
5 Outliers must be explained by biology	5 Shrew and grey whale examples
6 Potential implications for human longevity	6 Limited evidence
7 Call to contribute data	7 Most comprehensive study based on limited data
	8 Potential for more complex relationships than [2]

while the website conveys greater certainty with verbs like "conclude" and "are." The website allots substantial space to considering outliers and research implications, both of which point toward the possibility of greater longevity in humans. On the website, these moves regarding outliers and implications lead up to a call for more data—a clear call to action for citizen scientists to participate. This early draft of the manuscript, on the other hand, situates its exigence in the literature, following John Swales's (1990) Moves 1 and 2 in his "Create a Research Space" (CARS) model of scholarly introductions: "Establishing a Territory" and "Establishing a Niche." Using the literature to establish the "territory" of the relationship between heart rate and longevity, the manuscript then establishes a "niche" by pointing out a gap in this literature—the scant data the understood relationship is based on. The manuscript doubly secures this research niche by suggesting that the understood relationship may in fact be found to be more complex once additional data is examined. Notably, the manuscript's explicitly-stated exigence rests squarely in the scientific literature, with the text's relationship with citizen science and other publicly-oriented communication occluded. The scientific story and relationships embedded in the research article genre are not, at this point, challenged by these overt written choices. The activity related to citizen science, however, including its communication, does leave substantive and rhetorical traces on the manuscript.

As I followed the project, it became apparent that those traces stemmed in part from the creative, inventive work required to communicate with citizen scientists, work that served as something of a rehearsal for analysis and arguments that might eventually be used in professional science. Later in 2015, for example, Clay gave a publicly-oriented talk about the project, including solicitation for more citizen science participation, at a science café, a genre he explained he found useful for helping him "organize my thoughts about where to start, how to provide context" in a research article, including which ideas sound "ridiculous" and which work. While the event was intended to be for the public, Clay noted that a high proportion of the audience appeared to be retired scientists. Nevertheless, he invoked a public audience, as the genre and situation asked him to do so. While Clay found the makeup of the audience frustrating (since the point was to connect with a public audience), it reinforces the point made by Bucchi and others that these discourses and contexts routinely overlap, noting that on some issues experts in one specialty become members of the public on another. Clay and Brendan, in fact, both discussed how public writing could be used to reach other scientists for scientific purposes. Discussing, for example, a "cranky" audience member who stayed after the talk to take issue with some elements of his talk, Clay confessed that this exchange gave him useful feedback on a variable they would need to address more robustly in their research article. His experiences giving these sorts of talks and the inventional role they ended up playing for him in his scientific writing led him to express a preference for "giv[ing] a presentation or two on a project before I even start to write the paper," particularly "the introduction and the discussion." Increasingly, during our check-in interviews, Clay expressed an appreciation for how communicating with citizen scientists invited him to consider the "broader implications" of his work, implications that showed up in the introduction and discussion sections of research article manuscripts and that allowed him to target the higher tier journals ("high impact journals") that stress this. This may in part be accounted for by Clay's relative inexperience as a postdoc, but it demonstrates how interacting with the public can further both the development of scientific ideas and scientists' own development.

The recontextualization of material from citizen science communication to the team's scientific writing extended to other areas as well, including the visuals developed for research article manuscripts. While the team decided that the Heartbeats Project did not lend itself visually to videos, they did stress visual communication with citizen scientists in the form of clear, compelling charts, often placed sequentially to "tell a story." The project website, for example, featured a cleanly-designed scatter plot chart of all the species data submitted to the project, with a few evocative outlier examples highlighted through color and labeling (e.g., the grey whale, Etruscan shrew, humans). In his science café talk, Clay took this chart and duplicated it multiple times for consecutive slides, each time emphasizing a different species or group of species (carnivores, rodents, marsupials) to tell the story of the data visually, to "show motion." This analytical and rhetorical work then became a starting point for research article figures. The lab, in fact, routinely created figures for research articles that recontextualized material developed for citizen scientists and that would then be accessible to public audiences later, potentially being published in press releases and news articles, and shared on social media platforms. In Clay's words, why not "show a picture of the actual data that makes sense to people?" Why not "[build] the outreach figures ahead of time" and "[put] them into the paper?" While in his experience some journal editors resisted some of these visual innovations (one editor felt a panel of figures walking readers through the data just "showed the same information" redundantly and should be reintegrated into a single figure), other editors embraced these innovative visual strategies and readily used them.

Figure 1.2 maps the relationships between genres that the Heartbeats Project team used to communicate with citizen scientists and demonstrates how this communication served an inventional purpose for the team, helping them generate findings, arguments, and material to be repurposed.

Routinely participating in this network of genres gave the team a recurring reason to continue producing new material—new figures and analyses based on the data submitted—and a communicative reason to think more deeply

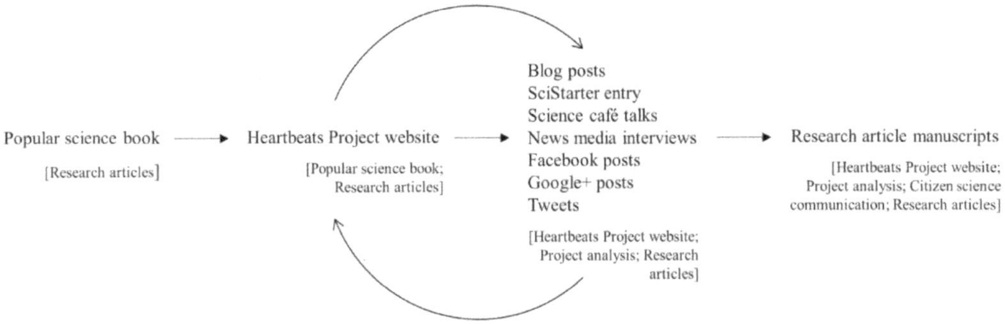

Figure 1.2 Citizen science communication as scientific invention [recontextualization in brackets].

at the level of implications (like human longevity) and values (like wonder). This activity was then recontextualized for scientific discourse, a movement that reverses the canonical model of public communication about science and bookends the research article, the dominant genre of professional science, with publicly-oriented genres. The Heartbeats Project demonstrates, then, one way citizen science can influence professional science. Beyond an opportunity to expand data collection or for public outreach, citizen science also offers scientists creative, inventional benefits, influencing the shape of the final research and the arguments made to situate it.

Shifting Networks, Genres, and Roles

The genres enlisted for the Heartbeats Project illustrate the shifted networks of genres that citizen science can prompt for scientists who choose to lead or engage with these projects (Figure 1.2). These are not the genres generally associated with scientific communication (e.g., research articles, literature reviews, research proposals, posters, abstracts, conference papers, grant proposals) or even with the occluded genres of scientific work (e.g., lab notebooks, emails, article reviews). Instead, these more closely resemble the genres enlisted to communicate with the public about published research articles (i.e., the traditional chain of genres in Figure 1.1 that amounted to important procedural knowledge for Clay). The timing of these genres, however, is changed and requires more direct communication between scientists and the public, a fact that may be both attractive and uncomfortable for some scientists, particularly those who feel they are not adequately trained or practiced in public science communication. For many, however, the challenges of citizen science will outweigh those misgivings as it presents itself as a fitting response to a commonly understood rhetorical situation, a situation that includes an exigence to share scientific research with the public and technological resources to do so widely and meaningfully. Clay, for example, noted the "huge disconnect" of having "the public pay for [science] through tax dollars" but having the final products in a form "the public can't read or have access to." This disconnect, in fact, is one that funding agencies like the National Science Foundation (NSF) have attempted to address by attaching requirements for public outreach in their grants and through the initiatives like the NSF's Public Access Repository. It is also part of the motivation behind the Open Science movement, whose proponents see openness and collaboration as values key both to cultivating relevant, rigorous science and to ensuring that this science is available to inform public decision-making. This latter concern encompasses perceived public hostility toward science and scientific findings and a general low level of scientific literacy (issues like vaccines and climate change are often cited). This complex of exigences (taxpayers, funding, collaboration, issues, literacy) have propelled a great deal of experimentation with both scientific genres and genre networks, including experiments with open access publishing, open data, open peer review and new genre elements like lay summaries, supplements, primers, and videos (Costello, 2016). Situated in this context, citizen science can be seen as a *response* to many of the pressures for change in scientific genre networks, holding potential for fostering public trust and literacy and, in Clay's words, affecting the "end game" (see Chapter 10, for a discussion of the deficit model and complications to the expert/public binary).

The Heartbeats Project, however, underscores how citizen science also *contributes* to the forces for change in scientific genres and genre networks. It is important to note that the larger scientific network of genres is generally more structured and regulated than other networks of genres, with multiple institutionalized forces acting conservatively to maintain stability—forces like university tenure and promotion practices, blind peer review, professional associations, funding and regulatory agencies, and others (Kelly & Maddalena, 2015, pp. 8–9). Interviews with members of the Heartbeats Project team, in fact, highlight the power of these forces and the limited value some scientists see in many of the genre innovations occurring at scientific journals (e.g., Clay's assessment of video supplements). Yet analysis of the Heartbeats Project's communication suggests that routinely engaging with citizen scientists influences scientists' inventional work, contributing incrementally to a movement toward accessibility and open science. Communicating with citizen scientists, for example, influenced the team's visual choices in a way that contributes to the accessibility of their research article manuscript. Shifting the network of scientific genres to include citizen science, then, promotes a degree of scientific genre change toward accessibility and toward addressing the "big picture" implications of scientific research.

The shifted timing and nature of public communication introduced by citizen science also contributes to the pressure for genre change in another important way, namely by altering the roles indexed in scientific genres. Genres provide a set of social rules, constraints, and resources for particular situations, including "reproducible speaker and addressee roles, social typifications of recurrent social needs or exigences, topical structures (or 'moves' and 'steps'), and ways of indexing an event to material conditions" (Miller, 1994, p. 71). Citizen science shifts the role of the scientist communicator from one communicating *to* the public to one communicating *with* the public. This collaborative relationship is part of the rhetorical power of citizen science; however, it also introduces a new set of relationships that have some similar precedents that writers can draw on (e.g., extended scientific collaborative teams) as well as introduce distinctive dynamics. These dynamics raise rhetorical problems not fully addressed by research article conventions. For example,

- At what point are citizen scientists coauthors and at what point is an acknowledgement sufficient? (See Chapter 2, for more about science authorship.)
- Is a data citation sufficient and what is the most accurate and effective way to cite data (Hunter & Hsu, 2015)?
- At what point are citizen scientists human subjects?
- Can citizen scientists play multiple roles?
- What ethical obligation do professional scientist authors have to make research articles based on citizen science projects available to citizen scientists? Is a press release or news article sufficient?
- Do citizen scientists have a right to published journal articles and to lay summaries tailored for them?
- Will the research article genre allow for a research exigence situated in citizen science or must the scientific story told in research articles continue to focus exclusively on the exigences presented by gaps in the scientific literature?

These are not simple questions to answer and they have not been answered often enough in the research article genre for them to be typified components. In the case of the Heartbeats Project, manuscript drafts do not list citizen scientists as co-authors, something Clay attributes to their fairly minor level of participation. Likewise, Brendan's suggestion during our interview to include a reward like a "keychain or t-shirt" to boost project participation does not invoke a collaborator role for citizen scientists. The reciprocity and obligations elicited by the Heartbeats Project is rather low. However, the set of factors that must be considered in order to answer questions related to roles and attribution draws attention to the fact that citizen science is not a single monolithic practice, but rather a set of practices with variations. The typified answers to this rhetorical problem will need to account for the various manifestations citizen science can take, something addressed by several typologies of citizen science practices (Bonney et al., 2009; Cooper, Dickinson, Phillips, & Bonney, 2007; Haklay, 2011; Wiggins & Crowston, 2011; Wilderman, 2007). Haklay's (2011) four-level typology of citizen science based on level of participation, for example, helps explain the Heartbeats Project team's decision not to include citizen scientists as coauthors:

1 "Crowdsourcing," with citizens as "sensors."
2 "Distributed Intelligence," with citizens as "basic interpreters."
3 "Participatory Science," with citizens participating in "problem definition and data collection."
4 "Extreme," with citizens collaborating on "problem definition, data collection and analysis." (p. 116)

In this typology, the Heartbeats Project fits at the level of "distributed intelligence," since participants must make several interpretive decisions about their data collection, including whether the heartrate data in a research article are valid (drugged animals, for example, should be excluded). While a step above the citizen scientist as "sensor," this level and type of participation raises only minimal questions about attribution and ethics. In contrast, projects like the Flint Water Study, a community-led citizen science project partnering with researchers at Virginia Tech University to investigate the contamination of Flint, Michigan's water supply (Maynard, 2016), raise more extensive rhetorical and ethical questions about the relationship between scientists and citizen scientists and the obligations this relationship raises. The obligation to communicate findings to citizen scientists and stakeholders in an accessible, actionable form, for example, is much greater and draws attention to how citizen science can encourage change at the level of genre and genre networks.

Conclusions: Implications for Scientific Communication

The Heartbeats Project case presents several findings relevant for those engaged in or teaching scientific communication in a shifting rhetorical landscape of science. While scientists might choose to involve citizen scientists in their projects for many reasons, including to expedite data collection, the Heartbeats Project shows that citizen science also offers scientists creative, inventional benefits that influence the shape of their final research and the arguments they develop to situate it (see Figure 1.2). Even in a project that did not yield widespread participation (participants numbered in the hundreds, not thousands) and where the communicative labor of working with citizens outweighed the yield in data, the professional scientists leading the project benefited from citizen science in other ways, namely rhetorical and scientific invention. Scientists considering organizing citizen science projects or engaging with existing projects should therefore realize that the benefits that arise from such engagement may extend beyond data and derive as much from their communication with citizen scientists, though this communication entails considerable work. With this in mind, scientists designing such projects might consider building in opportunities for citizen scientists to communicate with them beyond the submission of data.

In many of its forms, citizen science shifts the role of the scientist communicator from one of communicating *to* the public to one of communicating *with* the public. This collaborative relationship is part of the rhetorical power of citizen science, but it also raises rhetorical and ethical questions not fully addressed by research article conventions. These questions range from those about authorship and revised reader-writer roles, to those regarding the status of citizen scientists as collaborators to be acknowledged, human subjects to be protected, neither of these, or both. Ultimately, scientists designing citizen science projects need to consider the issue of *reciprocity* raised by their relationship with citizen scientists. For example, do scientists owe it to citizen scientists to "return" data and analysis to those who collected it? If so, in what form? Likewise, can citizen scientists expect scientists to advocate on their behalf with the data they have submitted, though scientists may see this as outside their purview or a contradiction in roles? These are evolving questions with evolving answers, but they point toward the pressure citizen science practices put toward accessibility, both the "technical" accessibility addressed by open access publishing and rhetorical accessibility at the level of genre. Professional scientists working with citizen scientists will need to consider these questions and many more specific to their projects and the communities they engage. Ultimately, however, the Heartbeats Project underscores that professional scientists working with citizen scientists are engaging in a rhetorical endeavor and that, like other rhetorical endeavors, influence is likely to move in both directions, with neither party left unchanged and the outcome not entirely predictable.

Acknowledgements

I am extremely grateful to the members of the Heartbeats Project team who consented to having me study their communication and most especially to Clay for his patience and generosity. I am also indebted to Carolyn Miller and to Stacey Pigg for their feedback on drafts of this chapter.

References

Audubon and Cornell Lab of Ornithology. (2016). *Citizen science reveals annual bird migrations across continents*. Retrieved March 22, 2016, from http://ebird.org/content/ebird/news/lasortemap/.

Austin, J. L. (1962). *How to do things with words*. New York: Oxford University Press.

Bakhtin, M. M. (1986). The problem of speech genres. In C. Emerson & M. Holquist (Eds.), V. W. McGee (Trans.), *Speech genres and other late essays* (pp. 60–102). Austin, TX: University of Texas Press.

Bazerman, C. (1994). Systems of genres and the enactment of social intentions. In A. Freedman & P. Medway (Eds.), *Genre and the new rhetoric* (pp. 79–101). Bristol, PA: Taylor & Francis.

Berkenkotter, C. (2001). Genre systems at work: DSM-IV and rhetorical recontextualization in psychotherapy paperwork. *Written Communication*, 18(3), 326–349.

Berkenkotter, C., & Huckin, T. N. (1995). *Genre knowledge in disciplinary communication: Cognition/culture/power*. Hillsdale, NJ: Lawrence Erlbaum.

Bonney, R., Ballard, H., Jordan, R., McCallie, E., Phillips, T., Shirk, J., & Wilderman, C. C. (2009). *Public participation in scientific research: Defining the field and assessing its potential for informal science education. A CAISE inquiry group report*. (p. 58). Washington, DC: Center for Advancement of Informal Science Education. Retrieved April 2, 2016, from http://eric.ed.gov/?id=ED519688.

Bucchi, M. (1998). *Science and the media: Alternative routes to scientific communications*. New York: Routledge.

Bucchi, M. (2004). Can genetics help us rethink communication? Public communication of science as a "double helix." *New Genetics and Society*, *23*(3), 269–283.

Bucchi, M. (2013). Style in science communication. *Public Understanding of Science*, *22*(8), 904–915.

Campbell, K. K., & Jamieson, K. H. (1978). Form and genre in rhetorical criticism: An introduction. In K. K. Campbell & K. H. Jamieson (Eds.), *Form and genre: Shaping rhetorical action* (pp. 9–32). Falls Church, VA: Speech Communication Association.

Cooper, C. B., Dickinson, J., Phillips, T., & Bonney, R. (2007). Citizen science as a tool for conservation in residential ecosystems. *Ecology and Society*, *12*(2), 11.

Costello, V. (2016, March 4). Lay summaries, supplements, primers: Scientists (and journals) strive to make science accessible to public (and each other). *PLOS Blogs*. Retrieved August 26, 2016, from http://blogs.plos.org/scicomm/2016/03/04/lay-summaries-supplements-primers-scientists-and-journals-strive-to-make-science-accessible-to-public-and-each-other/.

Devitt, A. J. (1991). Intertextuality in tax accounting: Generic, referential, and functional. In C. Bazerman & J. G. Paradis (Eds.), *Textual dynamics of the professions: Historical and contemporary studies of writing in professional communities* (pp. 336–357). Madison, WI: University of Wisconsin Press.

Fairclough, N. (1992). *Discourse and social change*. Cambridge, MA: Polity Press.

Freadman, A. (2002). Uptake. In R. M. Coe, L. Lingard, & T. Teslenko (Eds.), *The rhetoric and ideology of genre: Strategies for stability and change* (pp. 39–53). Cresskill, NJ: Hampton Press.

Gomm, R., Hammersley, M., & Foster, P. (2000). *Case study method: Key issues, key texts*. London: Sage.

Haklay, M. (2011). Citizen science and volunteered geographic information: Overview and typology of participation. In D. Sui, S. Elwood, & M. F. Goodchild (Eds.), *Finalizing a definition of "citizen science" and "citizen scientists"* (pp. 104–122). Berlin: Springer Science & Business Media.

Hilgartner, S. (1990). The dominant view of popularization: Conceptual problems, political uses. *Social Studies of Science*, *20*(3), 519–539.

Hine, C. (2015). *Ethnography for the Internet: Embedded, embodied and everyday*. London: Bloomsbury.

Hunter, J., & Hsu, C. H. (2015). Formal acknowledgement of citizen scientists' contributions via dynamic data citations. In *Digital libraries: Providing quality information* (pp. 64–75). Springer. Retrieved April 2, 2016, from http://link.springer.com/chapter/10.1007/978-3-319-27974-9_7.

Kelly, A. R., & Maddalena, K. (2015). Harnessing agency for efficacy: "Foldit" and citizen science. *Poroi*, *11*(1), 1–20.

Lewenstein, B. V. (1995). From fax to facts: Communication in the cold fusion saga. *Social Studies of Science*, *25*(3), 403–436.

Linell, P. (1998). Discourse across boundaries: On recontextualizations and the blending of voices in professional discourse. *Text: Interdisciplinary Journal for the Study of Discourse*, *18*(2), 143–158.

Luzón, M. J. (2013). Public communication of science in blogs: Recontextualizing scientific discourse for a diversified audience. *Written Communication*, *30*(4), 428–457.

Marcus, G. E. (1998). *Ethnography through thick and thin*. Princeton, NJ: Princeton University Press.

Maynard, A. (2016, January 27). Can citizen science empower disenfranchised communities? *Phys Org*. Retrieved April 3, 2016, from http://phys.org/news/2016-01-citizen-science-empower-disenfranchised.html.

McKinley, D. C., Miller-Rushing, A. J., Ballard, H. L., Bonney, R., Brown, H., Evans, D. M., ... Shanley, L. A. (2015). Investing in citizen science can improve natural resource management and environmental protection. *Issues in Ecology*, *19*. Retrieved February 27, 2016, from http://pubs.er.usgs.gov/publication/70159470.

Miller, C. R. (1984). Genre as social action. *Quarterly Journal of Speech*, *70*(2), 151–167.

Miller, C. R. (1994). Rhetorical community: The cultural basis of genre. In A. Freedman & P. Medway (Eds.), *Genre and the new rhetoric* (pp. 67–78). Bristol, PA: Taylor & Francis.

Myers, G. (2003). Discourse studies of scientific popularization: Questioning the boundaries. *Discourse Studies*, *5*(2), 265–279.

Silvertown, J. (2009). A new dawn for citizen science. *Trends in Ecology & Evolution*, *24*(9), 467–471.

Swales, J. M. (1990). *Genre analysis: English in academic and research settings*. New York: Cambridge University Press.

Swales, J. M. (2004). *Research genres: Explorations and applications.* New York: Cambridge University Press.

Wiggins, A., & Crowston, K. (2011). From conservation to crowdsourcing: A typology of citizen science. In *2011 44th Hawaii international conference on System Sciences* (pp. 1–10). IEEE. Retrieved April 2, 2016, from http://ieeexplore.ieee.org/xpls/abs_all.jsp?arnumber=5718708.

Wilderman, C. C. (2007). *Models of community science: Design lessons from the field.* Presented at the Citizen Science Toolkit Conference, Ithaca, NY: Cornell Lab of Ornithology. Retrieved April 2, 2016, from www.birds.cornell.edu/citscitoolkit/conference/toolkitconference/presentations.

2 Lines and Fields of Ethical Force in Scientific Authorship

The Legitimacy and Power of the Office of Research Integrity

Steven B. Katz and C. Claiborne Linvill

In 1992, *Technical Communication Quarterly* published a special issue dedicated to scientific misconduct. In that same year, the Office of Research Integrity (ORI) was formed in the Department of Health and Human Services (DHHS). Organizations like the National Academies of Science and Engineering, Public Health Service, and the Institute of Medicine also were working to codify and explain official definitions of "research misconduct." In the many years since, we've experienced an "increasing complexity of science" (Louis, Holdsworth, Anderson, & Campbell, 2008, p. 89): research teams have expanded, labs have become more multicultural, and technology has changed research and writing processes. Yet the definition of scientific misconduct has remained virtually unchanged: fabrication, falsification, and plagiarism (FFP). As we will see, this definition of scientific misconduct omits a host of authorship ethics affecting scientists, and thus the role of authorship in science and beyond.

For now, we can initially distinguish authorship ethics from other research ethics. Authorship ethics deal with the writing, communication, and publication of science (e.g., collaboration, the assigning of credit, the order of authors, peer review). Traditionally opposed to this is scientific research ethics, which deal with experiments (equipment, methodologies, data-gathering). But scientific research also includes ethics in *reporting* results, as we see in fabrication and falsification. Thus, even the simple (and false) distinction between authorship and research ethics begins to break down. To understand this is to begin to recognize the centrality of writing and authorship ethics throughout the scientific process, and an important argument for including them at least in the study if not the regulation of scientific ethics. Among the many authorship ethics touched on in this chapter, the ORI has chosen only to regulate "plagiarism," an extensive but not inclusive domain of authorship ethics.

DOI: 10.4324/9781315160191-2

This chapter delves into the wider implications of the ORI's (non) decision to exclude scientific authorship in its definition of scientific misconduct. Perhaps it should come as little surprise that the ORI, and other agencies whose deliberations and decisions it relies on, came to the conclusion that authorship ethics were not of concern to them, and that their focus should instead remain solely on fabrication and falsification, as well as plagiarism. In some ways, questions of authorship have been excluded from ethics in the West since the beginning of writing. In classical Greece, Plato (1956) and Aristotle (2007) prized "Ideal Forms" or observable "facts," respectively, and true content was much more important than language or rhetoric, which for these philosophers played little or no part in the discovery and dissemination of "Truth" or "scientific knowledge." This has been the mainstream view of scientists since the Newtonian revolution and the rise of empiricism: doing science happens outside and independently of language and authorship; science is objective; and writing is only important in the "conveyance" or "transmission" of results (Katz, 2009). Hence the mimetic (imitative) function of language to "copy" facts, and the ethical problem of copying other authors' ideas and words (plagiarism), become issues of scientific error and ethics based on the concept of "accuracy." But in this understanding of language as accurate and ethical transmission, the author's role in science is significantly reduced.

We won't have time in this chapter to explore the historically and philosophically contentious relationship between language, knowledge, and ethics in the multiple fields that study them. But we believe that the ORI's ongoing decision not to include authorship ethics in its definition of scientific misconduct provides a unique opportunity to interrogate the supposed, potential, and real power of ORI, and its concept of authorship ethics in relation to other scientific organs and institutions. We therefore will attempt to begin to reveal what we will call the "lines and fields of ethical force" in which the ORI, other scientific organizations and entities, and scientists themselves, are embedded and operate. The ultimate purpose of this chapter is to better understand the role of scientific authorship and to show how ethics in science should go beyond policing FFP.

A Historical Overview of the Office of Research Integrity

We begin our analysis of the Office of Research Integrity (ORI), and its role in the interplay of the lines and fields of ethical force in scientific authorship, by examining the history of the ORI, one of the nation's highest ethical bodies, and its decision to define research misconduct solely as FFP.

Research misconduct has only been formally defined and regulated in the United States on the federal level since the 1980s. Several government agencies and departmental committees developed during that time (and later combined) to regulate the work of scientists (ORI, n.d.-b). Major government funding agencies that oversee research include the National Institutes of Health (NIH), which is part of the U.S. Department of Health and Human Services (DHHS), and the National Science Foundation (NSF), which sets research guidelines for scientists and institutions that apply for their grants (NSF, 2002). But the top-most government services agency that regulates research under the umbrella of the DHHS is the Office of Research Integrity (ORI). The ORI developed when two branches of the Public Health Service (PHS), the Office of Scientific Integrity (housed in the NIH), and the Office of Scientific Integrity Review (housed in Office of the Assistant Secretary for Health), merged in May 1992 (ORI, n.d.-b).

Research misconduct became a public issue in 1981 during Congressional hearings that were held in response to some high-profile cases of unethical research (ORI, n.d.-b), most notoriously that of John R. Darsee, who published articles using fraudulent data and assigned false authorship credit (Jones, 2003). The hearings led to the passage of the Health Research Extension Act in 1985. This act required regulations regarding research fraud to be investigated and reported by federal government science agencies, which in turn led to: (1) the NIH publishing research guidelines in 1986 concerning funding, responsibilities, and conflicts of interest, codified in 1989 (ORI, n.d.-b) and (2) the NSF formally defining research misconduct in 1987 as "fabrication, falsification, plagiarism, or other serious deviation from accepted practices in proposing, carrying out, or reporting results from activities funded by NSF" (Boehm, 2012, p. 1). In 1989, the Public Health Service (PHS) published a similar definition, focused on "fabrication, falsification, and plagiarism" (Schechter & Schwartz, 1996). The National Academies of Sciences and Engineering, with the then-separate Institute of Medicine, and the Department of Health and Human Services' two subgroups—the ORI, and the Office of Science and Technology Policy (OSTP)—had each published definitions of research misconduct that, like all the previously published ones, stressed and defined the three tenants of fabrication, falsification, and plagiarism.

Over the past 30 years, the definitions of research misconduct have undergone some changes, but nothing drastic. The ORI published a definition in 2000 with the OSTP, then "revised" this definition in 2005 for the PHS after receiving numerous comments and complaints involving definitions of deception and falsification, as well as concerns about authorship and credit disputes. However, their 2005 revision is (as PHS acknowledges) a near carbon-copy of the original FFP definition. In fact, few of the definitions of scientific misconduct include anything beyond the three taboos of FFP. The DHHS, in their report "Public Health Service Policies on Research Misconduct," defines research misconduct as "fabrication, falsification, or plagiarism in proposing, performing, or reviewing research, or in reporting research results" (2005, p. 28386). The report defines fabrication as reporting made-up data or results; falsification as manipulating research "materials, equipment, or processes" or "changing or omitting data or results... such that the research is not represented in the research record" (also see Chapter 3); and plagiarism as the "appropriation of another person's ideas, processes, results, or words without giving appropriate credit" (misconduct does not include "honest errors" or "differences of opinion" [U.S. DHHS, 2005, p. 28386]). To summarize, as Francis Macrina succinctly puts it: "in science as in life, it is wrong to lie, cheat, or steal" (2014, p. 11).

Honest Disagreements?

Throughout the past decades and across all the federal agencies, definitions of research misconduct included at their core the three root words "fabrication, falsification, plagiarism." However, in 1995, the Commission on Research Integrity, also known as the Ryan Commission, was formed to make recommendations to the Secretary of Health and Human Services, the House Committee on Commerce, and the Senate Committee on Labor and Human Resources, about the definition of research misconduct and to address incidences of retaliation against whistleblowers who had reported misconduct (ORI, n.d.-b). The Ryan Commission recommended a new definition that went well beyond FFP: they ultimately defined research misconduct as "significant misbehavior that improperly appropriates the intellectual property or contributions of others, that intentionally impedes the progress of research, or that risks corrupting the scientific record or compromising the integrity of scientific practices" (1995, p. 15). This more socially broad and ethically aware definition by the Ryan Commission also includes examples of misconduct in the categories of misappropriation (breach of confidentiality, as well as plagiarism), interference (disrupting the work of another scientist), and misrepresentation (lying or withholding information, i.e., the more "interpersonal" side of data fabrication)—all dealing with authorship ethics and going further than FFP.

The Ryan Commission, then, recommended a definition of research misconduct that put two key authorship issues, intellectual property and intellectual contribution, at the forefront of the definition. Clearly, the Commission felt that authorship ethics were an integral aspect of scientific research misconduct. Despite this, their definition was *not* accepted in full by the then-Secretary of Health and Human Services, Donna Shalala. Instead, she implemented the FFP definition of research misconduct proposed by the National Science and Technology Council (ORI, n.d.-b). Even after the Ryan Commission's recommendations, the

DHHS's new Federal Research Misconduct Policy, published in December 2000, again defined research misconduct only as FFP (Ferguson, 2000). In fact, the Policy acknowledges the omission in its FAQs: "Are authorship disputes covered by this policy? Authorship disputes are not covered by this policy unless they involve plagiarism" (Ferguson, 2000).

Thus, governmental regulation and funding agencies charged with overseeing scientific ethics focused mainly on the data-gathering side of research misconduct (fabrication and falsification) and chose only to regulate and define one aspect of the authorship side of research (plagiarism), thus leaving out a multitude of other, nuanced, authorship ethics (see Tables 2.1 and 2.2). The ORI may have had internal questions about their narrow definition: they put out a call in 2002 for a survey to better understand scientists' research behaviors that might fall outside of the ORI's definition of research misconduct. However, two major scientific societies, the Federation of American Societies for Experimental Biology and the Association of American Medical Colleges, objected to the call for the survey (Martinson, Anderson, & de Vries, 2005). Filling in this research gap after the DHHS decided not to move ahead with their survey, Martinson, Anderson, and de Vries conducted a seminal survey in 2005 that was published in the interdisciplinary journal *Nature*. Martinson et al. (2005) asked scientists to self-report incidences of research misconduct (see Table 2.1).

The original table by Martinson et al. (2005) depicted the "top ten" sanctionable incidents of misconduct self-reported by entry-level and mid-career scientists. What emerged was that many of these incidents of scientific misconduct were authorship ethics—some of the very ones ORI choses to leave out in FFP. In their modification and reordering of the Martinson et al. table, Mogull and Katz (2012) reveal that in an expanded definition of authorship, most—if not all—of the 16 violations were concerned with writing and communication ethics, reordered here by percentages. It also contains the six misbehaviors *not* considered sanctionable by ethics compliance officers. These include not only authorship issues like "bad record keeping," which has the *highest* incidence of self-reportage, but also "assigning author credit" at 10%. Martinson et al. (2005) conclude: "Our findings suggest that U.S. scientists engage in a range of behaviors extending far beyond falsification, fabrication, and plagiarism" (p. 737). ORI's definition had been updated in 2005—theoretically in time to have read Martinson, Anderson, and de Vries' report—but ORI *did not* expand authorship beyond plagiarism in the three long-standing terms FFP. In fact, the ORI's new definition not only repeated previous terms, but expressly excluded authorship and credit disputes (U.S. DHHS, 2005). The ORI's actions seem clear: it chose not to regulate or define ethical issues concerning authorship, other than plagiarism, as being an essential part of scientific research.

Table 2.1 Research vs. Authorship Disclosures (Mogull & Katz, 2012). Originally Studied by Martinson, Anderson, and de Vries (2005)

Top 16 Behaviors Reported by Scientists (Early and Mid-Career)	*Percentage of Scientists Admitting to Engaging in Behavior in Previous 3 Years*	*Authorship Ethics Issue*	*Research Ethics Issue*
"Inadequate record keeping related to research projects"	27.5	✔	
"Changing the design, methodology, or results of a study in response to pressure from a funding source"	15.5	✔	✔
"Dropping observations or data points from analyses based on a gut feeling that they were inaccurate"	15.3	✔	✔
"Using inadequate or inappropriate research designs"	13.5		✔
"Overlooking others' use of flawed data or questionable interpretation of data"	12.5	✔	✔
"Withholding details of methodology or results in papers or proposals"	10.8	✔	
"Inappropriately assigning authorship credit"	10.0	✔	
"Circumventing certain minor aspects of human-subject requirement"	7.6		✔
"Failing to present data that contradict one's own previous research"	6.0	✔	
"Publishing the same data or results in two or more publications"	4.7	✔	
"Unauthorized use of confidential information in connection with one's own research"	1.7	✔	
"Using another's ideas without obtaining permission or giving due credit"	1.4	✔	
"Relationships with students, research subjects, or clients that may be interpreted as questionable"	1.4	✔	✔
"Not properly disclosing involvement in firms whose products are based on one's own research"	0.3	✔	
"Ignoring major aspects of human-subject requirements"	0.3		✔
"Falsifying or 'cooking' research data"	0.3	✔	✔

The Power and Influence of the ORI

Officially, the Office of Research Integrity "oversees and directs Public Health Service (PHS) research integrity activities" (ORI, n.d.-a). The PHS is comprised of 10 agencies, including the National Institutes of Health, the Food and Drug Administration, and the Centers for Disease Control and Prevention (ORI, n.d.-a). PHS provides billions of dollars in funding to support research at federal and non-government agencies, e.g., universities, hospitals, research institutions (ORI, n.d.-a). As the overseer of the integrity of research for the institutions PHS funds, ORI performs a variety of functions that affect a wide range of domains. In brief, the ORI: develops the policies and procedures to detect, investigate, and prevent research misconduct; implements programs to teach responsible conduct of research; reviews and monitors institutions' misconduct evaluations; evaluates policies and procedures of health and human services programs; and prevents retaliation against whistleblowers. It also recommends findings and actions when misconduct cases are referred to the Secretary of Health (ORI, n.d.-a).

The Strong Arm of the ORI

Clearly, ORI has many advisory roles, educational functions, and oversight responsibilities. But the ORI, through the DHHS, also exercises extensive punitive power for research misconduct in sciences funded by the Public Health Service (PHS). When an allegation of misconduct takes place at an institution receiving PHS funds, the ORI must be notified before the investigation begins, monitors the investigation, and is notified of the final results. The ORI then conducts its own independent investigation and if misconduct is confirmed, makes a recommendation for punishment to the DHHS and PHS (ORI, 2015b; ORI, 2016) and enforces it. Sometimes an incident is forwarded to the ORI directly, and the ORI determines if it warrants investigation by an institution (ORI, 2016). The ORI also can become more immediately involved in an investigation if there are circumstances such as a possible health hazard to humans or animals (ORI, 2015b). But most of the time ORI oversight is initiated and reactive.

ORI's real power is in its adjudication and sanctions. On the recommendation of the ORI, the punishments issued by the DHHS and PHS can be quite severe, including most notably "debarment from eligibility to receive Federal funds for grants and contracts, ... imposition of supervision on the respondent by the institution, ... [or] submission of a retraction of published articles by respondent" (ORI, n.d.-c). Here we see the real power of the ORI to reach across the entire federal government (to any and all funding agencies), reach individual institutions, and reach into any specified journal. To put this regulation into real-life application, the ORI has historically investigated and rendered rulings on about 12–16 cases each year (ORI, 2015a). These cases illustrate the real power the ORI possesses, as well as suggests intended or unintended consequences of its rulings, which in turn reveal the even greater potential of its power.

For example, Adam C. Savine, a former doctoral student in psychology at Washington University–St. Louis, was found guilty of falsifying data in three publications and six conference abstracts (National Institutes of Health, 2013). He entered a voluntary settlement for three years, in which he: (1) must have any research funded by the PHS supervised (and if he applies for any PHS funding, he has to include a supervision plan); (2) must have any institution that employs him and uses PHS funding certify to ORI that the research data are correct and based on legitimate experiments and are accurately reported; (3) cannot advise, serve on committees, boards, or other related oversight for PHS; and (4) have the senior authors request a retraction of his three articles (two from 2012, and one from 2010) (National Institutes of Health, 2013). This punishment is in line with most of other cases, although some are more lenient (requiring corrections instead of full retractions of papers, or have shorter punitive timeframes), while a few are more stringent, like in the case of Li Chen, who was found guilty of "recklessly" falsifying and fabricating data on four publications, one manuscript, and four grant applications, and without admitting fault, was disbarred from working with *any* U.S. government agency for three years (ORI, 2015c).

Punishment also is meted out for those found guilty of plagiarism, as in the 2013 case of Pratima Karnik, who received the same punishment as Mr. Savine but for two years instead of three, and without article retraction. It is perhaps worth noting, though, that hers is the only case from 2013 or 2014 regarding plagiarism as the *only* offense (ORI, 2015a). From 1992–2005, ORI found 162 instances of scientific misconduct, 19 of which were plagiarism cases, and 10 of which resulted in disbarment (Price, 2006). Price (2006) adds that "almost all of the 10 plagiarists debarred by ORI/PHS from federal funding also falsified and/or fabricated research material, thereby compounding the seriousness of their plagiarism" (p. 46). Only eight of the 162 cases were for plagiarism alone, implying that plagiarism in ORI cases is usually committed by scientists who also were willing to perform "research-related" acts of misconduct in data-falsification (cf. Masic, 2012).

The Supposed, Potential, and Limited Power of the ORI

Beyond the fact that scientists can be as ethically frail as the rest of us humans, what these cases illustrate well is how far the ORI's power stretches in all PHS cases where it is officially mandated to monitor and adjudicate incidents of scientific misconduct when an institution has finished its investigation. We also see here the strength of its punitive powers, the kind of moderate-to-severe punishments the ORI recommends to and enforces for the DHHS and PHS. We can easily understand how being banned from receiving funds, requiring oversight, and not being allowed to receive any U.S. funds for three years could in the short term affect the careers of anyone found guilty of FF or P by the ORI. There are even more heinous cases where violations of FFP actually set back research in a field, like the fabrication of stem cell data in AIDS research (Beardsley, 2006), where retractions of the unethical act, while necessary, will not mitigate the harm done to research, patients, and/or the research field.

But these cases also point to some important questions about potential future effects of the ORI's power. The ORI's case rulings online note only the punishment, but a closer reading of the rulings reveals other rhetorical clues: "Mr. Adam C. Savine, *former* doctoral student," and "Dr. Li Chen, *former* Postdoctoral Fellow" (National Institutes of Health, 2013; ORI, 2015c; emphasis ours). While the ORI ruling in each case of misconduct appears cut and dry, less obvious may be the likely impact of the ORI findings and the subsequent judgment and punishment on the arc of an individual's career. We need not speculate to observe, for example, that Mr. Savine was not allowed to graduate from his doctoral program (Bernhard, 2013) and that Dr. Chen was dismissed from Mount Sinai School of Medicine (Grant, 2014).

A crucial dimension of the ORI's power is "the range of motion" of ORI's reach, the long-term effects of its forensic judgments that extend far beyond the documented rulings and punishments. Findings of research misconduct by the ORI can result in scientists losing funds, equality with their peers, seniority, and respect within their universities; scientists also stand to lose the trust of society and the public as well (Committee on Assessing Integrity, 2002, p. 8). And even more, scientists stand to lose the one thing needed to practice science at all: their credibility in their field (Committee on Science, Engineering and Public Policy, 2009; Hoshiko, 1991; Macrina, 2014; Penrose & Katz, 2010).

However, this discussion of the ORI also begins to expose some of ORI's porous borders, if not the limitations of its grasp. Certainly, the ORI's power is limited—by Congressional statute, as an agency of the DHHS; by its self-definition; by its work with other agencies, organizations, and journals, discussed more below; and by the limitations of "ethics" itself (to stop bad acts *a priori*, given free will, or the inability to repair damage *a posteriori*).

By statute, the investigative authority of ORI is not legally activated until a case or a request is sent to them from other institutions, and then only those that fall under the purview of PHS (ORI, 2016). Further, "[t]he ORI has no direct involvement in the decision-making by an institution," whatever the ORI's separate finding. And, it is the DHHS "that takes final administrative actions against a respondent as well as the entity that oversees the appeal process through the HHS Departmental Appeals Board" (ORI, 2016). One little-noticed attribute and irony of the true power of ORI is its legal muscle to set the parameters of its ethical focus and authority. This includes the decision whether to deal with authorship ethics. The ORI delimits its own power by its restricted definitions of scientific misconduct that exclude scientific authorship ethics other than plagiarism. Certainly, plagiarism, which remains squarely in ORI's realm, is a problem, and has become more complicated—and thus further delimited—by the emergence of the internet, open-access journals and software, public cultural commons, and the crowdsourcing of peer review. But ORI's power to decide to focus only on plagiarism within the larger set of authorship ethics provides a window not only into the working of that power, or its view of authorship, but also its relationship to other scientific organizations and entities. Why might the ORI limit its power?

Perhaps simple practicality is one justification for the ORI dismissing most authorship issues from consideration. In her article published in the *Croatian Medical Journal* (1999), Mary Scheetz of the ORI's Division of Policy and Education suggests that the messy myriad of issues involved in authorship ethics may be too cumbersome and difficult for the ORI to fully evaluate. As we will discuss further below, it appears that ORI leaves the articulation and implementation of scientific authorship ethics to journals, which on its face makes sense, but provides other insights into how ORI works in the broader ethical context.

Mere "Authorship Disputes"?

There is more to the ORI's decision than practicality, however. Despite Martinson et al. (2005) finding that authorship ethics violations were numerous and self-reported more than research ethics violations, Scheetz (1999) argues that the ORI and the PHS cannot be directly involved in authorship disputes because so many of them are *not related to scientific misconduct*. That is, per ORI's FFP definition, authorship disputes are not a part of *research*, and *research is not authorship*. Plagiarism is the only authorship issue ORI will consider. As Scheetz wrote: "ORI receives many allegations that do not meet the Public Health Service definition of scientific misconduct.... [P]lagiarism allegations...are later determined to be authorship disputes" (1999, p. 323).

The separation of plagiarism from authorship disputes, and its inclusion as a scientific research misconduct, may puzzle us. However, Scheetz states that there may be several reasons for our confusion: "ORI's interpretation of plagiarism under the Public Health Service definition of scientific misconduct has a narrower scope than the term plagiarism as used more casually in the non-regulatory context" (1999, p. 323).

Authorship issues that may fall out of this narrower definition of plagiarism include: (a) "authorship or credit disputes between collaborators or former collaborators"; (b) "misappropriation of collaborators' ideas"; (c) "disagreements over who should be an author, or the order of authorship"; (d) "questions of whether consent must be obtained in order for a collaborator to publish independently from his or her research team"; and (e) "whether a member of a research team can publish conflicting analyses" (Scheetz, 1999, p. 324).

However, beyond plagiarism, no matter how it is defined or delineated, the question of authorship ethics—and of authorship itself—reveals another, deeper, philosophical issue. We suggest that the question is not only the potential plethora of authorship problems, but also an epistemological question of the relationship of authorship to research in science. Perhaps authorship issues and disputes are *not* related to scientific misconduct because authorship is not related enough to research, data gathering, or knowledge-making in science for Scheetz and the ORI. Here we see the relevance of the old debate stretching back to Plato, in which writing is not part of the discovery or creation of knowledge, but the mere conveyance of it. For Scheetz, plagiarism *qua* scientific misconduct includes "both the theft or misappropriation of intellectual property and the substantial unattributed textual copying of another's work. It does not include authorship or credit disputes" (Scheetz, 1999, p. 324).

But if authorship *is* related to scientific research, how are publication issues, interpersonal relations, other forms of communication, and writing itself related to scientific research? In this more rhetorical worldview, publication, the process of interpersonal communication, and writing (with all its "human problems") are not only the basis of the dissemination but the validation of knowledge as a social construction (Kuhn, 2012). The very act of writing itself is an "epistemic" activity that is the basis of science and of knowing: planning, formulating, calculating, proposing, producing, recording, reporting, disseminating, retesting, arguing, validating, accepting claims—are all central to scientific research as a human, social enterprise (cf. Latour & Woolgar, 1986). In excluding this expanded notion of authorship in scientific research, excepting plagiarism, the ORI's decision casts authorship only in its *negative* relationship to research and knowledge—as plagiarism.

Taking the other, rhetorical view of the centrality of authorship in the making as well as communicating of science itself, we begin to see the ripple of epistemological power in ORI's decision to focus solely on plagiarism as a singular, final criterion of research misconduct in science. Excluded forms of communication (see Table 2.1) and non-plagiaristic author ethics such as the responsibility of co-authors to develop knowledge and share with each other are how science—and scientific research—are done. The ORI does not consider such authorship ethics in its purview; it thus seems to downplay the role of authorship in science. Yet the ORI still in different ways (re)configures the wider ethical landscape in which scientists as authors work. It is to this wider landscape that we now turn.

Beyond the ORI

One Institution among Many...?

The ORI, a government agency in the DHHS whose cabinet secretary reports directly to the president of the United States, is a powerful regulatory institution for those receiving funds from its granting agencies in PHS, including the influential NIH. That the ORI is a powerful institution needs no further proof. But its relationships to other organizations and entities in science, such as universities, associations, and journals, and the effect on authors are not as obvious. Let's start with other organizations in science.

According to its self-description (ORI, 2016), the ORI first determines that institutions such as universities or other research facilities have regulations in place to set "local" policy, monitor the ethical conduct of research, and investigate alleged misconduct. In organizations such as universities, for example, where research is funded by PHS-related organizations, these criteria are met via the establishment of an Institutional Review Board (IRB). At each research organization that falls under ORI's purview, IRB offices constantly develop ethical modules and materials for researchers *in all disciplines* (including graduate students) involved with the study and/or use of human or animal subjects, and monitor them through the submission and approval of written protocols and consent forms.

These IRB offices thus enforce federal principles and ethical thresholds of violations based on the moral standards and codes first articulated in 1979 in *The Belmont Report* (National Commission). Beginning with the horrific findings of medical experiments revealed in the Nuremberg trials, the ethical scope and detail of *The Belmont Report* generally includes but extends well beyond the FFP strictures of ORI. While *The Belmont Report* focuses primarily on biological research (and medical research at that), many of the principles and discussion consist of or include communication issues, such as the interaction between the researcher and subject(s), the written protocols and consent forms that will govern experiments and how subjects are handled, and the privacy/publication of data that could cause harm to research subjects. Researchers must adhere to and/or update their protocols as necessary throughout a research project, from planning to publication (National Commission, 1979). Not too far off in the distance, ORI has the responsibility and power to monitor these research activities at institutions receiving PHS funding, and as we have seen, to intervene when requested by the institution and/or deemed absolutely necessary by ORI itself.

In fact, there is enormous incentive—ethical, legal, and financial—on these institutions to comply with federal regulations, and thus the ORI. IRB offices that serve entire universities are under enormous pressure. Researchers at scientific organizations that repeatedly are found to have committed scientific misconduct may find the effectiveness of their IRBs, their educational and regulatory measures, and even their PHS funding, called into question by the ORI (ORI, 2016). Other researchers at these institutions too will discover it is more difficult to receive federal funding. Like the reputation of individual researchers, the reputations of scientific organizations matter. There have even been calls for institutions to be held fiscally accountable for their researchers based on accusations of self-interest or financial gain (Glanz & Armendariz, 2017). However, efforts to monitor institutions are not without controversies about academic freedom, the self-regulation of scientists, and fiduciary issues based on the way funds are immediately dispersed throughout the institution, labs, and research teams (Schneider, 2015).

In describing the interrelation of the ORI with scientific institutions receiving federal funding from agencies under DHHS, we begin to see that ORI's overt power, self-stopped along the borders of other organizations, in some ways increases its influence over those organizations. At the policy level, the ORI (if sporadically) defines research misconduct (based on FFP) and provides guidelines for organizations to follow. However, the responsibility to develop efficient, well-organized, and well-run IRB offices, and carefully monitor the ethical dimensions of all research involving animal or human subjects, is placed squarely on the shoulders of the institution (Scheetz, 1999).

There are other ways ORI exercises a *passive* power as an observer. The ORI does not have a say in which proposals receive funding based on their merits, the way funding agencies do. But as a federal agency with oversight power, the ORI still sits as an ethical "gatekeeper" over research laboratories and universities, passively determining by regulation the outer perimeters of the FFP policies of organizations under its jurisdiction. The organizations, in turn, must oversee and regulate all their member scientists and researchers. The ORI, then, usually *does not* have to deal with the actual rules and procedures for implementing its FFP policies at each organization—the "details" of developing regulations and deadlines for educating its faculty, monitoring ethical compliance of individual researchers, carrying out investigations, and handling all the paperwork have been delegated to the institutions.

ORI's full oversight responsibilities and legal and punitive powers are only invoked when someone is found guilty of scientific misconduct by the IRB, and/or the infraction is serious enough to warrant further investigation by the ORI. It is at this stage that the *active* power of the ORI to prosecute and judge ethical breaches is initiated. We have seen that the ORI's docket of work is rather full. In addition, there are many more ethical infractions in science than most people think, and investigations can be quite involved and lengthy. As we've seen, the ORI's adjudication and penal decisions probably profoundly affect the personal and professional lives of specific scientists found guilty of infractions. Thus, the ORI's passive and active powers ripple through the conglomeration of IRB offices, training, rules, regulations, and other authorship activities at many specific institutions. As Foucault (1972) argued, all organizations and institutions themselves begin in writing—begin in the statutes and legal codes that create them, and the written laws govern those agencies, *ad infinitum*.

We turn now to a key entity in scientific communication, and the authorship landscape: research journals, the organizations that support them, and their relationship to the ORI and each other.

Journals as Ethical Entities

We have seen that ORI's definition of FFP limits the focus on authorship ethics to plagiarism. By being "forced" to create their own guidelines for authorship and ethics, journals and the scientific societies that produce them can be understood to fill the oversight void left by the ORI. Interestingly, the ethical as well as practical power of journals as the entities of organizations was originally derived from the real concern of protecting scientific authors (from each other) by publication, such as with the Royal Society of London, which awarded credit to papers that arrive at the journal first (Penrose & Katz, 2010, p. 10).

Authorship and publication issues that fall under the purview of journals include discussions of defining the author(s), credit allocation, best practices for author order, self-plagiarism (also known as redundant publication), cultural differences in authorship decisions, the role of journal editors, the selection and ethics of peer reviewers, etc. (Anderson, Kot, Shaw, Lepkowski, & De Vries, 2011; Jones, 2003; Louis et al., 2008; Pearson, 2006; Penrose & Katz, 2010). Of all these issues, the allocation of credit appears in almost all discussions regarding publication ethics. This makes sense considering the importance that credit plays in furthering science. Credit is necessary for both moving publication forward and for securing new research grants, on which an individual scientist's or a lab's future projects depend. Credit is also necessary for tenure, promotion, and professional advancement. These discussions help reveal the idea that publication ethics are more than "forms of research misconduct that can undermine the scientific literature" (Wager, Fiack, Graf, Robinson, & Rowlands, 2009, p. 348). Publication ethics, like authorship itself, therefore play a much more central role in the creation of scientific knowledge, as well as its later diffusion, than the ORI decision concerning authorship ethics as plagiarism might lead one to think.

During the early 1980s and through the decades that followed, federal oversight boards continued to publish definitions of research misconduct that did not include authorship issues beyond plagiarism. Seemingly to fill the vacuum, the editors of scientific and medical journals published their own rules concerning such ethics. In 1985, several journal editors began issuing statements and guidelines regarding authorship, including the International Committee of Medical Journal Editors (ICMJE). The ICMJE is a group of editors of various international medical journals and their related associations who meet solely for the purpose of establishing general publishing guidelines for all of their journals (ICMJE, 2014). Their published, formal authorship guideline is arguably the seminal one in science publication, having since been adopted internationally by "hundreds of biomedical journals," by U.S. medical schools and scientific societies, and even by two countries (Jones, 2003, p. 244).

ICMJE's "Recommendations for the Conduct, Reporting, Editing, and Publication of Scholarly Work in Medical Journals" (2015) contains sections specifically for the role of authors as well as for publication issues. This includes definitions of the roles of authors and contributors, as well as author responsibilities (in issues like conflicts of interest). In fact, "scientific misconduct" is only one section amidst extensive rules and discussions of scientific communication, and authorship issues specifically. Since no federal policy exists to tell scientists how to ethically allocate credit, the ICMJE's guidelines serve as one of the only—if not the only—set of rules regarding authorship that applies equally to various scientific disciplines and journals. Scientists can reference these recommendations to attempt to navigate the complicated waters of authorship and credit when they are working toward a publication in a biomedical field.

However, the guidelines published by the ICMJE certainly have not settled the issue of authorship disputes or put to rest other authorship-related ethical issues in science. In fact, a study published in *Academic Medicine* showed that the chairs of departments of medicine did not change their authorship habits from 1979–1990, despite publications of guidelines by the ICMJE and many journals (Shulkin, Goin, & Rennie, 1993, cited in Jones, 2003). While we have no way of knowing which behaviors didn't change, we can safely assume they include at least some of the incidences of misconduct exposed by Martinson et al. (2005). If this is the case, it follows that journals do *not* completely control the process of how authorship is determined—or do not do so following ICMJE's guidelines—and/or do not adequately educate their authors. One possible explanation for this omission that goes to our primary point is that even editors are not all fully invested in the idea of authorship ethics. In interviewing 231 editors of science and health care journals, Wager et al. (2009) found that editors are not very concerned about problems in publication ethics, as they believe that misconduct rarely occurs—*despite* the publication of the Martinson et al. (2005) survey four years before.

Ironically, and perhaps of great epistemological and ideological significance, it may be that scientific journal editors buy into the lesser role of authorship ethics that ORI also has carved out for them in the narrower confines of plagiarism. In looking at ethical organizations such as IRBs, entities such as research journals, and their interrelationship to the ORI and each other, we can begin to see the intersection of ethics in a landscape that in a limited way resembles Actor-Network Theory (see Latour, 2005, 2013)—but is applied here to elucidate ethics in scientific communication and authorship. We see this interactive landscape as constituted by "lines and fields of ethical force." Through their presence, agency, and continuous actions, the ORI and other organizations/entities together create powerful and sometimes intersecting or conflicting lines and fields of ethical force in which science—and authorship ethics—must be enacted, communicated, and legitimized.

Writing Within the Lines

Arguably, due to its overarching range and ability to ban a scientist from research, the ORI serves as the governing body with the most power to ethically influence scientists funded under the DHHS. However, as we have pointed out, scientists' actions are regulated by different organizations and entities whose powers converge in different ways, one of which is captured in Figure 2.1 (Linvill, 2012), with the ORI serving as perhaps the most omnipresent, if normally distant, influence. In fact, we believe that the perceived and real power of the ORI is amplified and echoed within this network of influencers. Yet each organization or entity can be understood to extend its own lines of power into scientists' career and work. Together, these sources and others (not shown or unknown) create fields of ethical force(s) in which scientists, organizations, and entities are embedded or emerge, exist, and move.

Figure 2.1 illustrates how ORI oversees (by administering investigations and prescribing punishment) the funding sources that undergird other institutions. The shared areas imply how, as the nation's largest oversight body, ORI's rules influence and directly or indirectly ("symbolically") affect and impinge on all aspects of a scientist's career, even when punishment is not a threat. Figure 2.1 also shows how the regulators of scientific entities most related to authorship—journals and the professional societies that produce them—fill the ethical space left by the ORI's commitment to FFP, and yet are *not* within the realm of the ORI's direct and/or actual influence except when the ORI demands a retraction.

Linvill (2012) found that scientists look to journals when making authorship decisions. The journals may not have the power to regulate authorship via the type of sanctions that ORI can impose, but journals and their editors do have the advantage of working directly with

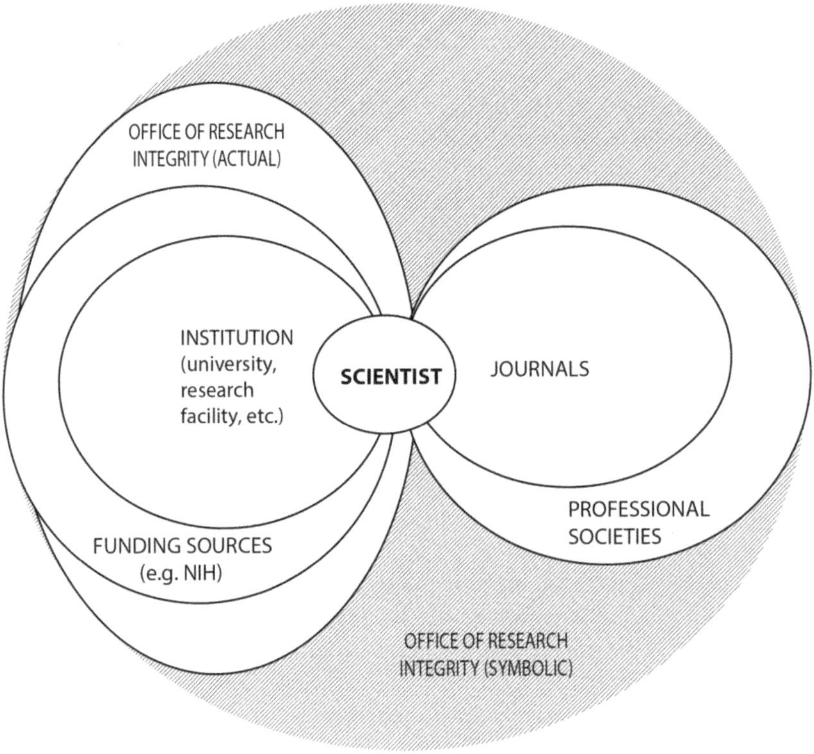

Figure 2.1 Lines and Fields of Ethical Force on Scientists (Linvill, 2012).

scientists and controlling what gets published. In addition, as scientists are immediately concerned with how to get published in a journal, they will often read and follow that journal's rules and instructions for authors. It would seem that the ability to write regulations, enforce codes, educate scientists about the process, and actually work with scientists as they make decisions as authors rests in the hands of journals and editors, rather than in a large organization such as the ORI. Yet authorship decisions, although important, can be scattered in the field among different resources and influencers—along different vectors of the lines and fields of ethical force.

We also might surmise that the ethics of the scientists themselves, as individuals with unique histories, are of prime if not primary importance (hence they are the central focus of Figure 2.1), even as they are intersected and interact with/in other ethical fields scientists are embedded in. Indeed, scientists work in professions and environments where their reputation is of the utmost importance and is also quite personal to them. But in Figure 2.1, we see that the ethics of individual scientists *are* touched and affected both by large governmental agencies like ORI, as well as other scientific organizations/entities. However, we see that the other scientific organizations/entities are in closer proximity to individual scientists, and thus in fuller force than the more vast and distant ORI. These organizations/entities also might appear to be self-regulating as well as single lines of force, until one sees them in the context of the *entire* ethical field, as in Figure 2.1.

In a previous section, we noted that the ORI has the power to choose what it will and will not investigate after the verdict of the institution; to decide where its power and influence will end, as provided by the federal law; and to punish offenders in ways that are highly significant and painful to scientists. But other organizations and entities have various powers to regulate and punish too. We can only begin to tease out some of these powers (and consequences) here. First is the power of journals to act as gatekeepers through managing peer review and publication; journals may reject manuscripts, require revision, and even black-ball authors. More foundationally, we see the power of funding agencies that also perform gatekeeping roles through peer review, ultimately controlling the purse strings of scientists and labs, which in turn directly influences and sets the agenda of the particular field(s) (Penrose & Katz, 2010, p. 175). Thus, the differential of organizations/entities when compared to the ORI may be *asymmetrical*. But other organizations and entities wield considerable power over scientists, and authorship ethics, as well.

One point that can be made here is that the power (and legitimacy) of the ORI, granted to it by law, does not occur in isolation, but also in relation to and tension with other ethical organizations and entities with which it interacts. Likewise, the other organizations and entities, particularly IRBs but also journals, have powers that are magnified by the existence and actions of the ORI. Thus, Figure 2.1 is the beginning of a mapping of power dynamics and differentials of ethics in science. In this interplay of lines and fields of ethical force, power rests not only with the ORI but with all the other organizations and entities as well. No matter the source, that power is amplified and redistributed in a constant interplay that begins to belie any one-dimensional notion of ethics as a choice between right and wrong. What is being amplified and redistributed, however, is not only power for power's sake, but power to accomplish work—the work of science, including authorship, even when the latter is under-recognized and its ethics may be obscured by the ORI's definitions. But authorship and its ethics are essential to the whole modern scientific enterprise. And those ethics are often opposed.

Tensions Among Ethical Fields of Force

What we begin to see in the ethical relationship of organizations and entities is both the outer limits of the ORI's influence, regulated by legislation and self-definition, and how this influence extends well beyond where it legally ends into the lines of force exerted by other institutions that control distribution and legitimacy in science. Furthermore, taken together, these lines of force are not only in relation but also in tension with each other, creating shifting shapes of influence at the local and individual levels, and beyond. What constitutes the ethical lines and fields of force, therefore, may be understood not only as spheres of power, but as tensions among ethics themselves. Table 2.2 illustrates the way authorship ethics (which includes scientific communication and publication ethics) can, in a limited two-dimensional way, be understood to be paired and in continuous conflict.

In fact, when we regard authorship and even writing in science as epistemic—as a way of discovering or making knowledge, rather than merely conveying it—another characteristic may be a non-bifurcation of moral conduct into the mutually exclusive categories of "true or false," "accurate or inaccurate," or "good or bad" ("bad" meaning *morally* bad behavior, not honestly inept, null, or erroneous research, following the distinction made by DHHS [2005, p. 28386]). Rather than mutually exclusive categories of ethics, ethical conflicts may appear to manifest

Table 2.2 Scientific Authorship: Sample Ethics in Continuous Tension

Authorship Ethic	*In Tension with*	*Authorship Ethic*
Collaboration	↔	Competition
Sharing	↔	Drive for originality
Fulfilling responsibilities	↔	Over-commitment
Order of authors	↔	Personal career/ambition
First author	↔	Et al.
Plagiarism	↔	Cultural Commons
Plagiarism	↔	Cultural norms
Proprietary	↔	Open source
Open access	↔	Profits
Ownership of information	↔	Freedom of speech
Sharing of data	↔	Need for secrecy
Social importance of science	↔	Period of research privacy
Objectivity in peer review	↔	Natural influence of one's own research
Reporting errata	↔	Ethos of journal
Institutional investigation	↔	Institutional reputation/ federal funding
Peer review as gatekeeping	↔	Communication technologies' speed
Peer review as gatekeeping	↔	Crowdsourcing involving audience
Maintaining professional ethos	↔	Public's right/need to know
Formal science	↔	Public access
Publicly funded science	↔	Privately funded science

themselves or be understood as continuous tensions between opposing but *equally accepted and even central* sets of conventions or values in authorship ethics (e.g., sharing *and* competition, collaboration *and* credit).

In the equally linear terminology of classical rhetoric (Aristotle, 2007), what Table 2.2 shows are *not* sets of contradictory values (being vs. not being), but rather sets of contraries (being and nothingness)—opposites that are understood to exist in the same class of things (the national security need for secrecy *and* the public's right to know). According to the *topoi* of "contradiction," proving one of the choices correct means that the other choice is wrong (a thing cannot in the same time or place be and not be). But in the *topoi* of "contraries," proving one of the choices is correct does not make the other choice wrong, since there may be other options in the same class (see Corbett, 1990, pp. 116–119).

Seeing authorship ethics as *tensions* within the lines and fields of ethical force rather than as distinct categories—as contraries rather than contradictions—may begin to move us away from the more limited notion of ethics as falsification, fabrication, and plagiarism, away from what might be regarded as a "false" epistemology (good/bad, accurate/inaccurate) upon which FFP is currently based. Instead of distinct categories (collaboration vs. competition), imagine these ethical lines of tension playing out in fields of three or more dimensions. The world of scientific authorship is a complex place where ethics turn into and out of themselves, forever oscillating. Authorship ethics by their very nature may present epistemological problems of indeterminacy and power that require individual and social recognition, if not regulation. Exploration and discussion will probably show that ethics are not so easily divided or classified, but rather are in both the abstract and in action, always immanent and emerging within fields of force(s), to use posthumanistic language (Braidotti, 2013). The tension between two or more "right" choices, while not exclusive to science, is possibly a dimension of ethics in scientific authorship that makes the study of them important. The values of scientific authorship may be central to the discipline of science as a profession. These ethical tensions thus can be seen to be integral to scientific writing *and research* when the practice of science itself is understood as a rhetorical and social endeavor, and scientific writing understood as a professional act.

Conclusions: The Implications for ORI as (Non) Locus

As an institution, the ORI has a symbolic as well as an actual (indirect and direct) role in regulating scientific research. ORI's responsibility is shared with other institutions in the Public Health Service (PHS), including 10 funding agencies (ORI, 2016). But the influence of the ORI, as an overseer of all PHS research, extends beyond these agencies to other scientific organizations such as universities and entities such as journals.

Earlier we concluded that the ORI was a magnifier of the power of other organizations and entities. We also concluded that the power of the ORI is very much dependent at least in part on the network of other institutions in which it is situated, through which it is constituted, and with which it interacts. The ORI would be a legal but seriously weakened organization without other ethical organizations and entities that complement it. Likewise, without the ORI, the power of the other organizations and entities in science would be seriously weakened: they would have nothing to point to in specific cases except ethical ideals, empirical facts, or reasoned dogma—no higher authority beyond science to which to appeal ethical problems.

It is in these relationships from which the lines and fields of ethical force emerge. In a sense, within this matrix, the ORI is decentered, and its power somewhat dispersed. So too are other organizations and entities, and even authors. However, because of the seeming adoption of FFP by nearly all organizations and entities in the network, the notion of author, and the role of authorship ethics, is diminished in science. Thus, in what we today would call "a belief in science," what ORI says (or doesn't say) about scientific authorship ethics not only establishes criteria for the validity of research, but also sets a standard for notions of authorship that ripple beyond science and throughout culture at large. By refraining from adding authorship ethics to its definition of research misconduct in science as FFP, the ORI continues the anti-rhetorical philosophy of Plato inherited by empirical science (cf. Bacon, trans. 1620). It separates the human act of authoring from the act of knowledge-making, which is presented (in publication) as an objective process in which language and authors play no significant part because they are supposed to remain detached.

An examination of tensions that exist in and emerge from the lines and fields of ethical force belies a bifurcated view of knowledge and authorship. In an *expanded* view of scientific authorship, one that would *include* falsification and fabrication (FF) as well as plagiarism (P), all science might be considered a process of authorship—of symbolization at every level of research, from "creating" and encoding substances in the lab, to organizing and arguing for the validity and significance of results to the field in papers and proposals and presentations. Even "raw data," isolated in the lab or found in the world, are already symbolized and processed by minds, experiments, senses, and machines (Latour & Woolgar, 1986). It is also through these processes that the status of other results and arguments constantly shift (Latour, 1987).

Conceived in relation to lines and fields of ethical forces constantly shifting in tension, the concept of writing, and thus of authorship ethics, emerge as their own force and are empowered. Authorship ethics, which might first appear as dualistic choices controlled by different organizations and entities—including the ORI—may be more like multidimensional planes where ethical institutions and other elements move in wrought and somewhat unstable planes. These planes, and the ethical tensions within them, reveal the higher function and value of scientific authorship. Sharing and competing, co-authoring and ordering, publishing and peer reviewing are not merely communication, but the activity of science itself.

Acknowledgment

Steve Katz wishes to thank Ann M. Penrose, his co-author of 3+ editions of *Writing in the Sciences: Exploring Conventions of Scientific Discourse,* whose thinking and many conversations were formative in the earlier development of the vision of ethics now articulated here.

References

Anderson, M., Kot, F. C., Shaw, M. A., Lepkowski, C. C., & De Vries, R. G. (2011). Authorship diplomacy. *American Scientist*, *99*, 204–207.

Aristotle. (2007). *On rhetoric: A theory of civic discourse* (2nd ed.). (G. A. Kennedy, Trans.). New York: Oxford University Press.

Bacon, F. (1620). *The new organon or true directions concerning the interpretation of nature.* Retrieved March 2, 2017, from www.constitution.org/bacon/nov_org.htm.

Beardsley, S. (2006, March 1). Down in flames: Can stem cell research recover from Woo Suk Hwang? *Scientific American.* Retrieved February 28, 2017, from www.sciam.com/article.cfm?id=down-in-flames.

Bernhard, B. (2013, March 7). Washington U student's mentor talks about discredited research. *St. Louis Post-Dispatch.* Retrieved March 2, 2017 from www.stltoday.com/lifestyles/health-med-fit/washington-u-student-s-mentor-talks-about-discredited-research/article_e2275d60-1ead-5906-851a-59c7a4daf6e5.html.

Boehm, A. F. (2012, December 1). *NSF OIG Form 34 (11/12) – (Replaces OIG 11/02).* Retrieved March 2, 2017, from www.nsf.gov/oig/_pdf/dearcolleague.pdf.

Braidotti, R. (2013). *The posthuman.* Cambridge: Polity Press.

Committee on Assessing Integrity in Research Environments. (2002). Integrity in scientific research: Creating an environment that promotes responsible conduct. [The National Academies Press OpenBook]. Retrieved March 2, 2017, from http://www.nap.edu/openbook.php?record_id=10430&page=R1.

Commission on Research Integrity, U.S. Department of Health and Human Services (CAIRE). (1995). *Integrity and misconduct in research.* Retrieved March 2, 2017, from https://ori.hhs.gov/sites/default/files/report_commission.pdf.

Committee on Science, Engineering and Public Policy, National Academy of Sciences, National Academy of Engineering, and Institute of Medicine of the National Academies. (2009). *On being a scientist: Responsible conduct in research* (3rd ed.). Washington, DC: National Academies Press. Retrieved March 9, 2017, from www.nap.edu/read/12192/chapter/1.

Corbett, E. P. J. (1990). *Classical rhetoric for the modern student* (2nd ed.). Oxford: Oxford University Press.

Ferguson, B. A. (2000, December 5). *Federal research misconduct policy.* Retrieved March 2, 2017, from http://ori.hhs.gov/federal-research-misconduct-policy.

Foucault, M. (1972). The discourse on language. In A. M. S. Smith (Trans.), *The archeology of knowledge and the discourse on language* (pp. 215–237). New York: Pantheon.

Glanz, J., & Armendariz, A. (2017, March 8). Years of ethics charges, but star cancer researcher gets a pass. *The New York Times*. Retrieved March 9, 2017, from www.nytimes.com/2017/03/08/science/cancer-carlo-croce.html?_r=0.

Grant, B. (2014, April 25). Gene therapy researcher faked data. *The Scientist*. Retrieved March 2, 2017, from www.the-scientist.com/?articles.view/articleNo/39825/title/Gene-Therapy-Researcher-Faked-Data/.

Hoshiko, T. (1991, October 28). Facing ethical dilemmas: Scientists must lead the charge. *The Scientist*. Retrieved March 2, 2017, from www.the-scientist.com/?articles.view/articleNo/12047/title/Facing-Ethical-Dilemmas—Scientists-Must-Lead-The-Charge/.

International Committee of Medical Journal Editors. (2014). *About ICMJE*. Retrieved March 2, 2017, from www.icmje.org/about-icmje/.

International Committee of Medical Journal Editors. (2015). *Recommendations for the conduct, reporting, editing, and publication of scholarly work in medical journals*. Retrieved March 2, 2017, from www.icmje.org/icmje-recommendations.pdf.

Jones, A. H. (2003). Can authorship policies help prevent scientific misconduct? What role for scientific societies? *Science and Engineering Ethics, 9*(2), 243–256. doi:10.1007/s11948-003-0011-3.

Katz, S. B. (2009). Biotechnology and global miscommunication with the public: Rhetorical assumptions, stylistic acts, ethical implications. In Grady & G. Hayhoe (Eds.), *Connecting people with technology: Issues in professional communication* (pp. 167–175). Amityville: Baywood.

Kuhn, T. (2012). *The structure of scientific revolutions* (4th ed.). Chicago, IL: University of Chicago Press.

Latour, B. (1987). *Science in action: How to follow scientists and engineers through society*. Cambridge: Harvard University Press.

Latour, B. (2005). *Reassembling the social: An introduction to actor-network theory*. Oxford: Oxford University Press.

Latour, B. (2013). *An inquiry into modes of existence: An anthropology of the moderns*. Cambridge, MA: Harvard University Press.

Latour, B., & Woolgar, S. (1986). *Laboratory life: The construction of scientific facts*. Princeton, NJ: Princeton University Press.

Linvill, C. C. (2012). *Do the ethics of authorship matter? A rhetorical and historical analysis of authorship ethics in science*. Presentation, Association of Teachers of Technical Writing, St. Louis, MO.

Louis, K. S., Holdsworth, J. M., Anderson, M. S., & Campbell, E. G. (2008). Everyday ethics in research: Translating authorship guidelines into practice in the bench sciences. *Journal of Higher Education, 79*(1), 88–112. Retrieved March 2, 2017, from www.jstor.org/stable/25144651.

Macrina, F. L. (2014). *Scientific integrity: Text and cases in responsible conduct of research* (4th ed.). Washington, DC: ASM Press.

Martinson, B. C., Anderson, M. S., & De Vries, R. (2005). Scientists behaving badly. *Nature, 435*(9), 737–738.

Masic, I. (2012). Plagiarism in scientific publishing. *Acta Informatica Medica, 20*(4), 208–213. doi:10.5455/aim.2012.20.208-213.

Mogull, S. A. & Katz, S. B. (2012). *Scientific communication ethics vs. ethics in science.* Presentation, Association of Teachers of Technical Writing, St. Louis, MO.

National Commission for the Protection of Human Subjects of Biomedical and Behavioral Research. (1979, April 18). *The Belmont report: Ethical principles and guidelines for the protection of human subjects of research.* Retrieved March 2, 2017, from www.hhs.gov/ohrp/regulations-and-policy/belmont-report/.

National Institutes of Health. (2013, March 13). *Notice number: NOT-OD-13-049: Findings of research misconduct.* Retrieved March 2, 2017, from https://grants.nih.gov/grants/guide/notice-files/NOT-OD-13-049.html.

National Science Foundation. (2002, March 18). *Part 689-Research misconduct.* Retrieved March 2, 2017, from www.nsf.gov/oig/_pdf/cfr/45-CFR-689.pdf.

Office of Research Integrity. (n.d.-a). *About ORI.* Retrieved March 2, 2017, from http://ori.hhs.gov/about-ori.

Office of Research Integrity. (n.d.-b) *About ORI – Historical background.* Retrieved March 2, 2017, from http://ori.hhs.gov/historical-background.

Office of Research Integrity. (n.d.-c). *Administrative actions.* Retrieved March 2, 2017, from https://ori.hhs.gov/administrative-actions.

Office of Research Integrity. (2015a). *Case summaries.* Retrieved August 25, 2016, from https://ori.hhs.gov/case_summary.

Office of Research Integrity. (2015b). *Sample policy and procedures for responding to research misconduct allegations.* Retrieved August 25, 2016, from http://ori.hhs.gov/sample-policy-procedures-responding-research-misconduct-allegations.

Office of Research Integrity. (2015c). *Case summary – Chen, Li.* Retrieved March 2, 2017, from https://ori.hhs.gov/chenli.

Office of Research Integrity. (2016, August 4). *Frequently asked questions.* Retrieved March 2, 2017, from http://ori.hhs.gov/content/frequently-asked-questions.

Pearson, H. (2006). Credit where credit's due. *Nature, 440*(7084), 591–592. doi:10.1038/440591a.

Penrose, A. M. and Katz, S. B. (2010). *Writing in the sciences: Exploring conventions of scientific discourse* (3rd ed). New York: Longman.

Plato. (1956). *Phaedrus* (W. C. Helmbold & W. G. Rabinowitz, Trans.). New York: Macmillan Publishing.

Price, A. (2006). Cases of plagiarism handled by the United States Office of Research Integrity 1992–2005. *Plagiary: Cross-Disciplinary Studies in Plagiarism, Fabrication, and Falsification*, 46–56. Retrieved March 2, 2017, from http://quod.lib.umich.edu/cgi/p/pod/dod-idx/cases-of-plagiarism-handled.pdf?c=plag;idno=5240451.0001.001.

Schechter, A. N., & Schwartz, J. P. (1996). Urgent call for comments on commission on research integrity's report. *The Catalyst*, *4*(4). Retrieved March 2, 2017, from https://archive.org/stream/nihcatalystpubl1996nati_2/nihcatalystpubl1996nati_2_djvu.txt.

Scheetz, M. D. (1999). Office of Research Integrity: A reflection of disputes and misunderstandings. *Croatian Medical Journal*, *40*(3), 321–25.

Schneider, L. (2015). What if universities had to agree to refund grants when there was a retraction. *Retraction Watch* [and commentaries]. Retrieved

March 2, 2017, from http://retractionwatch.com/2015/01/19/universities-agree-refund-grants-whenever-retraction/.

Shulkin, D.J., Goin, J.E., & Rennie, D. (1993). Patterns of authorship among chairmen of departments of medicine. *Academic Medicine, 68*, 688–692.

U.S. Department of Health and Human Services. (2005, May 17). Public health service policies on research misconduct (42 CFR parts 50 and 93). Federal Register Online. *Federal Register* 70.94. Retrieved March 2, 2017, from http://ori.hhs.gov/sites/default/files/42_cfr_parts_50_and_93_2005.pdf.

Wager, E., Fiack, S., Graf, C., Robinson, A. R., & Rowlands, I. (2009). Science journal editors' views on publication ethics: Results of an international survey. *Journal of Medical Ethics, 35*(6), 348–353. Retrieved March 2, 2017, from http://jme.bmj.com/content/35/6/348.full.

3 Science vs. Science Commercialization in Neoliberalism (Extreme Capitalism)

Examining the Conflicts and Ethics of Information Sharing in Opposing Social Systems

Scott A. Mogull

Science commercialization directly applies the latest findings from scientific research into new products and services that are then advertised and sold in a capitalist market economy. Since the 1980s, science commercialization has increasingly become a major goal of many scientific researchers, which corresponds with the rise of neoliberalism as an extreme form of capitalism (Boggio, Ballabeni, & Hemenway, 2016; Holloway, 2015). Notably, the social system of science commercialization is in opposition to the philosophical and ethical foundations of "open" science with the historical mission of pursuing scientific research as the general pursuit of knowledge and promoting the free information exchange for the improvement of society (David, 2005; Evans 2010). The contrast between scientific society and neoliberal science commercialization is evident in the observation by R. K. Merton, a prominent scholar of scientific society, who stated, "The pursuit of science is culturally defined as being primarily a disinterested search for truth and only secondarily, a means of earning a livelihood" (Merton 1957, 26). The prioritization between truth and profit has direct implications for technical communication in the context of science commercialization. In the social system of science commercialization, information secrecy and selective communication of partial, even biased, data are common practice. Furthermore, the social system of science commercialization leads to a state of ignorance among individuals outside of a particular commercial organization due to only investigating topics that support commercialization and selectively disclosing limited information about a scientific product (Evans, 2010; Fernandez Pinto, 2015). Generally speaking, organizations of science commercialization only provide information that positively supports a product (unless required by law or litigation to disclose additional information). Due to the secrecy and distortion (or bias) of technical communication in science commercialization, scholars are

DOI: 10.4324/9781315160191-3

concerned that future scientific progress will be impeded (David, 2005; Evans, 2010; McCain, 1991).

Although concerns of information sharing in science commercialization have been raised, the fields of technical communication and science and technology studies lack a detailed case study that illustrates the scope of the problem and begins to examine solutions. In this chapter, I explore the case of Treximet to illustrate how information is communicated to health care professionals and scientists in the context of pharmaceutical commercialization, a prominent form of neoliberal science commercialization. In this example, I find that the information revealed to these experts is not only selective but distorted to broaden the commercial market for a specific therapeutic drug treatment. This case study shows that information sharing in science commercialization is in direct contrast to the ethical foundations of information sharing in science and, moreover, echoes the ethical concern that the commercializing of scientific advancements, particularly by the pharmaceutical industry, is motivated by the desire for excessive profit through exploitative commodification rather than authentic human good (Angell, 2005; Avorn, 2003; Brody, 2014; Foucault, 2004; Hoedemaekers, 2001).

Information Sharing in Science Commercialization: The Case of Treximet

The case of Treximet versus Imitrex, two therapeutic drugs to treat migraine headaches researched and developed by the pharmaceutical company GlaxoSmithKline, illustrates the secretive, persuasive presentation of study data that leads to a definitive and consistent scientific conclusion promoted essentially as propaganda in an authoritarian style. In science commercialization, commercial organizations forcefully promote a singular, definitive conclusion of the data that lacks the nuances and discourse space for productive scholarly debate that are characteristic of research findings presented in scientific society. Such hallmarks of scientific commercialization are particularly concerning in the medical field because obfuscating detailed information that restricts or qualifies the appropriate use of a scientific product and deliberately misleading experts (specifically health care professional and scientists) in an overly false positive interpretation of scientific data leads to unnecessary pain and suffering by patients and impedes future research. More relevant to the discussion here, such information sharing practices of science commercialization violate the ethical foundation of information sharing in science and undermine future advancement. In analyzing this case, I examined the scientific data for Treximet provided by the commercial organization to technical audiences in the late 2000s when Treximet was launched in the U.S. market. This representative case is contextualized in the principles of scientific communication ethics, which provides a broader perspective of the long-term detrimental impact of science commercialization on science.

Business vs. Scientific Drivers

Science commercialization is not inherently contradictory to scientific ethics—quite contrary, often scientific research is prompted by a desire to improve the human condition and advance society (Gower, 1997). In particular, the underlying goal of scientific research is to apply new knowledge to develop innovations that achieve such humanistic outcomes. Yet, many scientists believe the reality of commercializing science in neoliberalism society conflicts with pursuit of science for the public good (Hemmungs Wirten, 2015; Small & Mallon, 2007). In practice, meaningful innovation of scientific and technical products leading to actual advancement is exceedingly rare; rather, the majority of new technical products, such as new pharmaceutical drugs, offer relatively little to no benefit or useful advantage over existing products (Angell, 2005; Goulding, 1983).

In the case of Treximet, its story can be criticized from the beginning. The timing of the initial market launch of Treximet, in 2008, when the new therapeutic drug was approved for the U.S. market to treat migraine headaches, was suspicious. Although the timing is not conclusive evidence in itself, this "scientific breakthrough" for the treatment of migraines coincided with the patent expiration of the company's first therapeutic drug for migraine headaches, Imitrex (Mogull & Balzhiser, 2015). A critic might interpret that such "serendipitous" circumstances indicate an economic or business motivation (drive) for the new therapeutic drug rather than a scientific breakthrough that would lead to improved care for individuals. To clarify the situation, the drug formulation of Imitrex tablets would no longer be protected from commercial competition upon patent expiration, and other pharmaceutical drug companies could manufacture and sell the same chemical formulation as a generic drug. Because the patent-protected monopoly of Imitrex was expiring for GlaxoSmithKline, the company faced reduced profits due to competition from lower-priced, chemically equivalent therapeutics from generic drug manufacturers (Caves, Whinston, Hurwitz, Pakes, & Temin, 1991). Financially, GlaxoSmithKline would benefit by converting Imitrex consumers to Treximet, which was a new patent-protected therapeutic drug for migraine headaches that could be sold at a premium price, rather than lose a source of revenue to other companies manufacturing a cheaper, generic version of Imitrex. Such business strategy of shifting consumers to purchase new patent-protected products upon expiration of patent-protected monopolies of older products is common practice for pharmaceutical and medical products (Bouchard, 2012), and such a strategy was inferred from GlaxoSmithKline's online, direct-to-consumer marketing of Imitrex and Treximet (Mogull & Balzhiser, 2015).

Further evidence supporting the assumption that business, not science, was primarily driving the launch of Treximet was the scientific review classification by the drug reviewers at the U.S. Food and Drug Administration (FDA), which classified Treximet as "a drug that appears to have therapeutic qualities similar to those of an already marketed drug" (FDA, n.d.). This review classification is notable because Treximet was *not* considered by the scientific reviewers at the FDA to be a technological breakthrough in the treatment of migraine headaches (for a review of the FDA drug classification system, see Sanborn, Goodwin, & Pessetto, 1991). Therefore, the technical evaluation as well as the timing of the market launch support the conclusion that the motivation for the commercialization of Treximet was primary a business-driven goal to generate profit rather than a scientifically driven goal to improve the treatment of migraines.

Scientific Evidence for Commercial Products

Despite business profit serving as the primary motivation for the commercialization of Treximet, the scientific evidence still must adequately support the innovation. Yet, the stringency of the scientific evidence necessary for commercialization varies significantly by industry and country. In the governmentally regulated pharmaceutical industry in the U.S., the FDA requires that clinical trial data show that a new therapeutic drug is more effective than a placebo for treating a defined medical condition. This relatively simple, but controversial, requirement means that new therapeutic drugs do not have to be compared to other drugs treating the same condition nor perform better than other drugs already available on the market (Angell, 2005). One caveat to this requirement is that pharmaceutical companies must provide the FDA with clinical trial data showing greater effectiveness of one therapeutic drug over another in order to make claims in the marketing literature of increased effectiveness over the other drug. Such clinical trial data are provided to the FDA in the New Drug Application (NDA) as part of the approval review process for a new therapeutic drug (Foote & Neumann, 2003). Specifically, the NDA statistical review, which may be accessed from the publically available FDA Drug Approval Database Drugs@FDA, provides the most technical data on the performance and safety of a new therapeutic drug such as Treximet.

Generally, access to the performance data of scientific products, including pharmaceutical drugs, is highly limited and restricted because organizations develop and analyze products in secrecy and only provide access to the original data when required by law (as in the case of the NDA) or through litigation. Yet, in the NDA for Treximet (referred to as "Trexima" in the documentation), clinical trial data were provided for the new migraine therapeutic drug along with results from a placebo and two other drugs: Imitrex (called "sumatriptan," the company's previous therapeutic drug) and naproxen, which is the chemical name for the over-the-counter medication "Aleve" (Trexima Statistical Review, 2005). These three drugs and placebo were evaluated in the standard clinical study trial protocol for measuring the effectiveness of migraine relief in adults at two hours following ingestion. The comparison of Treximet to Imitrex and naproxen in clinical trials, which was required to support marketing claims, showed difference effectiveness between sexes. In two clinical trials (see Figure 3.1), more female participants reported pain relief from Treximet/Trexima (59% and 67%) than for Imitrex/sumatriptan (50% and 55%), naproxen/Aleve (43% and 44%), or placebo (29% and 29%). In contrast, more male participants reported pain relief from Imitrex/sumatriptan (55% and 61%) than from Treximet (46% and 51%), naproxen (46% and 47%), or placebo (21% and 23%).

The interpretation of the clinical trial data divided by sex is noteworthy: Among the therapeutic drugs evaluated, Treximet was the most effective treatment of migraines in females. In contrast, Treximet was *not* the most effective treatment in males (Imitrex was more effective). In clinical practice, Treximet would likely be more effective than Imitrex for females and should be the first choice of therapeutic drugs for that group. In contrast, these data indicated that males would benefit from taking Imitrex/sumatriptan as a first choice of therapeutic drugs and that Treximet may be only marginally better than naproxen/Aleve in the population. As a second option for treatment, males might benefit more (considering both pain relief and expense) by trying the over-the-counter drug naproxen before trying the more costly Treximet to treat migraines.

Since the clinical trial data align partially, but not entirely, with the business objectives, I explored the way this technical information was communicated to experts (health care professionals and scientists). In

Gender	Trexima	Sumatriptan	Naproxen	Placebo
Male	21/46 (46%)	26/47 (55%)	19/41 (46%)	9/42 (21%)
Female	186/316 (59%)	156/315 (50%)	139/323 (43%)	100/340 (29%)

Gender	Trexima	Sumatriptan	Naproxen	Placebo
Male	24/47 (51%)	31/51 (61%)	23/49 (47%)	13/56 (23%)
Female	213/317 (67%)	169/310 (55%)	134/307 (44%)	89/304 (29%)

Figure 3.1 Clinical trial data by sex from two independent studies (top and bottom) from the Treximet New Drug Application (NDA) that was submitted to the FDA (Trexima Statistical Review, 2005, p. 22).

the analysis here, the two research questions that arose were: (1) how was the technical information represented and communicated to experts in the context of commercialization and (2) how would this communication be evaluated through the ethical lens of information sharing in scientific society?

Technical Communication to Health Care Professionals

In the clinical study data, Treximet was more effective for migraine pain relief than the alternative therapeutics evaluated for females but not males. Despite the fact that the company GlaxoSmithKline had clinical trial data that Treximet had different effectiveness in different treatment groups (females and males), no difference between the sexes, and, in fact, no negative performance data, was provided in the Treximet package insert (the technical documentation for pharmaceutical drugs provided to health care professionals that summarized the relevant performance, prescribing, and risk information). Rather, in the company-written package insert, GlaxoSmithKline (2012) pooled the data from the clinical trials that were previously separated by sex in the NDA submitted to the FDA. By pooling data into broader categories, the details of different effectiveness between different groups were sequestered, and Treximet appeared effective for "all participants" (see Table 3.1). In the pooled data presented in the package insert, Treximet appeared more effective for "all patients" (at 65% and 57%) than Imitrex/sumatriptan (55% and 50%), naproxen [sodium] (44% and 43%), or placebo (28% and 29%). Pooling the data from females and males masked the lower effectiveness of Treximet in males when compared to the other therapeutic drugs because the total number of females in the clinical trials (n = 2,532 or 87%) greatly outnumbered the total number of males (n = 379 or 13%).

Table 3.1 Clinical Trial Data Comparing Treximet to Two Other Therapeutic Drugs Imitrex/Sumatriptan and Aleve/Naproxen Sodium as Published in the Treximet Package Insert for Health Care Professionals

Treatment	*Study 1 Data ("All Patients")*	*Study 2 Data ("All Patients")*
Treximet	65% n = 364	57% n = 362
Imitrex/sumatriptan	55% n = 361	50% n = 362
Naproxen sodium	44% n = 356	43% n = 364
Placebo	28% n = 360	29% n = 382

Adapted from GlaxoSmithKline (2012).

In the package insert, the clinical trial data were presented through a rhetorically selected lens to deliberately mislead the audience—notably healthcare professionals—to a false interpretation that Treximet was more effective for "all patients" (exact phase used, which implied both sexes). Alternatively, a more detailed resolution of the data, which appeared in the Treximet NDA (see Figure 3.1), would have led healthcare professionals to a different qualitative conclusion regarding the most effective therapeutic drug to prescribe for migraine headaches based on the sex of the patient. Further reinforcing the false interpretation by healthcare professionals was the "gender statement" in the Treximet package insert, which stated

> In a pooled analysis of 5 pharmacokinetic studies, there was *no effect of gender* [emphasis added] on the systemic exposure of TREXIMET. In a study comparing the pharmacokinetics of sumatriptan in females and males, *no differences were observed between genders* [emphasis added] for AUC, C_{max}, T_{max}, and $T_{1/2}$.
>
> (GlaxoSmithKline, 2012, p. 4)

Notably, this statement reported that "no effect of gender" was detected and "no differences were observed between genders" in the standard studies for systemic exposure of the drug in females and males.

A rhetorical analysis of the gender statement revealed a technically accurate yet misleading report to healthcare professionals, which was consistent with and reinforcing of the presentation of pooled clinical trial data in the package insert. Of particular note, the measures of systemic exposure in the gender statement should not be conflated with effectiveness, which was measured in the clinical trial data. The similar terms, "exposure" and "effectiveness," have a nuanced difference in meaning. Effectiveness is the measure that a therapeutic drug resolves a medical condition (such as a migraine headache). In contrast, exposure is the amount of the therapeutic drug that circulates within an individual's body. I do not intend to suggest that healthcare professionals do not understand the difference between these terms, but rather that in the context of skimming a technically dense document, healthcare professionals would focus on key summary statements that inform practice and not have time to critically evaluate each claim. In this case, the qualitative conclusions in the gender statement, "no effect of gender" and "no differences between genders," along with the absence of any information or reference regarding the different effectiveness of Treximet between females and males (as shown in Figure 3.1), would lead healthcare professionals to a false overall conclusion that no difference between the sexes applies to all measures and features of Treximet. Here, the gender statement helps to mislead healthcare professionals to make a false interpretation that the biological activity of Treximet (at least in the conventional assays) was identical between females and males. Moreover, the phrasing of the gender statement and lack of any reports of nonstandard investigations excludes the awareness of healthcare professionals to any difference of Treximet between sexes whatsoever.

Critically, the company's motivation for such a positive gender statement and absence of negative information needs examination. For this discussion, the important point is that the technical information to healthcare professionals is rhetorically constructed to support business goals rather than to assist healthcare professionals in providing the best treatment to patients. The presumed effect of this biased technical documentation is that health care professionals, believing that they have seen clinical trial data—and that data was presented objectively—will confidently prescribe the patent-protected, high-priced Treximet as the most effective option (the first-choice therapeutic drug) for "all patients" suffering from migraine headaches when compared to the generic Imitrex/sumatriptan or "over-the-counter," non-prescription naproxen. Although Treximet does appear to be a valid first-choice therapeutic for female patients when compared to the other drugs, the implications for male patients are twofold: a less effectiveness treatment for migraine pain relief from Treximet coupled with an increased financial cost when compared to Imitrex/sumatriptan or naproxen.

In the case of Treximet, the distortion of technical data in the package insert increases the chance for ineffective treatment of males, which would result in increased pain, suffering, and expense for these patients. While a detailed analysis of the impact of ineffective treatment on patients is beyond the scope of this work, a few observations are worth noting. First, commercialization of science leads to underserved treatment of a minority population, which may have unique needs different from the majority of the consumer market, and any population may comprise a minority group (even traditionally privileged groups). Such data illustrate that the best treatment for any minority population with different treatment outcomes from the majority is subject to omission from technical documentation if such findings are contrary to the business drivers of the commercial sponsor. Second, patients who revisit their physicians because the initially prescribed Treximet did not provide sufficient pain relief would be prescribed the next-"best" alternative from the company's perspective. Interestingly, the effectiveness of these migraine treatments in the clinical trial data presented in the prescribing information (see Table 3.1) also corresponds to product profit for the company (from most to least): first the patent-protected Treximet, second the Imitrex (or generic sumatriptan), and finally the over-the-counter naproxen.

The major point of this discussion is that the distortion of technical information in the package insert impairs the ability of physicians to properly initially prescribe therapeutic drugs for migraines as well as address or comprehend inconsistent reports of treatment effects in patient follow-up visits—specifically in regards to conflicting reports of effectiveness from females and males. Initially, physicians are deliberately put into a state of ignorance to the different effectiveness of Treximet between females and males. Yet after a number of consistent reports from male patients, physicians may have a localized state of confusion in which an individual physician may internally question his or her clinical observations with the previous paradigm of Treximet effectiveness for "all patients" as promoted by the company. Ultimately, a physician may gather sufficient personal clinical data to solidity a pattern in his or her own mind that substitutes for the package insert. Yet, such a state of clarity may be unlikely because many physicians would be unable to gather sufficient local evidence from enough patients to clearly show the effectiveness of Treximet in minority population. Even if such clarity was achieved by an individual physician, the scale of such knowledge would be miniscule in comparison to the company-sponsored paradigm.

In the Treximet case, the technical information provided in the package insert reinforces the argument that the pharmaceutical industry is focused nearly exclusively on commercialization and contradicts two major claims of the industry—the false narrative of interest in improving patient treatment and, particularly relevant to this analysis, that pharmaceutical companies provide physician medical education rather than marketing (Angell, 2005). In science commercialization, the secrecy and distorted presentation of scientific data leads to an unusually consistent and definitive scientific conclusion that nearly exclusively supports the sales of the product. Treximet is a noteworthy case because the clinical study data showed a difference in the effectiveness by sex—with an advantage of the therapeutic drug in females but not males. Thus even without data distortion, Treximet would remain a valid primary therapeutic option for females, who represent the majority of migraine headache patients and therefore the largest market. Despite such a relatively large "honest" market potential, the distorted pooled clinical study data between sexes, which was the decision made by the company, maximized the size of the potential market and thus revenue.

Technical Communication to the Scientific Community

In science commercialization, commercial organizations have both the most detailed technical information about a technology as well as a vested interest in the commercial success of the product. This combination of knowledge and financial incentive for the direct application of scientific knowledge establishes the context for potential ethical conflict. In the context of science commercialization, the technology becomes the product rather than scientific knowledge, which, in contrast, is the product of scientific research (Knorr-Cetina, 1991; Latour & Woolgar, 1986). Thus in science commercialization, technical communication to the scientific community becomes a secondary priority that may be subject to obfuscation and distortion so that the information communicated outside of an organization aligns with the primary goal of selling product or technology.

Considering the case of Treximet in the context of scientific society, the scientific knowledge from the clinical study data may be used to advance the scientific understanding of the therapeutic drug and future treatments of migraine headaches. Complete, unbiased communication of technical information is an essential foundation of the scientific method that connects the investigations of individual research groups into a larger community investigating a scientific topic or problem (Mogull, 2017; Polanyi, 1962). Through the scientific system of open communication, current researchers can build on the knowledge and experience of others so that a scientific puzzle or issue is further developed by many different teams over generations of researchers (Committee on Responsibilities of Authorship in the Biological Sciences, 2003; Garfield, 1980; Montgomery, 1996; Polanyi, 1962; Wilson, 1998). The scientific method and formal communication process create a cycle that results in solving scientific puzzles at the fastest possible rate (for a key theoretical essay on the process of scientific problem solving, see Polanyi, 1962). In each iterative cycle of the scientific method, findings from published research papers, or theories, lead to more advanced research questions, or hypotheses, which are subsequently investigated through the next turn of the cycle (American Psychological Association [APA], 2010; Clapham, 2005; Committee on Science, Engineering, and Public Policy, 2009; Knottnerus & Tugwell, 2007; Mogull, 2017).

This practice of information sharing dates back to the foundations of modern scientific societies, such as the Royal Society of London, which is noteworthy for having established the principles of communication and scientific credit in the mid-1600s alongside publication of the first scientific journal, *Philosophical Transaction* (Bazerman, 2011). As established by the Royal Society of London in the late 1600s, the first scientist or team to communicate information by publishing research findings in a peer-reviewed scientific journal (n.b., not necessarily the first to make a discovery) is credited with the discovery and thus reaps the rewards associated with such contribution to society (Committee on Science, Engineering, and Public Policy, 2009). Importantly, this 350-year-old tradition of open information communication directly contributed to the prominence and advancement of Western scientific society—particularly when contrasted with other, more secretive societies (Wilson, 1998).

Due to the importance of technical communication in science, the journal article, or an original research article published in a peer-reviewed scientific journal, is the primary product and currency of science (Knorr-Cetina, 1991; Latour & Woolgar, 1986). In contrast, scientific discoveries that are not communicated do not advance science and thus do not confer value to the scientific society (APA, 2010; Clapham, 2005). The practices of information secrecy, which are common in science commercialization, do not confer value to the scientific society but do provide relatively short-term commercial value to organizations. Due to the conflict between information sharing and necessary science commercialization, the formal system of patenting technology was designed to balance the needs of both society and corporations by protecting the commercial sale of a technology for a period of time while enabling and promoting full disclosure of scientific information (Hemmungs Wirten, 2015; Hoedemaekers, 2001). As illustrated in the Treximet case study, commercial organizations both exploit patent protection and use trade secrets to prevent full disclosure of scientific information. By leveraging both systems, commercial organizations maximize their commercial advantage from a patent-protected monopoly for the current technology and selective disclosure of the technical information for future market opportunities of newer technologies. Through these dual actions, such science commercialization organizations are parasites on the scientific community and publically funded research.

In the following discussion, I examine the clinical trial data for Treximet communicated by GlaxoSmithKline-sponsored researchers to the scientific community through a journal article published in the medical journal *JAMA*. Notably, the clinical trial data of Treximet were reported consistently to scientists and healthcare professionals, which lacks the resolution of effectiveness by sex that was originally reported to the FDA. In a scientific journal article, GlaxoSmithKline-sponsored researchers provided data for migraine headache relief at two hours post treatment for Treximet (sumatriptan-naproxen sodium), the two other therapeutic drugs Imitrex (sumatriptan) and naproxen sodium (Aleve), and placebo. The data provided to scientists (see Table 3.2) is identical to the merged sex data communicated to physicians in the Treximet package insert (see Table 3.1). In the data table from the *JAMA* journal article, the combined number of all participants reporting migraine headache relief at two hours for study 1 and 2 again leads to the conclusion that Treximet (sumatriptan-naproxen sodium) was more effective (at 65% and 57%) than Imitrex (sumatriptan) (55% and 50%), naproxen sodium (44% and 43%), or placebo (28% and 29%). Additionally, the authors used the statistical *p* value to reinforce the perception of reliability to the data provided, which does indicate that the merged data of both sexes were statistically significant for the entire population although *p* values do not provide any measure to whether the groups compared were ethically selected.

Table 3.2 Excerpt of Treximet Clinical Data Detailing Research Funded by GlaxoSmithKline from a Peer-Reviewed Scientific Journal Article Published in the Medical Journal *JAMA*

	Number of Participants Reporting Headache Relief at Two Hours/Total Participants (Percent of Efficacy Population)				*Sumatriptan-Naproxen Sodium vs. Placebo*	*Sumatriptan-Naproxen Sodium vs. Sumatriptan*
	Sumatriptan-Naproxen Sodium	*Sumatriptan*	*Naproxen Sodium*	*Placebo*		
Study 1	237/364 (65%)	200/361 (55%)	157/356 (44%)	102/360 (28%)	$p < .001$	$p = .009$
Study 2	207/362 (57%)	182/362 (50%)	158/364 (43%)	109/382 (29%)	$p < .001$	$p = .03$

Adapted from Brandes et al. (2007).

Another point of interest that reinforces the use of scientific journal articles as commercialization marketing appears in the discussion section of the article. Interestingly, the authors of the Treximet journal article repeatedly used phrases such as, "The *superior efficacy* [emphasis added] of sumatriptan-naproxen sodium relative to sumatriptan monotherapy…" (Brandes et al., 2007, p. 1452). The authors' unusually phrased definitive conclusion in the scientific journal article is more commonly observed in marketing communications rather than in scientific discourse, which traditionally phrases conclusions more cautiously and employs hedging statements to qualify such claims (Hyland, 1996). Furthermore, the overly simplistic conclusion that Treximet is "superior" to the other therapeutic drugs for all patients lacks the intricacies and nuances, as well as the opposing views, that are hallmarks of science. In this case, presenting pooled data through a lens that distorts differences between females and males and implies a "superior" universal solution for both sexes conceals the important feature that the biochemical nature of migraines might be different in females and males. Obfuscating this key difference impedes the scientific community's comprehensive knowledge of migraines and may delay future discovery and commercialization of more effective drug treatments for both sexes.

In science commercialization, commercial organizations tend to withhold key information and resources that would facilitate research from others outside of the organization in order to maintain a competitive advantage (Campbell et al., 2002; Hilgartner & Brandt-Rauf, 1994). When key pieces of information are withheld, such as in this example that females and males respond differently to the therapeutic drug Treximet, scientists outside the organization are unable to build effectively on previous work and therefore must spend time and resources unnecessarily rediscovering previous findings (APA, 2010; Clapham, 2005; Knottnerus & Tugwell, 2007; Walport & Brest, 2011). Such withholding (or omission) of relevant data is considered a form of scientific misconduct (Clapham, 2005; Committee on Science, Engineering, and Public Policy, 2009; Knottnerus & Tugwell, 2007; Martinson, Anderson, & de Vries, 2005; Vision, 2010). Yet, the withholding of medical research data, particularly publically funded research data, has additional ethical implications due to the impediment of human care as well as the substantial public cost (both for treatment and further research) (Angell, 2005; Walport & Brest, 2011). Such issues were partially addressed in an article also published in *JAMA*, in which the author Chalmers (1990) stated, "Failure to report results in sufficient detail" is a form of scientific misconduct that "may either lead patients to receive ineffective or dangerous forms of care or result in a delay in recognizing that other forms of care are beneficial" (pp. 1405–1406). Particularly alarming is that researchers tend to withhold the most information in research areas that have the greatest impact for human health (for example, cancer) (Piwowar, 2011) because such areas also provide the greatest commercialization opportunities.

Conclusions: Reflections and Responses

In the Treximet case, the lens through which the data are presented and the definitive, conclusive rhetoric establish a consistent, yet distorted, paradigm among experts (both health care professionals and scientists) that maximizes the commercial benefit to the pharmaceutical company. Such scientific paradigms thus stall scientific understanding of a topic for a period of time and delay future research innovation (Kuhn, 2012). Moreover, the misleading conclusion that Treximet is superior for both sexes becomes increasingly ingrained in a current paradigm as both groups of experts repeat the distorted conclusion. This singular message of an authoritarian or propaganda nature is in contrast with the principles of scientific societies, which began and continue to function (partially) as free-thinking organizations outside of authoritarian control (Bazerman, 2011). Moreover, the authoritarian nature of science commercialization unifies a singular, definitive message, and the force and volume of this message by a company minimizes or even prevents the communication of dissenting viewpoints through legal action (such actions include confidentiality agreements, material transfer agreements, trade secrets protection, and potentially copyright protection) (Baycan & Stough, 2013; Caulfield, Harmon, & Joly, 2012; Durack, 2006; Evans, 2010; Kaiser, 1996; Murray & Stern, 2007). In science commercialization, the major counterbalancing force to one message or paradigm promoting a particular product is an opposing or competing product for the same economic market. However, such counterbalancing forces are restricted legally by patent protection of commercial scientific products, licensing, and even patent blocking, all of which prevent competition and eliminate the incentive for widespread communication of different messages. Notably, the economic resources that commercial institutions leverage to promote a singular message in science commercialization are in far excess of the relatively minor publicity of any potential dissenting voices that function within scientific society. Thus the only significant counterbalancing forces are other commercial organizations with equal resources who can send a counter message of equal coverage in various media—although presumably in an equally biased and distorted lens to promote their commercial product.

The scientific publishing process attempts to regulate ethical practice by limiting one's ability to publish research and thus reap the benefit of publication (i.e., science currency). One longstanding requirement, for example, is that journal editors only consider a manuscript for peer review if the author(s) verify that accepted protocols for treatment of human and animal subjects were followed and the research methods were approved by an ethics committee (International Committee of Medical Journal Editors [ICMJE], 2016). More recently, and in direct response to abuses by commercial organizations (such as in the Treximet example), many editors of peer-reviewed journals are requiring authors of journal articles to identify the original clinical trial design through preclinical trial registration in online databases such as clinicaltrials.gov (ICMJE, 2016) and to commit to sharing data in database repositories following publication (BMJ, n.d.). Additionally, the U.S. Congress (Zarin, Tse, & Sheehan, 2015) and major grant-funding agencies such as the U.S. National Institutes of Health and Wellcome Trust are requiring authors to register clinical trial design (Claufield et al., 2012). Such measures by these large organizations of substantial power are intended to counterbalance certain exploitation of biased and incomplete technical communication in science commercialization by commercial organizations. But such measures, while substantial and important in intent, are incomplete solutions. For example, in the case of Treximet, the clinical study was registered in clinicaltrials.gov and referenced in the *JAMA* journal article, but the details provided in clinicaltrials.gov were not the same as what was provided in the NDA. The clinical study design provided by the pharmaceutical company in clinicaltrials.gov stated that both sexes (male and female) qualified for participation (notably lacking any indication that the data for each sex would be independently analyzed). Furthermore, the Treximet clinical trial registration was dated October 14, 2005, which was several months *after* the NDA statistical review was published, which was dated August 5, 2005, but well before publication of the journal article in 2007.

More on point, such online repositories are insufficient in isolation to counterbalance omission and data distortion. In the case of Treximet, the original clinical trial data have been publically available from Drugs@FDA, a database hosted by the FDA. Yet, this data did not prevent or change publication of pooled data in either the package insert or scientific journal article. While such public databases might—in theory—be used by careful scientific reviewers of package inserts at the FDA or by careful scientific journal editors or manuscript reviewers, access to the clinical trial data—as history has shown—is inadequate to counter the propaganda techniques in science commercialization. Rather, a more active role is required of the scientists and health care professionals.

References

American Psychological Association. (2010). *Publication manual of the American Psychological Association* (6th ed.). Washington, DC: American Psychological Association.

Angell, M. (2005). *The truth about the drug companies: How they deceive us and what to do about it.* New York: Random House.

Avorn, J. (2003). Advertising and prescription drugs: Promotion, education, and the public's health. *Health Affairs (Project Hope)*, W3-104.

Baycan, T., & Stough, R. R. (2013). Bridging knowledge to commercialization: The good, the bad, and the challenging. *The Annals of Regional Science, 50*(2), 367–405.

Bazerman, C. (2011). Church, state, university, and the printing press: Conditions for the emergence and maintenance of autonomy of scientific publication in Europe. In B.-L. Gunnarsson (Ed.), *Languages of science in the eighteenth century* (pp. 25–44). Berlin: De Gruyter.

BMJ. (n.d.). *Research.* Retrieved March 10, 2017, from www.bmj.com/about-bmj/resources-authors/article-types/research.

Boggio, A., Ballabeni, A., & Hemenway, D. (2016). Basic research and knowledge production modes: A view from the Harvard Medical School. *Science, Technology, & Human Values, 41*(2), 163–193.

Bouchard, R. A. (2012). *Patently innovative: How pharmaceutical firms use emerging patent law to extend monopolies on blockbuster drugs.* Oxford: Biohealthcare Publishing (Oxford Limited).

Brandes, J. L., Kudrow, D., Stark, S. R., O'Carroll, C. P., Adelman, J. U., O'Donnell, F. J., ... Lener, S. E. (2007). Sumatriptan-naproxen for acute treatment of migraine: A randomized trial. *JAMA, 297*(13), 1443–1454.

Brody, H. (2014). Economism and the commercialization of health care. *Journal of Law, Medicine & Ethics, 42*, 501–508.

Campbell, E. G., Clarridge, B. R., Gokhale, M., Birenbaum, L., Hilgartner, S., Holtzman, N. A., & Blumenthal, D. (2002). Data withholding in academic genetics: Evidence from a national survey. *JAMA, 287*(4), 473–480.

Caulfield, T., Harmon, S. H. E., & Joly, Y. (2012). Open science versus commercialization: A modern research conflict? *Genome Medicine, 4*(2), 17.

Caves, R. E., Whinston, M. D., Hurwitz, M. A., Pakes, A., & Temin, P. (1991). Patent expiration, entry, and competition in the US pharmaceutical industry. *Brookings Papers on Economic Activity: Microeconomics*, 1–66.

Chalmers, I. (1990). Underreporting research is scientific misconduct. *JAMA, 263*(10), 1405–1408.

Clapham, P. (2005). Publish or perish. *BioScience, 55*(5), 390–391.

Committee on Responsibilities of Authorship in the Biological Sciences, National Research Council. (2003). *Sharing publication-related data and materials: Responsibilities of authorship in the life sciences.* Washington, DC: National Academies Press.

Committee on Science, Engineering, and Public Policy—National Academy of Sciences, National Academy of Engineering, and Institute of Medicine of the National Academies. (2009). *On being a scientist: A guide to responsible conduct in research* (3rd ed.). Washington, DC: National Academies Press.

David, P. A. (2005). From keeping 'nature's secrets' to the institutionalization of 'open science.' In R. Ghosh (Ed.), *Code: Collaborative ownership and the digital economy* (pp. 85–108). Cambridge, MA: MIT Press.

Durack, K. T. (2006). Technology transfer and patents: Implications for the production of scientific knowledge. *Technical Communication Quarterly, 15*(3), 315–328.

Evans, J. A. (2010). Industry collaboration, scientific sharing, and the dissemination of knowledge. *Social Studies of Science, 45*(5), 757–791.

Fernandez Pinto, M. (2015). Tensions in agnotology: Normativity in the studies of commercially driven ignorance. *Social Studies of Science, 45*(2), 294–315.

Foote, M., & Neumann, T. K. (2003). A primer of drug development. *Journal of the American Medical Writers' Association, 18*, 61–66.

Foucault, M. (2004). The crisis of medicine or the crisis of antimedicine? *Foucault Studies, 1*, 5–19. (Original work published 1974).

Garfield, E. (1980). Has scientific communication changed in 300 years? *Essays of an Information Scientist, 4*(8), 394–400.

GlaxoSmithKline. (2012). *Treximet prescribing information.* Retrieved March 10, 2017, from www.gsksource.com/pharma/content/dam/GlaxoSmithKline/US/en/Prescribing_Information/Treximet/pdf/TREXIMET-PI-MG.PDF.

Goulding, I. (1983). New product development: A literature review. *European Journal of Marketing, 17*(3), 3–30.

Gower, B. (1997). *Scientific method: An historical and philosophical introduction.* London: Routledge.

Hemmungs Wirten, E. (2015). The patent and the paper: A few thoughts on late modern science and intellectual property. *Culture Unbound, 7*(4), 600–609.

Hilgartner, S., & Brandt-Rauf, S. I. (1994). Data access, ownership, and control: Toward empirical studies of access practices. *Knowledge, Creation Diffusion, Utilization, 15*(4), 355–372.

Hoedemaekers, R. (2001). Commercialization, patents and moral assessment of biotechnology products. *Journal of Medicine and Philosophy, 26*(3), 273–284.

Holloway, K. J. (2015). Normalizing complaint: Scientists and the challenge of commercialization. *Science, Technology, & Human Values, 40*(5), 744–765.

Hyland, K. (1996). Writing without conviction? Hedging in science research articles. *Applied Linguistics, 17*(4), 433–454.

International Committee of Medical Journal Editors. (2016). *Recommendations for the conduct, reporting, editing, and publication of scholarly work in medical journals.* Retrieved March 10, 2017, from www.icmje.org/icmje-recommendations.pdf.

Kaiser, M. (1996). Toward more secrecy in science? Comments on some structural changes in science—and on their implications for an ethics of science. *Perspectives on Science, 4*(2), 207–230.

Knorr-Cetina, K. (1991). Epistemic cultures: Forms of reason in science. *History of Political Economy, 23*(1), 105–122.

Knottnerus, J. A., & Tugwell, P. (2007). Communicating research to the peers. *Journal of Clinical Epidemiology, 60*(7), 645–647.

Kuhn, T. S. (2012). *The structure of scientific revolutions: 50th anniversary edition* (4th ed.). Chicago, IL: The University of Chicago Press.

Latour, B., & Woolgar, S. (1986). *Laboratory life: The construction of scientific facts.* Princeton, NJ: Princeton University Press.

Martinson, B. C., Anderson, M., & de Vries, R. (2005). Scientists behaving badly. *Nature, 435*(7043), 737–738.

McCain, K. W. (1991). Communication, competition, and secrecy: The production and dissemination of research-related information in genetics. *Science, Technology, & Human Values, 16*(4), 491–516.

Merton, R. K. (1957). Priorities in scientific discovery: A chapter in the sociology of science. *American Sociological Review*, 22(6): 635–659.
Mogull, S. A. (2017). *Scientific and medical communication: A guide for effective practice.* New York: Routledge.
Mogull, S. A., & Balzhiser, D. (2015). Pharmaceutical companies are writing the script for health consumerism. *Communication Design Quarterly*, *3*(4), 35–49.
Montgomery, S. L. (1996). *The scientific voice.* New York: Guilford Press.
Murray, F., & Stern, S. (2007). Do formal intellectual property rights hinder the free flow of scientific knowledge? An empirical test of the anti-commons hypothesis. *Journal of Economic Behavior & Organization, 63*(4), 648–687.
Piwowar, H. A. (2011). Who shares? Who doesn't? Factors associated with openly archiving raw research data. *PLoS One, 6*(7), e18657.
Polanyi, M. (1962). The republic of science: Its political and economic theory. *Minerva, 1*(1), 54–73.
Sanborn, M. D., Goodwin, H. N., & Pessetto, J. D. (1991). FDA drug classification system. *American Journal of Health-System Pharmacy*, *48*(12), 2659–2662.
Small, B., & Mallon, M. (2007). Science, society, ethics, and trust: Scientists' reflections on the commercialization and democratization of science. *International Studies of Management & Organization, 37*(1), 103–124.
Trexima Statistical Review. (2005). *Statistical review and evaluation: Clinical Studies.* Retrieved March 10, 2017, from www.accessdata.fda.gov/drugsatfda_docs/nda/2008/021926s000_StatR.pdf.
U.S. Food and Drug Administration. (n.d.). *Drugs@FDA frequently asked questions: What do the chemical type and review classification codes stand for?* Retrieved March 10, 2017, from www.fda.gov/Drugs/InformationOnDrugs/ucm075234.htm#chemtype_reviewclass.
Vision, T. J. (2010). Open data and the social contract of scientific publishing. *BioScience, 60*(5), 330–331.
Walport, M., & Brest, P. (2011). Sharing research data to improve public health. *The Lancet, 377*, 537–539.
Wilson, E. O. (1998). *Consilience: The unity of knowledge.* New York: Vintage.
Zarin, D. A., Tse, T., & Sheehan, J. (2015). The proposed rules for U.S. clinical trial registration and results submission. *The New England Journal of Medicine, 372*(2), 174–180.

4 Visualizing Science

Using Grounded Theory to Critically Evaluate Data Visualizations

Candice A. Welhausen

In late 2015, healthcare providers in northeastern Brazil discovered that cases of a rare condition that causes severe fetal brain abnormalities had increased dramatically among newborns. This condition—microcephaly—was believed to be associated with exposure to Zika, a mosquito-borne virus that can cause mild flu-like symptoms.

Transmitted by *Aedes albopictus* and *Aedes aegypti* (the latter being the same mosquito species that inundated port cities in the United States with yellow fever throughout the nineteenth century), Zika is not new. Indeed it was discovered in Uganda in the mid-twentieth century. Yet, the virus had previously failed to garner serious attention among researchers because its symptoms (if any) are generally minor (Maron, 2016). However, in early 2016 as the outbreak escalated in Brazil, the Centers for Disease Control and Prevention (CDC) issued a travel advisory for pregnant women to countries in the Western Hemisphere where cases had been identified (CDC, 2016a). By April, 346 "travel associated" cases had been confirmed in the continental United States (CDC, 2016b) with that number expected to grow into the summer and early fall (Korte, 2016). Soon thereafter the agency confirmed the link between Zika and microcephaly (CDC, 2016c), adding that the virus might also be responsible for some serious neurological conditions in adults (Cha & Sun, 2016). As then CDC principal deputy director Dr. Anne Schuchat put it: "Everything we look at with this virus seems to be a bit scarier than we initially thought" (Korte, 2016, para. 2). As Zika's "pandemic potential" (Lucey & Gostin, 2016, p. 865) grew, public health agencies and news organizations created thematic maps targeted to non-expert audiences that communicated visual risk information about the spread of the virus.

In this chapter, I use grounded theory to analyze a collection of these maps to theorize the creation of these scientific visuals during a specific type of crisis and emergency risk scenario: an emergent public health threat. Originally developed by Glaser and Strauss (1999), this approach is "a general methodology, *a way of thinking about and conceptualizing data*" (Strauss & Corbin, 1994, p. 275) that uses an inductive, iterative process. This process involves collecting data and then coding,

DOI: 10.4324/9781315160191-4

categorizing, and analyzing that data. Unlike many hypothesis-driven empirical methods, grounded theory "calls for a lack of preconceptions," as Ross characterizes it (2013, p. 98; also citing Lewis & Whitely, 1992). This method is designed to lead to constructing a theory, which should explain all of the observed data, and can in turn be tested when new data are available. A grounded theory approach usually starts with data collection and/or a particular research question; a theory can then potentially emerge through this "constant comparative method," as Glaser and Strauss (1999) define it.

Grounded theory has gained substantial traction in the field of professional and technical communication (McNely, Spinuzzi, & Teston, 2015), and a number of studies that specifically focus on visual communication have employed this approach (e.g., Cooke, 2003; Portewig, 2008; Teston, 2012). Further, research on public health threats has been conducted retrospectively in the field—that is, after the immediate hazard has been resolved (e.g., see Ding, 2009, 2013; Welhausen, 2015a).

This chapter theorizes the data visualization strategies used to construct visual risk communication targeted to public audiences in the early stages of a pandemic. More specifically, I use grounded theory to tease out trends and patterns in the ways that such information is conveyed. Graphics like figures, tables, and illustrations are fundamental to advancing arguments in scientific disciplines (Perini, 2005; see also Chapter 5). Further, research in technical and professional communication has theorized the persuasiveness of scientific drawings and illustrations (Buehl, 2014; Reeves, 2011; Richards, 2009) and statistical graphics (Brasseur, 2004; Dragga & Voss, 2001; Kimball, 2006; Welhausen, 2015b). However, less is known about the design strategies used to communicate visual risk information during an epidemic's early stages and the ways that these choices might shape risk perception.

Literature Review: Disease Maps and Visualizing Quantitative Risk Information

Thematic maps have long been created to document the spread of infectious and communicable diseases. Today, public health researchers construct these visuals to advance a hypothesis about the spatial and temporal distribution of an outbreak (or potential outbreak) in order to better control its spread. Yet because disease maps are frequently compiled from large collections of data, they are often not seen as arguments themselves, but as objective scientific information that is simply rendered into a visual form. Much like the process that Latour and Woolgar (2013) describe wherein scientific information becomes solidified into knowledge in laboratory-based research settings, data visualizations too quickly become disconnected from the individual records that comprise a particular data set.

Data visualizations are not the data they represent. As Richards (2009) points out (drawing from Latour and also citing the influence of Gaston Bachelard on Latour's argument): "Data can undergo dozens, hundreds, even thousands of reifications before they assume their final form in a published visual representation" (p. 186). The process of visualization renders abstract, aggregated information into a concrete representation that emphasizes certain features while also de-emphasizing others.

Quantitative information about the spread of Zika was disseminated to public audiences during 2016 primarily through thematic maps created by public health agencies and news organizations (using data collected by these agencies). Maps control geographic space in powerful ways (see Crampton, 2010; Harley, 2009; Pickles, 2004; Wood, 1992). Disease maps in particular invoke "metaphorical control" by dividing the space into "diseased and not diseased" areas, which often results in specific public health actions (see Welhausen, 2015b). For instance, in all likelihood the CDC used the quantitative information shown in many of the maps included in this chapter to assess Zika-associated risks for pregnant women. The agency then issued a travel advisory targeted to this audience for specific countries.

Maps are also commonly used to communicate visual risk information to public audiences. Yet, while studies in risk communication have investigated how non-experts construct knowledge from many commonly used graphical genres (see reviews by Lipkus & Hollands, 1999; Ancker, Senathiraja, Kukafka, & Starren, 2006), less emphasis has been directed toward disease maps.

Methods

To explore the design choices that creators of data visualizations used to communicate visual risk information about the spread of Zika to non-experts, I posed the following research question: How is quantitative information about risk visually communicated during an emergent public health threat? I then reviewed 32 thematic maps published between mid-January and mid-April of 2016 by the CDC, World Health Organization (WHO), and mainstream English-language news organizations such as *The New York Times* (see Table 4.1 for a list of the maps). In order to include a wide range of maps in this study, I conducted an online search using only the key word "Zika." I reviewed the articles and reports that resulted from this search, and compiled a collection of data visualizations showing the current/potential spread of the virus. I did not include visualizations of conditions related to Zika like the microcephaly epidemic in northeastern Brazil. Of the maps in my sample, the first five listed in Table 4.1 were published directly on the CDC or WHO websites, respectively. The remaining were created using data from CDC, WHO, other public health agencies, or individual researchers.

Table 4.1 List of Maps

Map #s	*Source*	*Title of Article or Report*	*Title of Graphic*	*Last Updated*
1	*CDC* (a)	Zika travel information	Countries and territories reporting active mosquito transmission of Zika virus	4/18/16
2	*CDC* (b)	Zika virus disease in the United States, 2015–2016	Same as report	4/12/16
3	*WHO* (a)	Zika situation report	Global status of Zika virus	4/7/16
4	*WHO* (a)	Same as above	Distribution of Zika virus in the Americas, 2015–2016	4/7/16
5	*WHO* (b)	Same as above	Countries and territories with autochthonous of Zika virus circulation 2007–2016	2/5/16
6	*The Washington Post* Sun (2016)	CDC confirms Zika virus causes microcephaly, other birth defects	Estimated range of the mosquitoes behind the Zika virus (Source: CDC)	4/13/16
7	*The Washington Post* Cha & Sun (2016)	What is Zika and what are the risks as it spreads	The spread of the Zika virus	2/4/16
8	*NYT* McNeil, Saint Louis & St. Fleur (2016)	Short answers to hard questions about Zika virus	Zika Virus (Sources: CDC; Pan American Health Organization)	3/18/16
9	*NYT* Same as above	Same as above	None (Source: Kraemer et al. (2015a); Simon Hay, University of Oxford)	3/18/16
10	*NYT* Same as above	Same as above	Areas pregnant women should avoid (Sources: CDC; Pan American Health Organization)	3/18/16
11	*NYT* Same as above	Same as above	How Zika spread around the world (Source: WHO)	3/18/16
12	*NYT* Peçanha & Wallace (2016)	Air travel between U.S. and Zika-affected areas	Risk of local Zika transmission	2/6/16

(*Continued*)

Map #s	*Source*	*Title of Article or Report*	*Title of Graphic*	*Last Updated*
13	*The Huffington Post* Schumaker (2016)	The latest on Zika: Argentina's mosquito problem deepens	Where there's active Zika transmission	2/18/16
14	*The Huffington Post* Almendrala (2016)	An illustrated guide to the Zika outbreak	How Zika spread from a Ugandan forest to the Americas	1/22/16
15	*The Huffington Post* Same as above	Same as above	Wider mosquito ranges, more tropical diseases (Sources: CDC; Kraemer et al. (2015b))	1/22/16
16	*The Indian Express* Full citation information is missing.	Zika virus outbreak: All you need to know	Spread of the Zika virus Two maps entitled: 1 Zika affected countries; 2 Range of the Aedes mosquito (Sources: CDC; Kraemer et al. (2015b))	2/4/16
17	*The Associated Press* Cheng, Satter, & Goodman (2016)	Health officials want more Zika samples, data from Brazil	Where Zika virus is active (Source: CDC)	2/4/16
18	*Nature* Butler (2016)	Zika virus: Brazil's surge in small-headed babies questioned by report	Zika in the Americas (Sources: European Centre for Disease Prevention and Control (Zika-transmission data); Kraemer et al. (2015a))	2/3/16
19	*BBC News* (2016)	Same as above	Zika virus past and present (Source: CDC)	4/13/16
20	Same as above	Same as above	How Zika virus spread from Africa (Source: Lancaster University)	4/13/16
21	Same as above	Same as above	Countries with cases of Zika virus in the Americas (Source: WHO, 4/7/16)	4/13/16
22	Same as above	Same as above	Global distribution of Aedes mosquitoes Two maps entitled: 1 Aedes aegypti mosquito; 2 Aedes albopictus mosquito (Source: Kraemer et al. (2015b))	4/13/16

23	*CNN* CNN Health Staff (2016)	Map: Tracking Zika virus	Countries with locally transmitted cases of Zika virus (Sources: CDC; Pan American Health Organization)	4/14/16
24	*The Economist* See *The Economist* in References	Zika fever virus chequers	No separate title for map (Source: Dr K. Khan St. Michael's Hospital, Toronto)	1/23/16
25	*Science News* Rosen (2016)	Rapid spread of Zika virus in the Americas raises alarm	Risk of local Zika transmission (Original Source: Bogoch et al. (2016))	2/20/16
26	*CBS* Welch (2016)	Zika virus and pregnancy: What women need to know	Zika transmission risk in the United States (Original Source: Bogoch et al. (2016))	2/16/16
27	*CBS* See CBS/AP in References	What Americans don't know about Zika could hurt them	No title (Sources: "Image based on data mapped by Olga Wilhelmi, NCAR GIS Program")	4/8/16
28	*ABC News* Mohney (2016)	Experts explain key to stopping Zika virus from spreading in the US	Zika virus travel alerts (Source: CDC)	2/27/16
29	Same as above	Same as above	The mosquitos that spread Zika virus Two maps entitled: 1 Approximate distribution of Aedes aegypti mosquitoes 2 Approximate distribution of Aedes albopictus mosquitos (Source: CDC)	2/27/16
30	*USA Today* Szabo (2016)	Five things to know about the Zika virus	Zika virus (Source: Pan American Health Organization)	1/15/16
31	*USA Today* Ungar (2016)	Zika confirmed in people in Indiana and Ohio, among other states	Zika cases in travelers	2/9/16
32	Breitbart See AFP in References	Zika virus found in Danish tourist returning from South America	The Zika virus (Source information not legible on graphic even at full zoom)	1/27/16

Grounded Theory Approach

Grounded theory methodology, Creswell (2014) explains, is generally characterized by the following: "generating categories of information (open coding), selecting one of the categories and positioning it within a theoretical model (axial coding) and then explicating a story from the interconnection of these categories (selective coding)" (p. 196). I conducted the open coding phase by writing brief observations/descriptions of the maps I collected until I noticed preliminary patterns in the overall design. I formulated several broad categories including *genre* and *color choice*, which I integrated into my analysis as I continued to describe and categorize the visual features of the maps.

I moved into the second stage when I began to narrow and adapt these categories, going back to visuals I analyzed earlier and updating my observations. For instance, I quickly noticed that all of the graphics in my sample were maps. However, I initially kept the term *genre* as a preliminary category. I later changed *genre* to *perspective* to organize my description of the geographic area shown in each map. I then further narrowed this category into *global* and *continent*, for example, which also continued to evolve into *hemispheric* because many maps show the entire (or nearly the entire) Western Hemisphere.

My last step was to connect the final categories I developed to an existing theoretical framework (as applicable) in order to generate a theory about the visual construction of these maps, which I describe in the Discussion. Throughout my process, I followed many of the key characteristics of a grounded theory approach described by Charmaz (2014, see p. 7; also citing Glaser, 1978; Glaser & Strauss, 1967; Strauss, 1987) including "the constant comparison method" (see Glaser & Strauss, 1999) and reviewing relevant scholarship on risk communication theory after conducting my analysis.

In order to quantify visual trends and patterns in the maps I assembled and analyzed, I developed the following final categories, which are explained in more detail in the remainder of this section: perspective (global, hemispheric, continental, country/national), color, high or low context, and design aesthetic.

Perspective refers to the geographic coverage of each map, which I classified as follows:

- *Global:* shows all of the continents or a clear majority of the continents.
- *Hemispheric:* shows the majority of the Western Hemisphere.
- *Continental:* shows an entire continent and/or portions of continents (e.g., South America) but not most of the Western Hemisphere.
- *Country/national:* shows an entire country or nation (e.g., the United States and its territories).

Arguably all thematic maps use the *perspective* that Kimball (2006) has defined as an "aerial position of power" (p. 378), giving viewers an omniscient point of view over the space depicted. Yet the exact geographic area that a map shows is also significant, particularly in a disease map, because this coverage communicates the extent of the "diseased" and "non-diseased" space and consequently how serious the spread of the epidemic is (or may be). For instance, maps that show the entire globe signify to viewers that the risk is a worldwide threat, whereas maps that show a country signify that the risk is contained within that specific region.

Under the broad category of *perspective*, I also observed whether the map showed (1) the current and/or potential geographic range of the mosquitoes (*Aedes aegypti* and *Aedes albopictus*) that transmit the virus (e.g., *mosquito-related*), (2) geographic areas with current and/or past "active" virus transmission (e.g., *Zika-related*), and (3) the map's primary rhetorical purpose—that is, whether it was designed to show temporal, spatial, and/or comparative relationships. Thematic maps always communicate spatial information, which is usually situated within a particular timeframe. Maps that show outbreaks of infectious disease also visually compare "diseased" and "non-diseased" space. Yet the ways that these relationships might be visually prioritized were worth noting because of the overall rhetorical effect on viewers. More specifically, I sought to determine: is the most important feature of this map the spatial, temporal, or the comparative relationships? Are all three equally important? What relationships are emphasized by the creator's design choices?

Color describes whether the map uses a warm color scheme such as red, orange, and/or yellow, or a cool color scheme such as green, blue, and/or purple. Many maps used blue to show the ocean, which I did not take into account when classifying the color scheme. Rather *color* specifically refers to the hue used to visualize information about the current or potential spread of Zika. I also coded the overall level of lightness, brightness, and saturation of the color's hue—that is, if the color choices were bright and highly saturated, or muted, flat, or pastel.

We perceive warm colors as advancing or popping out towards us, whereas cool colors appear to recede away from us. This effect may also explain the metaphoric associations that we tend to assign to color choice. For instance, warm colors are often interpreted as urgent, serious, and attention-inducing. Red and orange in particular are commonly used in Western cultures to alert and warn readers, while cool colors usually elicit the opposite response. Amare and Manning (2013) have argued for a broader understanding of color that focuses on the overall emotional effect on viewers rather than narrowly focused, individualized meanings that are often highly context-specific. Thus, whether a map uses warm or cool colors can profoundly affect how its visual risk information is perceived.

Other visual characteristics such as the lightness/darkness, brightness, and saturation of a color's hue can intensify the effect of color choice. For instance, in his work on social semiotics, Machin (2007) proposes that lightness, brightness, and saturation communicate the level of seriousness, intensity, and emotional weight of a color. According to his theoretical framework, darker and more saturated colors are perceived as more serious, emotional, and dramatic; light, bright, and flat colors are perceived in the opposite sense.

High or *low context* is a dimension of intercultural communication (Hall, 1976) that I have argued can also be applied to visual modes, specifically data visualizations of epidemic disease (Welhausen, 2015a). More to the point, I have proposed that these graphics are often high context because they frequently lack explicit explanatory information—language-based or visual—that tells readers how to interpret the graphic. Indeed many creators of data visualizations today may expect that readers will generally already have the requisite cultural knowledge to understand the visual information conveyed because the conventions that these forms tend to use have long been established (see Kostelnick, 2004).

For this portion of my analysis, I classified each visual as *high* or *low* context to indicate the extent to which readers were expected to understand the map based on its visual features alone (perspective, use of labels, and color and shading, for instance). Unlike the previous two categories in my analysis—*perspective* and *color*—*high* and *low* are not absolute categories. Rather, visual communication (and indeed language-based communication) may have features of both depending upon the context. Further, images that I classified as *low* context may still have included limited explanatory information. However, in order to identify potential patterns, I classified visuals as *low* context if there was any textual and/or visual content in the article or report where the map was published *beyond* the basic features that are conventional to this genre that appeared to clarify what the map shows or otherwise explicitly explain the map. Explanatory information might also take the form of supplemental visuals included within the map itself or the article where the map was published like drawings or photographs (along with captions) that further elucidate some aspect of the map. In sum, I sought to determine: does the map stand alone? Or are readers being told how to interpret the map? I categorized a map as *high context* when readers were expected to infer meaning *entirely* from the visual representation.

Design Aesthetic: Finally, I categorized the overall design aesthetic of each map as *clean* or *cluttered* by critiquing its effectiveness. Here I used Williams' (2014) well-known principles of design (based on Gestalt theory of perception): contrast, repetition, alignment, and proximity. I then combined my evaluation of each map using Williams' categories with Kostelnick's theoretical work on what he has described as a "modernist design aesthetic" (e.g., 1990, 1998, 2007). This particular style,

Kostelnick (1998) explains, is characterized by "functional simplicity" and a "clean, minimalist" appearance (p. 476). Data visualizations, in particular, he argues achieve the "clarity" associated with a modernist design aesthetic by using "high-contrast displays guided by perceptual principles" (2007, p. 283).

Much like *low* or *high context*, whether a design looks *clean* or *cluttered* is much more fluid than my binary classification suggests. However, in order to quantify my findings, I used the following categories to evaluate the holistic effect of a map's design scheme—that is, how its execution of Williams' principles of design functions as a gestalt:

- *Clean*: strong execution of the principles in ways that align with a "modernist design aesthetic." These maps tend to use a sans serif typeface, effective figure/ground contrast primary in terms of color choice, and strong proximity in terms of figure legends and labels.
- *Cluttered*: ineffective execution of one or more design principles in a way that negatively affects the overall design aesthetic. A map might fall into this classification if there was insufficient white space, if the designer included too much visual or textual information (too many labels, for instance), or if the design used a color scheme that created too much contrast or appeared visually jarring (as was the case for several maps that used warm, bright, saturated colors).

After completing my analysis I found that only contrast and proximity were applicable; I did not use alignment or repetition. Contrast emerged as the most important principle because all of the maps in my sample relied heavily on color to communicate information about visual risk. Proximity applied to maps that included labels for countries as well as multiple variables in the legend.

Results

I used the coding process I describe previously to identify and quantify the visual trends and patterns described in this section. Next, I will discuss these findings using the categories I outlined previously in the Methods.

Perspective

The primary visual content of the maps I analyzed fell into one of two categories: (1) those that show the geographic distribution of the virus, that is, locations with past and/or current "active" Zika transmission (i.e., *Zika transmission-related*) and (2) those that show the current and/or potential geographic range of the mosquitoes (*Aedes aegypti* and *Aedes albopictus*) that transmit the disease (i.e., *mosquito-related*).

The majority—24 maps—showed the former, while eight showed the latter, and two maps (maps 18 and 32) showed both. The most common map types were those that showed a *global* (14) or a *hemispheric* (12) *perspective,* while five were *national* (United States), and only one was *continental.*

As previously stated, thematic maps show spatial relationships. Indeed each map in my analysis conveyed either *Zika transmission-related* or *mosquito-related* visual information within a particular geographic space. I also coded whether each map communicated comparative relationships. Because the primary goal of these maps was to show either *Zika transmission-related* or *mosquito-related* visual information in contrast to unaffected geographic space, each map also compared space affected (or potentially affected) by the Zika virus to space not affected.

Finally, one might expect that all of the maps included in this analysis would also show temporal relationships because thematic maps show how select variable(s) occupy a defined geographic space at particular point(s) in time. Yet this was not the case. For instance, map 1 (see Figure 4.1) shows "reported active transmission" presumably on the date that the map was published; however, the exact time frame is not clear. Of the 32 maps I analyzed, I categorized half as showing temporal relationships because the exact timeframe shown was conveyed. Of these, the overt purpose of several maps

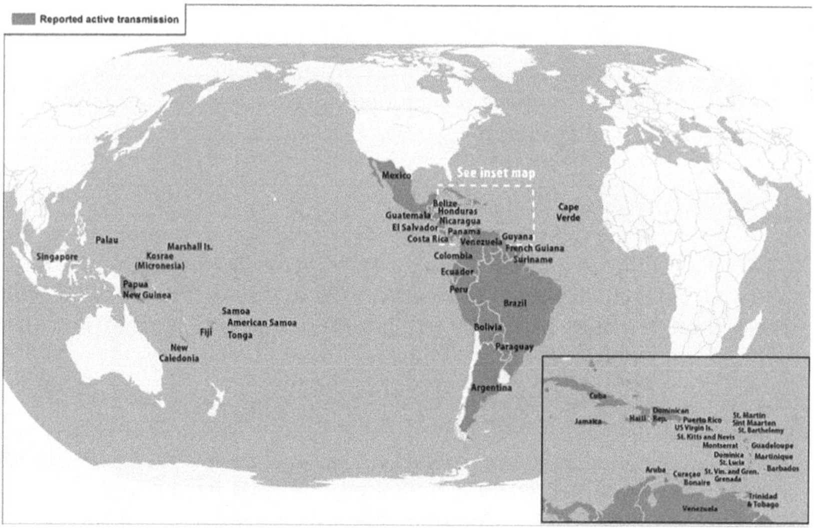

Figure 4.1 Global map showing "reported active Zika transmission" (CDC, 2016a). Color figure in plate section.

(maps 7, 11, and 20, for example) was specifically to show how Zika has spread east since it was discovered.

Color

The majority of maps (19) used a warm color scheme, most frequently shades of orange and/or yellow (some maps used both). Six maps (maps 5, 16, 18, 22, 24, 27) used bright, highly saturated shades of red. Eleven maps used cool colors with the majority of these using shades of blue and/or purple; three maps used shades of green (maps 16, 20, and 32). There was a balance between graphics that used bright, highly saturated colors and those that used muted or pastel shades. Two maps used both a warm and cool color scheme: map 18 and map 22 (see Figure 4.2). In these maps warm colors indicated high levels of risk (probability of the mosquitoes that spread the disease appearing in the region), while cool colors indicated low to minimal risk.

High Versus Low Context

I classified 13 graphics as *high* context and the remainder (19) as *low* context. Several of the maps also included embedded illustrations (maps 13, 14, 15 and 32), which I classified as *low* context. *High* context graphics were more likely to be created directly by CDC and WHO (maps 1–5) and probably also used by expert audiences. For instance, map 5 (Figure 4.3 shown on next page) includes no explanatory information. *Low* context graphics, conversely, were more likely to appear in publication venues targeted to non-expert readers (*Huffington Post*, *CBS*, *ABC*, and *USA Today*).

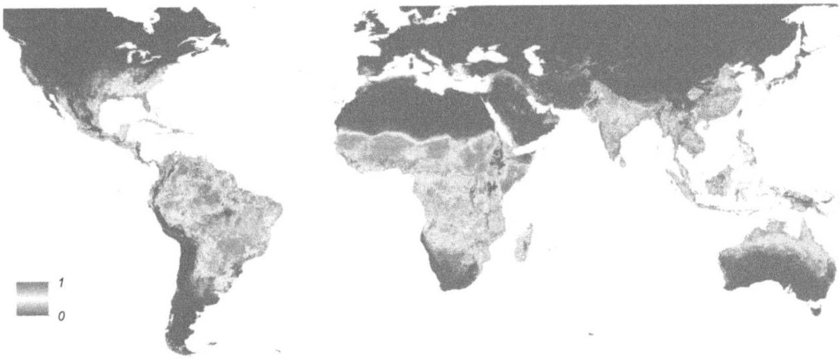

Figure 4.2 "Global distribution of Aedes mosquitos" (Kraemer et al., 2015a). Color figure in plate section.

Figure 4.3 "Countries and territories with autochthonous of Zika virus circulation 2007–2016." Color figure in plate section.

Source: Reprinted from Zika situation report. February 5, 2016. Neurological syndrome and congenital anomalies, WHO, 2016.

Design Aesthetic

Finally, I categorized each map as *clean* or *cluttered.* As I forecast in the Methods, this criterion was somewhat difficult to apply because some of the visuals executed some of the principles of design more effectively than others. For instance, I classified map 21 as *clean* because the map used strong contrast and mostly good proximity with almost adequate white space. However, the detail of the Caribbean islands in the upper right-hand corner was cluttered because the designer attempted to label all of the islands.

Of the maps in my sample, I classified four as *cluttered* and the remainder as *clean.* Interestingly, I noted that three of the four I classified as cluttered were created by CDC and WHO, respectively (maps 3, 4, and 5—see Figure 4.3). All of the maps I classified as *cluttered* used too much contrast primarily in terms of color (e.g., aesthetically this design choice was visually overwhelming) and/or insufficient white space. The maps I classified as *clean* consistently used strong figure/ground contrast (primarily in terms of color choices) and effective proximity (i.e., countries were labeled consistently and labels were grouped where necessary; many used a bold, sans serif typeface).

Discussion: Toward a Theory of Visual Risk Communication in Crisis and Emergency Risk Scenarios

Risk Perception Theory and Hazard + Outrage

Since its origins in the 1980s, risk communication research has recognized fundamental differences in the ways that experts and non-experts perceive risks. Experts tend to evaluate risk in terms of "qualitative and quantitative measures" (Dransch, Rotzoll & Poser, 2010, p. 296). Conversely, non-experts tend to consider what risk communication expert Peter Sandman (2014a) has described as *hazard* + *outrage*—that is, how dangerous the risk is likely to be as well as how much anxiety it invokes and not necessarily how likely it is that the hazard will occur.

Non-experts tend to assess risks in terms of characteristics like *voluntariness*, *controllability*, and *familiarity* that have been defined by risk perception theory (i.e., the psychometric paradigm; see Covello, Peters, Wojtecki, & Hyde, 2001, p. 385; Fischhoff, Slovic, Lichtenstein, Read, & Combs, 1978; Sandman, 1987; Slovic, 1987). From this perspective, risks that are seen as *voluntary*, *controllable*, and *familiar*—like driving a car, for example—are generally less threatening than those that are not well *understood*, associated with *uncertain* outcomes, and/or have high *catastrophic potential*. For instance, whenever we travel—regardless of the mode—we *voluntarily* assume risks. Yet because we believe we have more *control* when we drive and because driving is usually more *familiar*, driving usually feels less "risky" than flying.

Media Coverage During Crisis and Emergency Risk Scenarios: Analysis of Study Results

Outbreaks of epidemic disease often constitute what Sandman (2014b) has described as a *high hazard/high outrage* (crisis and emergency risk communication) situation in which both perceived danger and anxiety are high. Such scenarios often warrant risk communication that "help(s) people bear their feelings (their outrage) and cope effectively with serious hazards" (Sandman, 2004, para. 2), provides timely, accurate, and action-oriented information (Reynolds, Seeger, & CDC, 2014), and recognizes what is not known, avoids making promises, and validates the public's concerns (Sandman, 2004). Risk communication "should make the public aware of its vulnerability to a particular risk and inform it about the most effective protection measures" (Dransch et al., 2010).

Arguably, media coverage of Zika during the timeframe this study was conducted in 2016 adhered to these guidelines. For example, information about the virus's spread (like the maps discussed in this chapter) was publicized after the threat was discovered (Pearson, 2016), and pregnant women were advised not to travel to affected countries

(LaMotte & Goldschmidt, 2016). Yet media coverage has also been shown to over-emphasize evolving public health threats like infectious and communicable disease with less attention directed toward "chronic risks" like not exercising and smoking (Bomlitz & Brezis, 2008, para. 1). Reporting in such scenarios has also been criticized for "emphasiz[ing] the more sensational aspects of a crisis" including "wrongdoing, blame, and danger" (Reynolds et al., 2014, p. 36). Indeed, this characterization may describe news organizations' initial communication approach. As Ungar (1998) argues in his analysis of the 1995 Ebola outbreak, during a "hot crisis" reporting strategies tend to move from "alarming" to "reassuring" as concern and worry begins to increase among public audiences (p. 36). Ungar's observations also align with Sandman's (2014b) strategy of "precaution advocacy"—that is, "trying to arouse concern and motivate preparedness even though there's nothing bad on the immediate horizon" (para. 8)—in the early stages of a potential pandemic.

The results of the study in this chapter suggest that visual communication too may follow a similar pattern. In my analysis below I complete the final stage of my grounded theory approach, linking the results I discussed in the previous section to several of the characteristics defined by risk perception theory.

Perspective

The maps in this study situate the risk of Zika in space (and often time), allowing viewers to compare specific "diseased" and "non-diseased" areas rather than the number of cases (bar charts) and/or the number of cases over time (line graphs). The geographic space shown in these maps (indeed in any disease map) conveys to viewers the current level of containment or lack thereof over the potential spread of the disease.

The maps in this study overwhelmingly depict Zika as a global and hemispheric level threat, which increases the perceived level of hazard. Most maps (24) also show areas of active transmission. Further, eight show the potential range of the mosquitos that transmit the disease, visually reinforcing lack of containment, *control*, and consequently *uncertainty*, all of which usually increase risk perception.

Color

The maps in this study also tend to use warm (19) rather than cool colors (10), which convey urgency, alarm, and danger. Indeed the warm colors used in many of the maps combined with a global or hemispheric perspective (for instance, maps 3, 5, 7, 8) together visually signify that Zika is an urgent, global (or hemispheric)-level threat. Map 25 (see Figure 4.4), for instance, illustrates this effect.

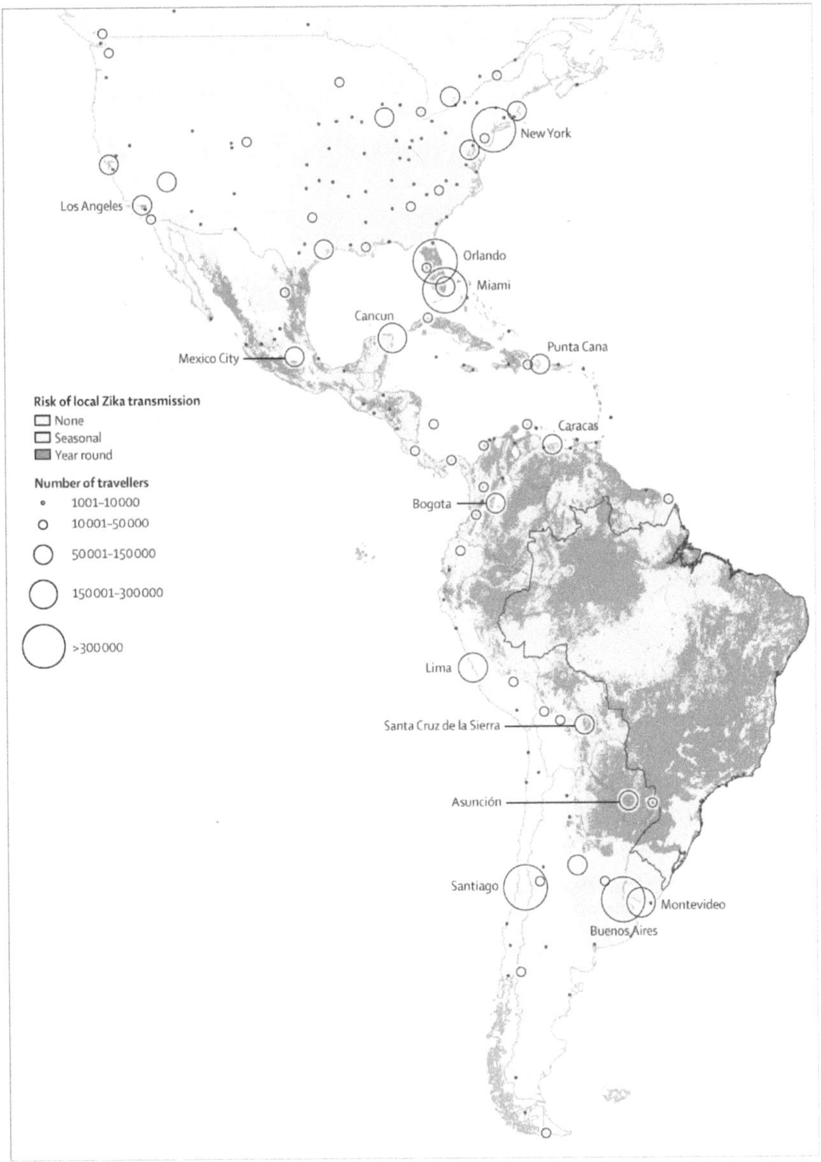

Figure 4.4 "Final destinations of travellers departing Brazil by potential for autochthonous Zika transmission." Color figure in plate section.

Source: Reprinted from *The Lancet*, 387, Bogoch et al., "Anticipating the international spread of Zika virus from Brazil," 335–336, Copyright (2016), with permission from Elsevier.

Together, these design choices show lack of containment and *control*, reinforcing *dread* and *catastrophic potential* by visually escalating the perception of the threat. Muted warm colors, which are used in some of the maps (7 and 8, for example), can mitigate this message. However, the brighter and the more saturated the color, the more pronounced the effect as shown in Figure 4.3, for instance.

Low Context

Explanatory textual and visual information is included in most of the maps in this study to explain the risk more directly and explicitly. Providing this additional information can downplay an alarming visual message conveyed through perspective and color by promoting *trust in institutions* (the news organization that creates the maps and the public health agencies who collect the data) and *understanding* (by explaining the virus's *natural origin*, for instance). However, much of this additional contextual information is language-based, and if viewers do not read the article, they may rely on the visual message alone to assess the threat. Further, if an article admonishes readers to not become alarmed, but the map communicates the opposite message (like Figures 4.3 and 4.4), the visual message may override the textual message.

Design Aesthetic

The layout choices in most of the maps adhere to the visual conventions of a clean design aesthetic, which simplifies, streamlines, and abstracts the quantitative information shown. Clean design may promote *understanding* and *trust in institutions*, and may downplay *uncertainty* because the visual information promotes "clarity." But a minimalist aesthetic may also visually reinforce that Zika is an urgent, global (or hemispheric)-level threat because it may boost the credibility of the visual message. For instance, maps 9 and 12 have particularly strong aesthetic appeal. Map 9 uses shading to show the potential geographic range of *Aedes aegypti*. Areas in the mosquito's range are a deep, light brick shade of red, which fades to a light greyish beige in areas where the mosquito is not present. Map 12 shows "risk of local Zika transmission" in matte orange; mountains along with other natural features look hand-drawn, emphasizing *natural origin*.

Conclusions: Visualizing a Pandemic

All of the design choices that I explore in this study—perspective, color, high/low context, design aesthetic—contribute to the ways that Zika risk is framed in these maps and subsequently perceived by viewers. Perspective and color may carry the most weight of the overall visual message by establishing the geographic scope of the threat as well as whether viewers should be alarmed (warm colors) or reassured (cool colors). Low context and a clean design aesthetic may then either reinforce or downplay these dominant visual choices. A low-context textual message that overtly emphasizes alarm may visually bolster this message, while a clean design aesthetic may strengthen the credibility of the organization delivering the message.

Grounded theory has been met with a number of criticisms, which its creators have responded to (see Glaser & Strauss, 1999). I employ this inductive approach because it offers one strategy for lending insight into the visual reporting strategies used to communicate quantitative risk information about emergent public health threats like Zika. Such data are usually collected by public health agencies. However, they are often disseminated to public audiences through news coverage (Reynolds & Seeger, 2005), which profoundly shapes how viewers perceive risks (Slovic, 1986). As mentioned, media coverage has been critiqued for gravitating toward sensationalism. Yet such criticisms also seem to assume that when outrage (anxiety) escalates to panic, as was the case in Europe and the United States when Ebola struck West Africa in 2014 (see Higgins, 2014), the media is primarily responsible.

However, not all infectious and communicable diseases are perceived as equally dangerous or threatening, which also shapes how risk messages about a particular disease are perceived. The 2014–2016 Ebola outbreak was the worst to date. But the virus did not pose a danger to people in the United States (Fox, 2016). In contrast, Zika continued to gain ground in 2016. As of late August 2016 the virus was widespread throughout the Western Hemisphere, with rapid transmission in Puerto Rico (CDC, 2016d) and locally acquired cases diagnosed in two Florida neighborhoods (CDC, 2016b). Yet Americans remained generally unconcerned (Byrnes, 2016) by what then CDC Director Tom Frieden described as an "invisible crisis" (Fox, 2016, para. 7). This profound difference in the public response can be attributed, in part, to fundamental differences in the ways that Zika and Ebola align with many of the dimensions of risk perception theory.

Contracting either disease is *involuntary* (e.g., people do not agree to be exposed like they agree to travel-associated risks, for example), which is a factor that usually increases anxiety. But Zika may seem more *familiar* and *controllable* because mosquitoes are already common in the areas shown on the maps, and most people who live in these areas have been bitten by mosquitoes before with no severe adverse consequences. Further, they can also exert some *control* over exposure by engaging in preventative behaviors like using insect repellants.

Ebola, on the other hand, is not at all *familiar.* Consequently, exposure is seen as much less *controllable. Uncertainty* may also be high for both diseases because the long-term effects are unknown. However, perhaps the most important difference is *dread* and *catastrophic potential.* Ebola causes severe symptoms and has around a 50% mortality rate (WHO, 2016c). In sharp contrast, people with Zika often exhibit minor, if any, symptoms (CDC, 2016e). It is important to acknowledge that *dread* and *catastrophic potential*, however, are probably very high among pregnant women and those wanting to become pregnant who live in areas that could be exposed.

Ultimately, no risk communication strategy is neutral. Any visual or language-based choice that a risk communicator makes influences how non-experts will perceive the risk. But Ebola is intrinsically far more threatening than Zika. Therefore, many of the visual risk communication strategies that I identify in this study—maps that show a global or hemispheric perspective, warm colors, and clean design aesthetic—are likely to be perceived very differently for a disease like Ebola than a disease like Zika. Indeed, a similar visual communication strategy may have exacerbated the full-blown panic that ensued during the 2014–2016 Ebola outbreak (Welhausen, 2015a).

Consequently, when risk communicators decide whether visual risk communication should seek to "alarm" or to "reassure," they should also assess how the characteristics of the disease align with the dimensions of risk perception theory. More specifically, when possible, they should solicit information from the intended audience about how viewers may *already* perceive the risk. A visual risk communication strategy that does strongly reinforce lack of containment and *controllability* may be needed if the Zika outbreak worsens to persuade Americans to engage in behaviors that might reduce their risk, one of the key features of effective risk communication that Rohrmann (1992) identifies. However, should the risk change—for example, if cases of microcephaly begin to increase in the United States (escalating perceived *dread* and *catastrophic potential*)—then risk communicators would want to modify their overall risk message to emphasize containment (and reassurance) to mitigate increasing panic. The heuristic that I develop in this chapter offers a framework that creators of visual risk communication can draw from in determining how to design visual risk communication in order to achieve these objectives.

References

AFP. (2016, January 27). Zika virus found in Danish tourist returning from South America. *Breitbart*. Retrieved March 8, 2017, from www.breitbart.com/london/2016/01/27/zika-virus-found-in-danish-tourist-returning-from-south-america/.

Almendrala, A. (2016, January 22). An illustrated guide to the Zika outbreak. *The Huffington Post*. Last updated: January 22, 2016. Retrieved March 8, 2017, from www.huffingtonpost.com/entry/guide-to-zika-virus_us_56a272afe4b076aadcc675b6.

Amare, N., & Manning, A. (2013). Teaching form and color as emotion triggers. In E. R. Brumberger & K. M. Northcut (Eds.), *Designing texts: Teaching visual communication* (pp. 181–195). Amityville, NY: Baywood Publishing, Inc.

Ancker, J. S., Senathiraja, Y., Kukafka, R., & Starren, J. B. (2006). Design features of graphics in health risk communication: A systematic review. *Journal of the Medical Informatics Association, 13*(6), 608–618. Retrieved March 8, 2017, from www.ncbi.nlm.nih.gov/pmc/articles/PMC1656964/.

BBC News. (2016, August 31). *Zika outbreak: What you need to know*. Retrieved March 8, 2017, from www.bbc.com/news/health-35370848.

Bogoch, I. I., Brady, O. J., Kraemer, M. U., German, M., Creatore, M. I., Kulkarni, M. A., ... Watts, A. (2016). Anticipating the international spread of Zika virus from Brazil. *The Lancet, 387*(10016), 335–336. Retrieved March 8, 2017, from doi:10.1016/S0140-6736(16)00080-5.

Bomlitz, L. J., & Brezis, M. (2008). Misrepresentation of health risks by mass media. *Journal of Public Health, 30*(2), 202–204. Retrieved March 8, 2017, from doi:10.1093/pubmed/fdn009.

Brasseur, L. (2004). *Visualizing technical information: A cultural critique*. Amityville, NY: Baywood.

Buehl, J. (2014). Toward an ethical rhetoric of the digital scientific image: Learning from the era when science met Photoshop. *Technical Communication Quarterly, 23*(3), 184–206.

Butler, D. (2016). Zika virus: Brazil's surge in small-headed babies questioned by report. *Nature, 530,* 13–14. Retrieved March 8, 2017, from doi:10.1038/nature.2016.19259.

Byrnes, J. (2016, August 3). Poll: Vast majority not concerned about Zika. *The Hill*. Retrieved March 8, 2017, from http://thehill.com/blogs/blog-briefing-room/news/290216-poll-vast-majority-not-concerned-about-zika.

CBS/AP. (2016, April 8). What Americans don't know about Zika could hurt them. *CBS News*. Retrieved March 8, 2017, from www.cbsnews.com/news/what-americans-dont-know-about-zika/.

CDC. (2014). *Crisis and emergency risk communication, 2014 edition*. Retrieved August 4, 2017, from https://emergency.cdc.gov/cerc/resources/pdf/cerc_2014edition.pdf.

CDC. (2016a). National center for emerging and zoonotic infectious diseases. Division of global migration and quarantine. *Travelers' health. Zika travel information*. Retrieved March 8, 2017, from wwwnc.cdc.gov/travel/page/zika-travel-information.

CDC. (2016b). National center for emerging and zoonotic infectious diseases. Division of vector-borne diseases. *Zika virus. Case counts in the US*. Retrieved August 24, 2016, from www.cdc.gov/zika/geo/united-states.html.

CDC. (2016c). *CDC newsroom. CDC concludes Zika causes microcephaly and other birth defects*. Retrieved April 13, 2017, from www.cdc.gov/media/releases/2016/s0413-zika-microcephaly.html.

CDC. (2016d). *CDC newsroom. Zika infections increasing rapidly in Puerto Rico*. Retrieved August 1, 2016, from www.cdc.gov/media/releases/2016/p0729-zika-infections-puerto-rico.html.

CDC. (2016e). National center for emerging and zoonotic infectious diseases. Division of vector-borne diseases. *Zika virus. Symptoms, testing, and treatment*. Retrieved June, 21, 2016, from www.cdc.gov/zika/symptoms/.

Cha, A. E., & Sun, L. H. (2016, February 4). What is Zika? And what are the risks as it spreads? *The Washington Post*. Retrieved March 8, 2017, from www.washingtonpost.com/news/to-your-health/wp/2016/01/21/zika-virus-faq-more-than-a-million-infected-globally-a-dozen-in-the-united-states/.

Charmaz, K. (2014). *Constructing grounded theory: A practical guide through qualitative analysis* (2nd ed.). Thousand Oaks, CA: Sage Publications.

Cheng, M., Satter, R., & Goodman, J. (2016, February 4). Health officials want more Zika samples, data from Brazil. *The Associated Press*. Retrieved March 8, 2017, from http://midco.net/news/read/category/world/article/the_associated_press-health_officials_want_more_zika_samples_data_from-ap.

CNN Health Staff. (2016, April 12). Map: Tracking Zika virus. *CNN*. Retrieved March 8, 2017, from www.cnn.com/2016/01/27/health/map-zika-virus-transmission/index.html.

Cooke, L. (2003). Information acceleration and visual trends in print, television, and web news sources. *Technical Communication Quarterly*, *12*(2), 155–182.

Covello, V. T., Peters, R. G., Wojtecki, J. G., & Hyde, R. C. (2001). Risk communication, the West Nile virus epidemic, and bioterrorism: Responding to the communication challenges posed by the intentional or unintentional release of a pathogen in an urban setting. *Journal of Urban Health: Bulletin of the New York Academy of Medicine*, *78*(2), 382–391.

Crampton, J. (2010). *Mapping: A critical introduction to cartography and GIS*. London: Wiley-Blackwell.

Creswell, J. W. (2014). *Research design: Qualitative, quantitative, and mixed methods approaches* (4th ed.). Thousand Oaks, CA: Sage Publications.

Ding, H. (2009). Rhetorics of alternative media in an emerging epidemic: SARS, censorship, and extra-institutional risk communication. *Technical Communication Quarterly*, *18*(4), 327–350.

Ding, H. (2013). Transcultural risk communication and viral discourses: Grassroots movements to manage global risks of H1N1 flu pandemic. *Technical Communication Quarterly*, *22*(2), 126–149.

Dragga, S., & Voss, D. (2001). Cruel pies: The inhumanity of technical illustrations. *Technical Communication*, *48*(3), 265–274.

Dransch, D., Rotzoll, H., & Poser, K. (2010). The contribution of maps to the challenges of risk communication to the public. *International Journal of Digital Earth*, *3*(3), 292–311.

Fischhoff, B., Slovic, P., Lichtenstein, S., Read, S., & Combs, B. (1978). How safe is safe enough? A psychometric study of attitudes towards technological risks and benefits. *Policy Sciences*, *9*(2), 127–152.

Fox, M. (2016, July 4). 'Invisible' Zika virus epidemic frustrates health officials. *NBC News*. Retrieved March 8, 2017, from www.nbcnews.com/storyline/zika-virus-outbreak/invisible-zika-virus-epidemic-frustrates-health-officials-n602811.

Glaser, B. G. (1978). *Theoretical sensitivity*. Mill Valley, CA: Sociology Press.

Glaser, B. G., & Strauss, A. L. (1967). *The discovery of grounded theory: Strategies for qualitative research*. New Brunswick, NJ: Aldine Transaction.

Glaser, B. G., & Strauss, A. L. (1999). *The discovery of grounded theory: Strategies for qualitative research*. New Brunswick, NJ: Aldine Transaction.

Hall, E. T. (1976). *Beyond culture*. Garden City, NY: Anchor Press/Doubleday.

Harley, J. B. (2009). Maps, knowledge, and power. In G. Henderson & M. Waterstone (Eds.), *Geographic thought: A praxis perspective* (pp. 129–148). New York: Routledge.

Higgins, A. (2014, October 17). In Europe, fear of Ebola exceeds the actual risks. *The New York Times* (Europe). Retrieved March 8, 2017, from www.nytimes.com/2014/10/18/world/europe/in-europe-fear-of-ebola-far-outweighs-the-true-risks.html.

Kimball, M. (2006). London through rose-colored graphics. *Journal of Technical Writing and Communication, 36*, 353–381.

Kraemer, M. U., Sinka, M. E., Duda, K. A., Mylne, A. Q., Shearer, F. M., Barker, C. M., ... Hendrickx, G. (2015a). The global distribution of the arbovirus vectors Aedes aegypti and Ae. albopictus. *Elife*, *4*, e08347. Retrieved March 8, 2017, from https://elifesciences.org/content/4/e08347.

Kraemer, M. U., Sinka, M. E., Duda, K. A., Mylne, A., Shearer, F. M., Brady, O. J., ... Coelho, G. E. (2015b). The global compendium of Aedes aegypti and Ae. albopictus occurrence. *Scientific Data*, *2*. Retrieved March 8, 2017, from www.nature.com/articles/sdata201535.

Korte, G. (2016, April 11). 'Scarier than we initially thought': CDC sounds warning on Zika virus. *USA Today*. Retrieved March 8, 2017, from www.usatoday.com/story/news/politics/2016/04/11/scarier-than-we-initially-thought-cdc-sounds-warning-zika-virus/82894878/.

Kostelnick, C. (1990). Typographical design, modernist design aesthetics, and professional communication. *Journal of Business and Technical Communication*, *4*(1), 5–24.

Kostelnick, C. (1998). Conflicting standards for designing data displays: Following, flouting, and reconciling them. *Technical Communication, 45*(4), 473–482.

Kostelnick, C. (2004). Melting-pot ideology, modernist aesthetics, and the emergence of graphical conventions: The statistical atlases of the United States, 1874–1925. In C. Hill & M. Helmers (Eds.), *Defining visual rhetorics* (pp. 215–242). Mahwah, NJ: Lawrence Erlbaum Associates.

Kostelnick, C. (2007). The visual rhetoric of data displays: The conundrum of clarity. *IEEE Transactions on Professional Communication*, *50*(4), 280–294.

LaMotte, S., & Goldschmidt, D. (2016, January 16). CDC issues travel warning for pregnant women due to Zika virus. *CNN*. Retrieved March 8, 2017, from www.cnn.com/2016/01/15/health/zika-pandemic-travel-warnings-cdc/.

Latour, B., & Woolgar, S. (2013). *Laboratory life: The construction of scientific facts*. Princeton, NJ: Princeton University Press.

Lewis, L. F., & Whitely, A. (1992). Initial perceptions of professional facilitators regarding GDSS impacts: A study using the grounded theory approach. In *Proceedings of the Twenty-Fifth Hawaii International Conference on System Sciences* (Vol. 4, pp. 80–89).

Lipkus, I. M., & Hollands, J. G. (1999). The visual communication of risk. *Journal of the National Cancer Institutes Monographs, 25*, 149–163.

Lucey, D. R., & Gostin, L. O. (2016). The emerging Zika pandemic: Enhancing preparedness. *JAMA, 315*(9), 865–866.

Machin, D. (2007). *Introduction to multimodal analysis* (1st ed.). New York: Oxford University Press.

Maron, D. F. (2016, May 24). How Zika spiraled out of control. *Scientific American*. Retrieved March 8, 2017, from www.scientificamerican.com/article/how-zika-spiraled-out-of-control1/.

McNeil, D. G., Saint Louis, C., & St. Fleur, N. (2016, July 29). Short answers to hard questions about Zika virus. *The New York Times*. Retrieved March 8, 2017, from www.nytimes.com/interactive/2016/health/what-is-zika-virus.html.

McNely, B., Spinuzzi, C., & Teston, C. (2015). Contemporary research methodologies in technical communication. *Technical Communication Quarterly*, *24*(1), 1–13.

Mohney, G. (2016, January 27). Experts explain key to stopping Zika virus from spreading in the US. *ABC News*. Retrieved March 8, 2017, from http://abcnews.go.com/Health/experts-explain-key-stopping-zika-virus-spreading-us/story?id=36553111.

Pearson, M. (2016, February 2). Zika virus sparks 'public health emergency.' *CNN*. Retrieved March 8, 2017, from www.cnn.com/2016/02/01/health/zika-virus-public-health-emergency/.

Peçanha, S. and Wallace, T. (2016, February 6). Air travel between U.S. and Zika-affected areas. *The New York Times*. Retrieved March 8, 2017, from www.nytimes.com/interactive/2016/02/06/science/air-travel-from-countries-affected-by-zika.html.

Perini, L. (2005). The truth in pictures. *Philosophy of Science*, *72*(1), 262–285.

Pickles, J. (2004). *A history of spaces. Cartographic reason, mapping and the geo-coded world*. London: Routledge.

Portewig, T. C. (2008). The role of rhetorical invention for visuals: A qualitative study of technical communicators in the workplace. *Technical Communication*, *55*(4), 333–342.

Reeves, C. (2011). Scientific visuals, language, and the commercialization of a scientific idea: The strange case of the prion. *Technical Communication Quarterly*, *20*(3), 239–273.

Reynolds, B., & Seeger, M. W. (2005). Crisis and emergency risk communication. *Journal of Health Communication, 10*, 43–55.

Reynolds, B., Seeger, M. W., & CDC. (2014). Crisis + emergency risk communication. 2014 edition. Publication ID 221793. Retrieved June 21, 2017, from https://stacks.cdc.gov/view/cdc/25531

Richards, A. R. (2009). Argument and authority in the visual representations of science. *Technical Communication Quarterly*, *12*(2), 183–206.

Rohrmann, B. (1992). The evaluation of risk communication effectiveness. *Acta Psychologica, 81*, 169–192.

Rosen, M. (2016, January 22). Rapid spread of Zika virus in the Americas raises alarm. *Science News*. Retrieved March 8, 2017, from www.sciencenews.org/article/rapid-spread-zika-virus-americas-raises-alarm.

Ross, D. G. (2013). Common topics and commonplaces of environmental rhetoric. *Written Communication*, *30*(1), 91–131.

Sandman, P. M. (1987). Risk communication: Facing public outrage. *EPA Journal*, 21–22. Retrieved March 8, 2017, from www.psandman.com/articles/facing.htm.

Sandman, P. M. (2004). *Crisis communication: A very quick introduction*. Retrieved March 8, 2017, from www.psandman.com/col/crisis.htm.

Sandman, P. M. (2014a). *Dr. Peter M. Sandman introduction to risk communication and orientation to this website.* Retrieved March 8, 2017, from www.psandman.com/index-intro.htm.

Sandman, P. M. (2014b). *Dr. Peter M. Sandman. Pandemic flu and other infectious diseases index.* Retrieved March 8, 2017, from www.psandman.com/index-infec.htm.

Schumaker, E. (2016, February 18). The latest on Zika: Argentina's mosquito problem deepens. *The Huffington Post.* Retrieved March 8, 2017, from www.huffingtonpost.com/entry/zika-virus-new-evidence-microcephaly_us_56c5e785e4b08ffac127d59d.

Slovic, P. (1986). Informing and educating the public about risk. *Risk Analysis, 6*(4), 403–415.

Slovic, P. (1987). Perception of risk. *Science, 236*(4799), 280–285.

Strauss, A. L. (1987). *Qualitative analysis for social scientists.* New York: Cambridge University Press.

Strauss, A., & Corbin, J. (1994). Grounded theory methodology: An Overview. In N.K. Denzin & Y. S. Lincoln (Eds.), *Handbook of qualitative research.* (273–285). Thousand Oaks, CA: Sage Publications.

Sun, L. H. (2016, April 13). CDC confirms Zika virus causes microcephaly, other birth defects. *The Washington Post.* Retrieved March 8, 2017, from www.washingtonpost.com/news/to-your-health/wp/2016/04/13/cdc-confirms-zika-virus-causes-microcephaly-other-birth-defects/.

Szabo, L. (2016, January 15). Five things to know about the Zika virus. *USA Today.* Retrieved March 8, 2017, from www.usatoday.com/story/news/2016/01/15/five-things-know-zika-virus/78853408/.

Teston, C. (2012). Moving from artifact to action: A grounded investigation of visual displays of evidence during medical deliberations. *Technical Communication Quarterly, 21*, 187–209.

The Economist. (2016, January 23). *Zika fever. Virus chequers.* Retrieved March 8, 2017, from www.economist.com/news/science-and-technology/21688849-newly-emerging-disease-threatening-americas-virus-chequers.

Welch, A. (2016, February 16). Zika virus and pregnancy: What women need to know. *CBS News.* Retrieved March 8, 2017, from www.cbsnews.com/news/zika-virus-and-pregnancy-what-women-need-to-know/.

Welhausen, C. A. (2015a). Visualizing a non-pandemic: Considerations for communicating public health risks in intercultural contexts. *Technical Communication, 62*(4), 244–257.

Welhausen, C. A. (2015b). Power and authority in disease maps: Visualizing medical cartography through yellow fever mapping. *Journal of Business and Technical Communication, 29*(3), 257–283.

WHO. (2016a). *Zika situation report. Zika virus, microcephaly and Guillain-Barré.* Retrieved March 14, 2017, from www.who.int/emergencies/zika-virus/situation-report/7-april-2016/en/.

WHO. (2016b). *Zika situation report. Neurological syndrome and congenital anomalies.* February 5, 2016. Retrieved March 14, 2017, from www.who.int/emergencies/zika-virus/situation-report/5-february-2016/en/.

WHO. (2016c). *Ebola virus disease.* Retrieved March 8, 2017, from www.who.int/mediacentre/factsheets/fs103/en/.

WHO. (2017). *The history of Zika virus*. Retrieved March 8, 2017, from www.who.int/emergencies/zika-virus/history/en/.

Williams, R. (2014). *The non-designer's design book* (4th ed.). Berkeley, CA: Peachpit Press.

Wood, D. (1992). *The power of maps*. London: Guilford Press.

Ungar, S. (1998). Hot crises and media reassurance: A comparison of emerging diseases and Ebola Zaire. *British Journal of Sociology*, 36–56.

Ungar, L. (2016, February 9). Zika confirmed in people in Indiana and Ohio, among other states. *USA Today*. Retrieved March 8, 2017, from www.usatoday.com/story/news/nation/2016/02/09/zika-confirmed-oh-elsewhere-among-travelers-affected-areas/80052798/.

5 The Tree of Life in Popular Science

Assumptions, Accuracy, and Accessibility

Han Yu

In American popular culture, the tree of life is not an uncommon metaphor: it is generally taken to signify the progress of life, sometimes with a religious or spiritual undertone. The concept takes on a narrower, though no less significant, meaning in biological sciences where Charles Darwin is credited as one of the first to apply "tree thinking" to natural studies. His 1859 *On the Origin of Species* contains elaborate trees (Figure 5.1) that explain how hypothetical species or varieties of species, over thousands of generations, diverge into different units of life or go extinct.

In modern biology, the term "tree of life" is still used, especially in popular communication (see Zimmer, 2016), though a more accurate term to refer to these visual representations is "phylogenetic diagrams." Other

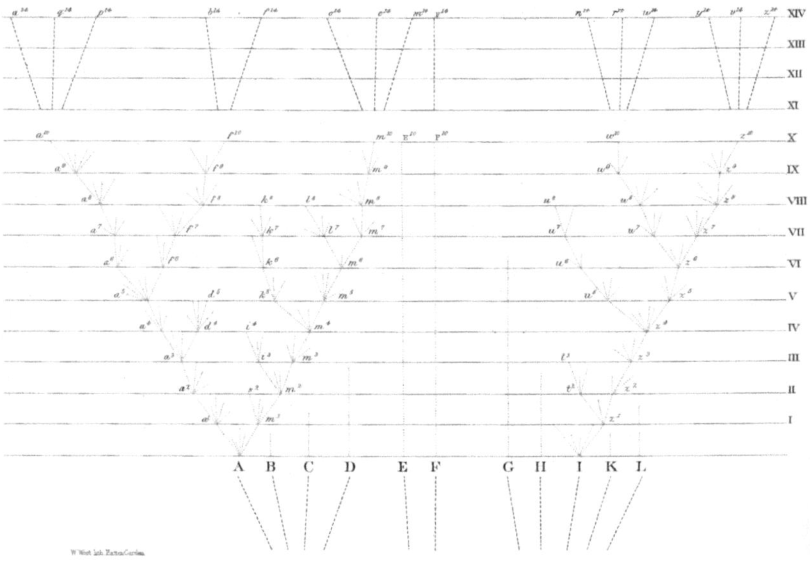

Figure 5.1 Charles Darwin's tree of life in *On the origin of species* (Darwin, 1859).

DOI: 10.4324/9781315160191-5

related terms include "phylogeny," "evolutionary tree," and "cladogram." These terms are sometimes used synonymously, sometimes with subtle differences and not always consistently ("Reading Trees," 2016). In this chapter, "phylogenetic diagram" is used to encompass these variations.

Most succinctly, a phylogenetic diagram is "a diagram that depicts the lines of evolutionary descent of different species, organisms, or genes from a common ancestor" (Baum, 2008a, p. 190). Despite their long history, phylogenetic diagrams became an essential tool to modern biology only in the last 20 or so years (Baum, Smith, & Donovan, 2005; Catley, 2006; Soltis & Soltis, 2003). By putting organisms into appropriate evolutionary and genetic contexts, these diagrams provide scientists with valuable clues for targeted research. In the study of HIV/AIDS, for example, they allow researchers to "identify the source of the virus," "detect viral recombination," and "track viral evolution" (Baum et al., 2005, p. 979). Moreover, by reconstructing the evolutionary history of genes, they allow researchers to understand changes, variations, and relationships at the genetic level (Bluis & Shin, 2003; Soltis & Soltis, 2003).

From the standpoint of education and public communication, phylogenetic diagrams facilitate the understanding of evolution at and beyond the level of individual species. By revealing the evolutionary patterns of diverse organisms and situating that process in geologic time, they present evolution not as a linear process but as a complex, branching event. In North America, where misconceptions about evolutionary mechanisms abound and public acceptance of the common-descent principle of evolution is low, phylogenetic diagrams can be an especially useful tool for public communication (Baum et al., 2005; Gregory, 2008; Scott & Giusti, 2006).

Last and probably most importantly, phylogenetic diagrams promote the tree-thinking heuristic advocated by contemporary scientists, philosophers, and educators (Baum et al., 2005; Gregory, 2008; O'Hara, 1988). This heuristic, enabled by billion-year-old historical as well as modern molecular evidence, allows humans to appreciate the mechanisms that created life on Earth, to realize the fragility of the planet's biodiversity, and to nurture a responsible "stewardship ethic" for maintaining the planet's environment and ecology (Catley, 2006).

Despite phylogenetic diagrams' merits for public science communication, the vast majority of current studies, as detailed below, approach them from the standpoint of formal education and focus on biology textbooks and classroom learning. Very few (MacDonald & Wiley, 2012) explicitly examine phylogenetic diagrams as they are used in popular science communication. Given the diagrams' ability to depict evolution (arguably one of the most important concepts in modern science) and to foster responsible ethics toward the world around us, it is important that we consider how phylogenetic diagrams are deployed in popular media and their potential effects.

This is what this chapter intends to do. It examines the use of phylogenetic diagrams in U.S. popular science magazines, discusses the evolutionary mechanisms these diagrams portray, and considers their potential impact on publics' uptake of evolution. Based on these findings, the chapter also suggests heuristics on how to present phylogenetic diagrams in popular science communication.

Basics of Phylogenetic Diagrams

Figure 5.2 dissects a simple phylogenetic diagram. It features a common ancestral root from where related branches diverge. The direction from the root to branch tips indicates the historical order of branching events. All branches end in *terminals,* which represent taxa that co-exist at a given time; the taxa can be individual organisms, groups of organisms, or genetic materials. Because these entities co-exist, the branches all terminate at the same height; extinct taxa, however, would terminate at a lower height. Related species or genes have recent common ancestors, which are signaled by *nodes.*

As a whole, phylogenetic diagrams portray evolutionary relationships through hierarchically nested units called *clades.* Within each clade are relatively closely related organisms that have split from a common ancestor/node and share derived characteristics (known as *synapomorphies*) (Baum et al., 2005; Gregory, 2008). For example, in Figure 5.2, B and C share a most recent node, node 1, and are more closely related to each other than they are to A or D. Thus, B, C, and node 1 constitute a clade. At the next level, B, C, and A are more closely related to each other than they are to D because they share the next most recent node, node 2. Thus, B, C, A, and node 2 constitute the next level of clade, with the B+C+node1 clade nesting within it.

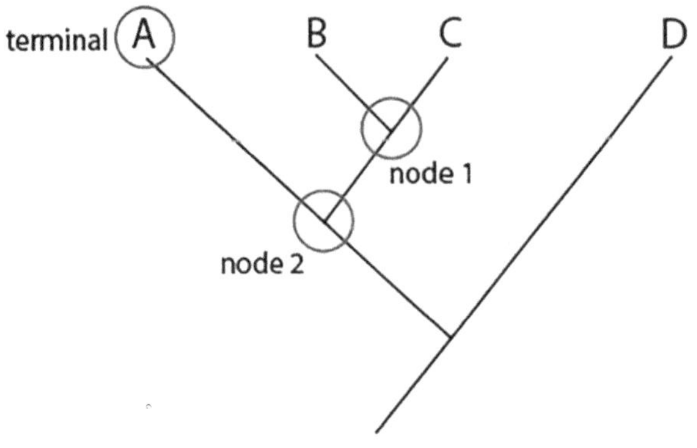

Figure 5.2 A simple phylogenetic diagram.

Current Research on Phylogenetic Diagrams

In current research, scholars draw upon perceptual and cognitive theories (such as Gestalt principles and visual metaphors) and use controlled experiments and visual analyses to explore how phylogenetic diagrams are constructed, presented, perceived, and processed.

One prevailing finding of these studies is that phylogenetic diagrams are prone to misinterpretation (Baum, 2008a; Baum et al., 2005; Gregory, 2008; Halverson, 2010; Meisel, 2010; Novick & Catley, 2007). Contributing to this reality is these diagrams' flexible (or one may say inconsistent) visual representations. To start, a given phylogenetic diagram can assume a variety of shapes and orientations as long as it maintains the same topology (pattern of branching). For example, Figure 5.2 seen earlier is known as a ladder: it contains a continuous, diagonal main branch. Side branches then split off from this main branch. This same diagram can be presented in a number of alternative displays shown in Figure 5.3. Figures 5.3a and 5.3b are known as a tree display: they explicitly depict "stacked," nested clades. The branches may be straight or curvy and extend vertically or horizontally. Figures 5.3c and 5.3d are, respectively, a radial and a circular display, formed by "bending" branches into spokes or curves. For readers who are not familiar with phylogenetic diagrams, it is counterintuitive to have to accept that these apparently dissimilar images all convey the same information.

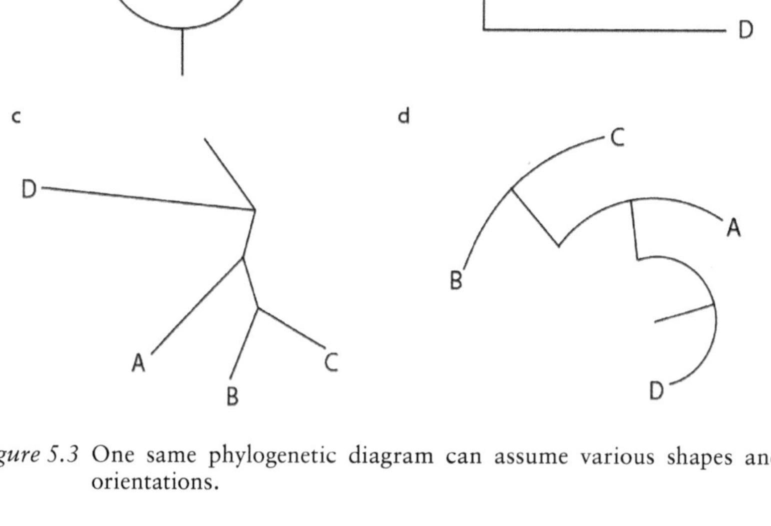

Figure 5.3 One same phylogenetic diagram can assume various shapes and orientations.

Not only are the shape and orientation of a diagram irrelevant, branch lengths and specific node positions generally do not matter either as long as they give rise to the same branching pattern. The two diagrams in Figure 5.4 thus express the same information as those in Figures 5.2 and 5.3. Then again, in some cases, branch lengths do matter and are employed to "depict either the amount of evolution occurring in a particular gene sequence or the estimated duration of branches" (Baum, 2008a, p. 190). This exception, however, may not be indicated in the diagram, and readers are expected to be able to discern it from "the context" of the diagrams (Baum, 2008a, p. 190).

Although Figures 5.2, 5.3, and 5.4 are informationally equivalent, they are not equally accessible from a perceptual and cognitive standpoint. For example, Novick and Catley (2007) found that the ladder display is more likely to create misinterpretation than the tree display. In their study, students saw the continuous, diagonal main branch in the ladder as a single event—due to the influence of the Gestalt principle of continuation—when in fact, that branch is "segmented" by side branches and thus represents multiple branching events. As another example, the circular display makes branches difficult to discern and compare by curving them in different directions (Kjærgaard, 2011).

Even more frustrating is the finding that unclear or incorrect diagrams abound (Catley & Novick, 2008; MacDonald & Wiley, 2012). A common mistake, for example, is to attach taxa not only at the terminal but in the middle of a branch. This design, from a visual perspective, suggests that the middle organism had become or turned into the terminal organism over time. This reading reinforces a common misconception about evolution known as anagenesis, which holds that as an old species undergoes change, it transforms into a new species; commonsensical as it may sound, anagenesis is supported by little evidence and cannot explain the net increase of species over time (Novick, Shade, & Catley, 2010). Precisely because phylogenetic diagrams are supposed to provide evolutionary evidence, incorrect and unclear designs exacerbate misconceptions.

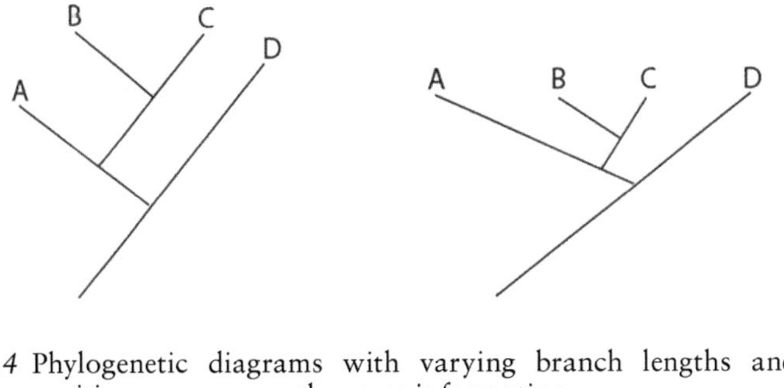

Figure 5.4 Phylogenetic diagrams with varying branch lengths and node positions can convey the same information.

In addition to studying how phylogenetic diagrams convey evolutionary evidence, current research considered the broader, social-cultural messages embedded in this visual artifact. For example, a bottom-to-top diagram whose branches do not end at the same level may trigger cognitive and metaphorical understanding of "up" as "more" and "better" while "down" as "less" and "worse" (Catley & Novick, 2008; Catley, Novick, & Shade, 2010; Lakoff & Johnson, 2003; Tversky, 1997; Tversky, Kugelmass, & Winter, 1991). Such diagrams thus suggest a progressive, teleological understanding of evolution in which "primitive," "lower," and "simpler" organisms progress, purposefully and as if by design, into "advanced," "higher," and "complex" ones. Not infrequently, humans or some humans (often Caucasians) are suggested to be the endpoint of this progression (Scott, 2010; Scott & Giusti, 2006).

Horizontal diagrams are not immune to such readings. With Figure 5.3b, for example, terminal taxa are aligned vertically to the side. Therefore, the top-most branch may be equated to evolutionary superiority, while the lower branches represent different levels of primitive states (Phillips, Novick, Catley, & Funk, 2010). A ladder display can be just as problematic. Because the English language is written and read from left to right, there is "a strong tendency by English speakers to portray temporal concepts from left to right" (Tversky et al., 1991, p. 529). This left-to-right habit, coupled with the "up-is-good-and-down-is-bad" metaphor, means that the organism occupying the top-right terminal of a diagonal ladder may be seen as the evolutionary endpoint (Gregory, 2008).

Of course, none of these readings is correct. All extant organisms, having evolved from a common ancestor on independent paths, are similarly advanced by virtue of existing today. Even though some organisms may be more morphologically complex, they are equally distant in evolutionary time from that common ancestor (Baum, 2008b). Indeed, it is dubious whether a complex morphology is biologically superior; bacteria, despite their simple morphology, are far greater in number than any other species on Earth and are much more adaptable, diverse, and resilient (Gould, 1994; Zimmer, 2016). Still, a progressive reading of evolution persists in popular culture, especially when the evolution of human is concerned (Meisel, 2010; O'Hara, 1988; Scott, 2010; Scott & Giusti, 2006). Such a reading is not only scientifically incorrect, it gives rise to an anthropocentric or Eurocentric worldview that underlies diverse modern problems from environmental degradation to racism.

Phylogenetic Diagrams in U.S. Popular Science Magazines

The above literature, insightful as it is, largely focuses on the use of phylogenetic diagrams in educational contexts: the visual examples being discussed are sourced from textbooks, and the audiences being considered are students. In following such a research path, we miss the opportunity to examine how phylogenetic diagrams are presented to public audiences and whether and how these images may be revised to dispel misconceptions about evolution and cultivate a tree-thinking heuristic among the publics. The following study addresses this gap by examining the use of phylogenetic diagrams in U.S. popular science magazines.

Research Methods

Popular science magazines were chosen as the source for this study because these publications, compared with other popular media such as newspapers and general magazines, have more in-depth reports on scientific topics and use more as well as more detailed visual representations. Pragmatically, digital archives of magazines retain full images and allow the kind of visual analysis pursued in this study, whereas newspaper archives often omit images (possibly to save space in order to archive larger numbers of issues). Four popular science magazines were used as data source: *American Scientist* (1913–present), *Popular Science* (1872–present), *Science News* (1922–present), and *Scientific American* (1845–present). These titles represent the most well-known popular science magazines in the U.S. and boast viewer bases in the millions (see, e.g., *Scientific American media kit*, 2016; Society for Science & the Public, 2016).

To locate phylogenetic diagrams from these publications, keywords (*cladogram*, *evolutionary tree*, *family tree*, *phylogenetic tree*, *phylogenetic diagram*, *phylogeny*, and *tree of life*) were searched for in respective magazine archives. These keywords had been generated by examining a small sample of articles from these magazines that contain phylogenetic diagrams as well as by drawing upon previous literature. All search results were reviewed, and articles that do not contain phylogenetic diagrams despite using the keywords were eliminated. This process resulted in 145 articles, which contained a total of 218 phylogenetic diagrams.

All diagrams were then coded in qualitative data analysis software NVivo. The coding scheme, developed based on Catley and Novick (2008) and MacDonald and Wiley (2012), is detailed in Table 5.1. The scheme focuses on the visual representations of the diagrams in order to explore their design implications and complications. Some of the coding categories, as noted in Table 5.1, may overlap.

Findings

Darwin's tree of life appeared in the 1850s, and other well-known trees, notably those by German biologist Ernst Haeckel, followed shortly after. However, nineteenth-century U.S. popular science magazines published only a few phylogenetic diagrams (three in total). Even in the

Table 5.1 Coding Scheme with Descriptions and Examples

Category	*Sub-Category*	*Description*	*Examples*
Display (may overlap with *orientation* unless otherwise noted)	*Ladder*	Employs a continuous, diagonal main branch. Branching events may appear curved or straight. A correct display that observes phylogenetic principles.	
	Almost a ladder	Looks like a ladder but has irregularities (e.g., inconsistent main branches). An incorrect display that violates one or more phylogenetic principles.	
	Tree	Branches are “stacked”. Branching events may be curved, straight, or slanted. A correct display that observes phylogenetic principles.	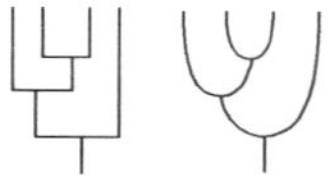
	Almost a tree	Appears like a tree but has irregularities (e.g., branches not ending at the same height when they should). An incorrect display that violates one or more phylogenetic principles.	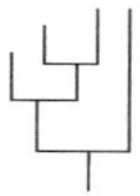
	Actual tree	Resembles a real tree. Has a main trunk with branches growing in different directions and ending at different heights. An incorrect display that violates phylogenetic principles.	

Orientation (may overlap with *display* unless otherwise noted)	*Top down*	Root is positioned at top, and branches extend downward. Can apply to any displays except for "true" ladders.	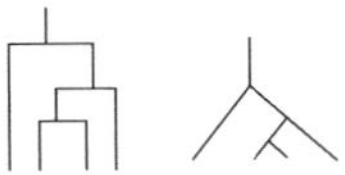
	Bottom up	Root is positioned at bottom, and branches extend upward. Can apply to any displays except for "true" ladders.	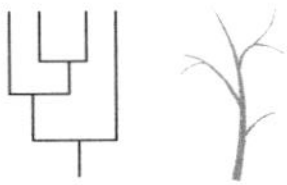
	Left to right	Root is positioned to the left, and branches extend toward right. Can apply to any displays except for "true" ladders.	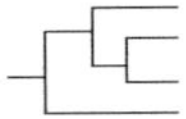
	Right to left	Root is positioned to the right, and branches extend toward left. Can apply to any displays except for "true" ladders.	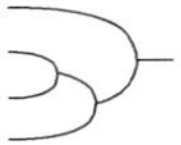
	Up to right	Root is positioned at bottom left, and a continuous main branch extends toward top right. Can apply to ladder and almost-a-ladder displays.	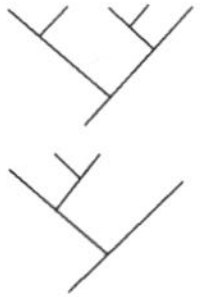
	Down to right	Root is positioned at top left, and a continuous main branch extends toward bottom right. Can apply to ladder and almost-a-ladder displays.	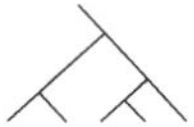
	Radial or *circular*	Root is positioned in the center, and branches extend in a radial or circular orientation. Can apply to tree and almost-a-tree displays.	

(Continued)

Category	Sub-Category	Description	Examples
Taxa image		Uses pictorial images to represent taxa. Can apply to any displays and orientations.	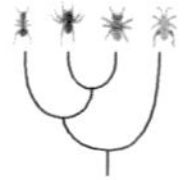
Non-terminal taxa		Taxa appear in the middle of a branch. Can apply to incorrect displays and any orientations.	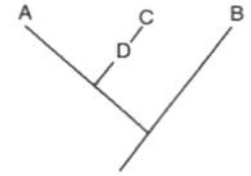
Correct uneven branch		Branches end at different heights to depict extinct organisms or quantify evolutionary change. Can apply to any displays and orientations.	
Synapomorphy		Labels synapomorphies, those derived characteristics shared within a clade. Can apply to any displays and orientations.	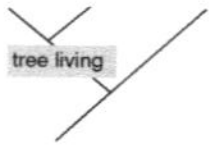
Time		Labels time. Can apply to any displays and orientations.	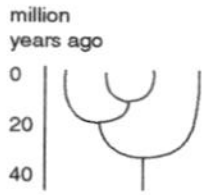
Human	*As endpoint*	Human or their *Homo* relatives occupy a position of visual prominence. Can apply to any displays and orientations.	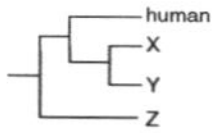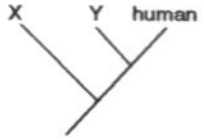
	Not as endpoint	Human or their *Homo* relatives do not occupy a position of visual prominence. Can apply to any displays and orientations.	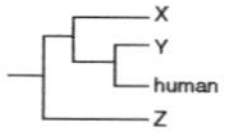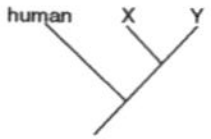

early- to mid-twentieth century, there were relatively few of them: only 31 were published from the 1900s to 1960s. It was only toward the end of the twentieth century and in the twenty-first century that phylogenetic diagrams became more popular: 118 were published from the 1970s to 1990s, and in the first 15 years of the twenty-first century, 66 were published. This growing trend corresponds to the scientific community's recent emphasis on tree thinking.

Diagram Explanation

Although the phylogenetic diagrams found in this study almost always use captions, these captions are often limited to commenting on what is ostensibly seen in the diagram: for example, what organisms are depicted or which organisms appear related. There is, by contrast, very little explicit explanation on how a phylogenetic diagram is constructed or interpreted: for example, what do the nodes stand for, or what is the concept of nested relationship or clade? Of the 145 articles found, only 25 contain any such explanations, most of which are provided through captions, and some are contained in the articles' body content. A similar lack of phylogenetic diagram explanation was found in biology textbooks and museum displays (Catley & Novick, 2008; MacDonald & Wiley, 2012). It seems, then, that the scientists and science communicators behind these publications assume audiences, including public audiences, to already know about and can readily interpret phylogenetic diagrams. Or, as Welhausen (Chapter 4) would put it, readers are expected to be well versed in the context of these images. However, given the design complications and misinterpretations involved in phylogenetic diagrams, such assumptions seem ill founded.

The design of branch length is a case in point. As mentioned earlier, all branches in a phylogenetic diagram should end at the same height to depict co-existing organisms. Sometimes, however, a diagram may contain extinct species, which would require branches to terminate earlier, or a diagram may demonstrate quantitative data such as the amount of genetic change, which would also result in branches of varying lengths. In this study, 27 such diagrams were found. Most of them (78%) do not acknowledge or explain their exceptional design, or the explanation may be obscure—as in "The horizontal component of separation represents evolutionary distance between organisms" (Kabnick & Peattie, 1991, p. 42). Because of this, it is difficult for readers to differentiate exceptional but correct diagram features from incorrect ones, which adds to the confusion in how these diagrams are constructed and interpreted.

As another example, branch thickness occasionally presents an issue. In phylogenetic diagrams, all branches should have the same width, but in this as well as other studies (Catley & Novick, 2008; MacDonald & Wiley, 2012), some diagrams feature branches of varying thickness.

Without explicit explanation, it is unclear whether this uncommon feature represents population size, number of species, geographical distribution, or other information (Catley & Novick, 2008).

Display Types

As noted in Table 5.1, five display types were coded in this study. Two of them are correct displays that observe phylogenetic principles: ladders and trees. The others (almost a ladder, almost a tree, and actual trees) fail one or more principles such as properly depicting nested relationships, branching events, or extant organisms. Figure 5.5 summarizes the distribution of these displays. As it shows, trees are the most common choice (30%), while ladders come in last at 15%. This means that only 45% of the phylogenetic diagrams found in this study are actually correct, while the majority (55%) are not.

Specifically, 39% of the diagrams are almost a tree or almost a ladder. These diagrams may have taxa in the middle of a branch, position extant organisms at different levels, or have "side branches" that split off without forming nested relationships. In Catley and Novick's (2008) words, these display types are especially insidious: looking very much like ladders or trees but violating certain rules, they reinforce deep-seated misconceptions about evolution. For example, 16% of the diagrams place taxa in the middle of a branch, which, as mentioned earlier, suggests evolution as anagenesis, a linear process of chimpanzees turning into humans (Shtulman, 2006).

The other incorrect display, actual trees, accounts for 16% of the total diagrams. Some of these are drawn realistically to resemble a real tree;

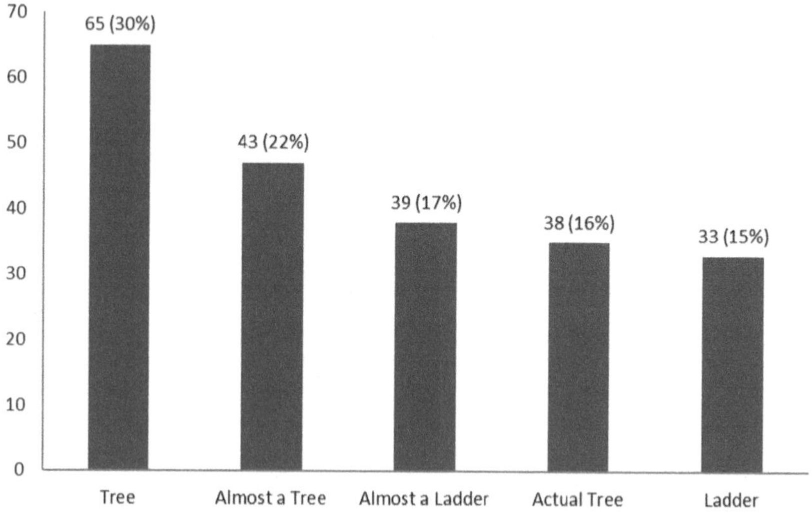

Figure 5.5 Display choices: A majority of the displays are incorrect.

others are schematic renditions of a real tree. Frequently, these displays show "lesser" life forms (e.g., bacteria) at lower branches and "higher" life forms (e.g., animals) at higher branches. "These diagrams are the epitome of a teleological view of evolution; they overtly suggest the notion of direction and progress" (Catley & Novick, 2008, p. 983). Despite evidence to the contrary, this progress-driven concept about evolution continues to be prevalent in popular belief (Catley & Novick, 2008; Scott, 2010), and the many actual tree displays in mainstream popular science magazines do nothing to deter it.

The finding that 55% of phylogenetic diagrams published by leading U.S. popular science magazines are incorrect is bleak, to say the least. Some readers may wonder if this is caused by earlier publications being less careful, but that is not the case: of the 66 diagrams published in the last 15 years, 46% are incorrect. Compared with biology textbooks, which have a 72% overall accuracy rate for phylogenetic diagrams (Catley & Novick, 2008), popular science magazines do much worse. If it is somewhat understandable that formal education materials are more vigorously vetted, it is deeply disappointing that popular science magazines, a medium for everyday citizens to gain exposure to science, are not held to more stringent standards. Moreover, as Catley and Novick's (2008) study shows, textbook quality is not consistent among school levels. While college textbooks' phylogenetic diagrams are 65%–83% correct, middle school and high school textbooks fair much worse at 35%–53%, making them no better than popular science magazines. By contrast, professional journals are the least tolerant of erroneous displays (Catley & Novick, 2008). With these data, one is tempted to conclude that scientists and science communicators consider it more important to "get the science right" when the target audience is more scientifically informed, more educated, or at least promised to become more educated. The needs of the "lay" or "low-level" readers, by comparison, fell to the wayside.

These findings have serious implications when we consider that 66% of the U.S. population does not have a bachelor's degree (National Center for Education Statistics, 2016; data reflect that of 2014). In other words, there is a substantial number of people who would not benefit from the apparently more serious focus on evolutionary science in college. For these American adults, their sources of scientific information are not rigorously peer reviewed journals or vetted textbooks but informal channels such as books, magazines, newspapers, TV, radio, the Internet (including online newspapers and magazines), and museums (Falk, Storksdieck, & Dierking, 2007; National Science Board, 2014). Ensuring correct visual representations in popular media should therefore be a priority for science communicators. This is especially so in the area of evolutionary science, given the widespread misconceptions and mistrust of Darwinian evolution in the U.S.

Orientations

As noted in Table 5.1, seven diagram orientations were coded in this study: vertical (top to bottom and bottom to top), horizontal (left to right and right to left), diagonal (up to right and down to right), and radial and circular. Figure 5.6 summarizes the distribution of these orientations. As the figure shows, by far the most common choice is the bottom-up orientation where the root of the tree is at the bottom and the branches extend upward. This finding is not surprising given our physical and cognitive familiarity with the vertical and especially the bottom-up orientation: humans are vertically oriented, and most life forms grow upward (Tverskey, 1997). However, when used to graph evolution history, this orientation, as mentioned before, risks creating the visual metaphor of evolution as a progressive event from "lower" and "primitive" organisms to "better" and "advanced" ones. As a sharp contrast, the other vertical choice, the top-bottom orientation, is used by only seven diagrams. This is despite the fact that this latter choice is equally capable of depicting nested relationships and can probably better connote evolution as a process of "descending" rather than progressive enhancements.

The second leading orientation choice is the left-to-right orientation where the root is positioned to the left and branches extend toward right. This orientation results in vertically placed taxa labels and thus similarly risks the top branches being perceived as "superior." At the same time,

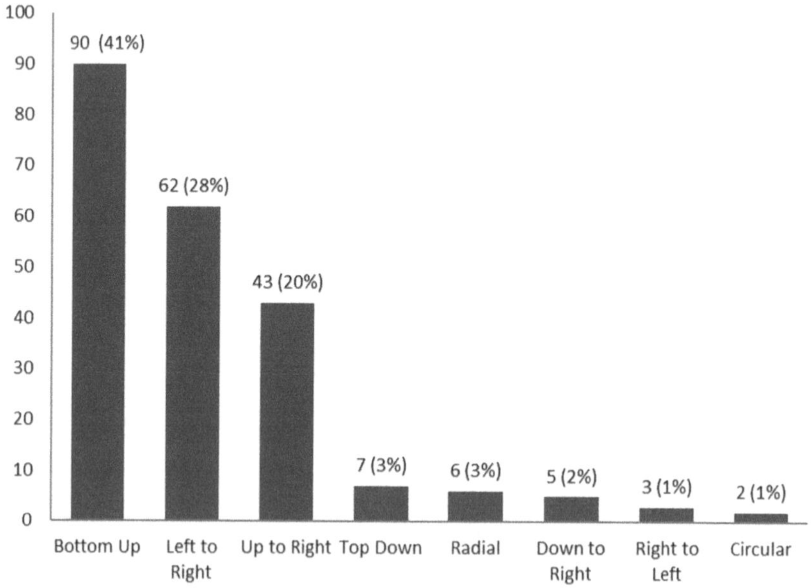

Figure 5.6 Orientation choices: Orientations that encourage a progressive reading of evolution dominate.

this orientation has a particular advantage in accessibility. Recall that in a phylogenetic diagram, time is read in the direction from root to tip; therefore, in a left-to-right diagram, time proceeds from left to right. This sequence corresponds to English speakers' tendency to conceive temporal events as happening from left to right (Tversky et al., 1991). Phylogenetic diagrams with other orientations disrupt this reading sequence and are more likely to lead to misinterpretation because readers may ignore the actual root-tip orientation and mistakenly perceive taxa on the left to happen before those on the right (Gregory, 2008).

The third leading orientation choice is the up-to-right orientation where the root is positioned at bottom left and a continuous main line extends toward top right. The prevalence of this choice is unfortunate from both a semiotic and an accessibility standpoint. Semiotically, an up-to-right ladder can signal evolution as progressive enhancements due to the combined effect of the left-to-right reading and "up-is-better" metaphor. From an accessibility standpoint, an up-to-right ladder is particularly difficult to comprehend. As Novick and Catley (2007) found, due to the influence of the Gestalt principle of continuation, students tend to incorrectly assume that the continuous main branch in the ladder is a single evolutionary event. Furthermore, because English readers are conditioned to read from left to right, they encounter the ladder's nested relationships in a reverse order: the least related clades are encountered first (Novick, Stull, & Catley, 2012). Figure 5.7a demonstrates this process. When reading this ladder from left to right, readers will first encounter C and A, even though A and B share a more recent ancestor and are more related.

The counterpart to this design, the down-to-right ladder (Figure 5.7b), presents a different story. Because of their left-to-right reading habit, English readers will process this ladder from top to bottom, which helps to discourage the "up-is-better" metaphor and a progressive reading of evolution. Moreover, when readers process these ladders from left to right, they generally encounter nested relationships in their right order (Novick et al., 2012). For example, with Figure 5.7b, readers will first

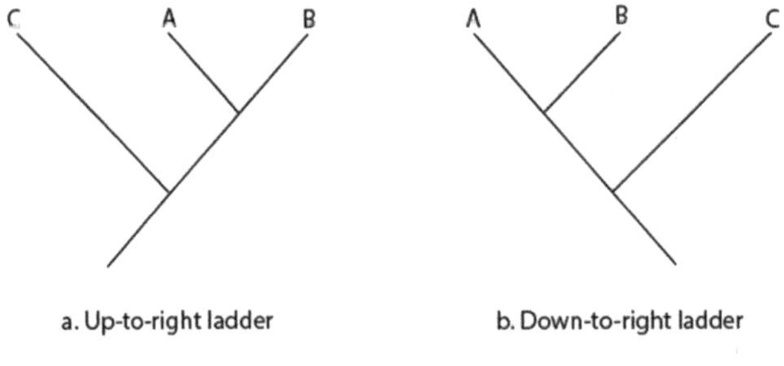

Figure 5.7 Up-to-right ladders are less accessible than down-to-right ladders.

encounter A and B, which are more closely related. Proceeding from this first clade, they will encounter C, who shares a more distant common ancestor with A and B and is less related. In spite of these advantages, down-to-right ladders are rare in popular science magazines: only five were found in this study.

Altogether, diagram orientations that promote an incorrect, progressive reading of evolution dominated; those that discourage such a view numbered only in single digits. This is not to say that these other orientations are all necessarily superior, because multiple factors complicate the design of a phylogenetic diagram. For example, although circular diagrams can disrupt common orientation biases and discourage a hierarchical reading of evolution, they are difficult to access given their lack of an obvious orientation. Such diagrams may have a place in professional publications that need to fold the branches in order to efficiently present large amounts of information to an informed audience (see Figure 5.8), but in popular communication, they are more likely to overwhelm than inform. The same may be said of radial designs. Then again, these designs *may* be appropriate for public audiences if they are able to convey general impressions of data, such as the complex radial tree published in a recent *New York Times* article that reveals latest biodiversity findings (Zimmer, 2016). The tree contains many data points, terminologies, and technical details but also a visually obvious message: bacteria occupy much of Earth's biodiversity, while humans and other "higher" lives fit on a small twig. That twig is positioned, most likely deliberately, at the bottom of the tree.

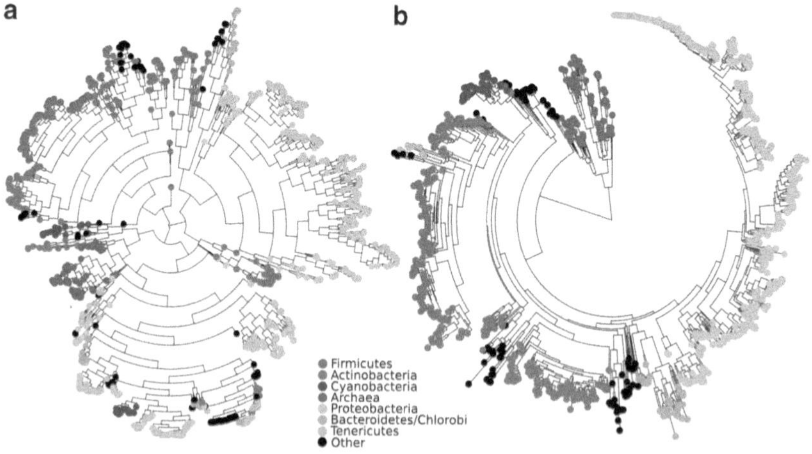

Figure 5.8 Circular designs pack in an enormous, some may say excessive, amount of information (Segata & Huttenhower, 2011, p. 7). Color figure in plate section.

Placement of Humans

Because the meaning of a phylogenetic diagram depends on its branching pattern, not the position of individual branches, one can usually rotate the branches of a given diagram without affecting its meaning. Figure 5.7, in demonstrating two orientations of a ladder, already shows this possibility. From Figure 5.7a to Figure 5.7b, branch C is flipped horizontally, but the two diagrams portray the same nested relationship: A and B form a more related clade, and then A, B, and C form the next clade. Similarly, Figure 5.9 shows how branches in a tree display can be rotated. From Figure 5.9a to Figure 5.9b, branches A and B are flipped, and then branch D is rotated from top to bottom. The two diagrams, however, still portray the same evolutionary relationship: A and B form a most closely related clade; A, B, and C form the next clade; and A, B, C, and D form the most distant clade.

Because branch positions are not fixed, which organism happens to occupy a top or a right-hand branch should be random. However, as Sandvik (2009) showed, such is not the case when the human species is graphed. In his study of phylogenetic diagrams in biology textbooks, Sandvik (2009) found that humans are consistently placed at the top-right corner of the diagram, a position that represents the metaphorical height and endpoint of evolution, a position where nature seemed destined to progress toward all along. This same tendency exists in popular science magazines, as shown by this study. Of the 54 diagrams found in this study that contain humans (*Homo sapiens*) or their recent *Homo* relatives (e.g., *Homo erectus*), 39% position the *Homo* species at the top-most branch or the top-right corner. Although 39% may not seem a majority, no other species portrayed in the many diagrams is consistently granted this visual prominence.

Some of these human-centered diagrams take on an incorrect display type (e.g., almost a tree), but others assume correct display types that observe phylogenetic principles (i.e., trees and ladders). These latter ones are especially problematic by being "technically correct." In Sandvik's (2009) words, it is "exactly because the graphical representation of the results is irrelevant to their correctness that scientists are entirely free to choose whatever ... ordering of taxa they like" (p. 438). What became

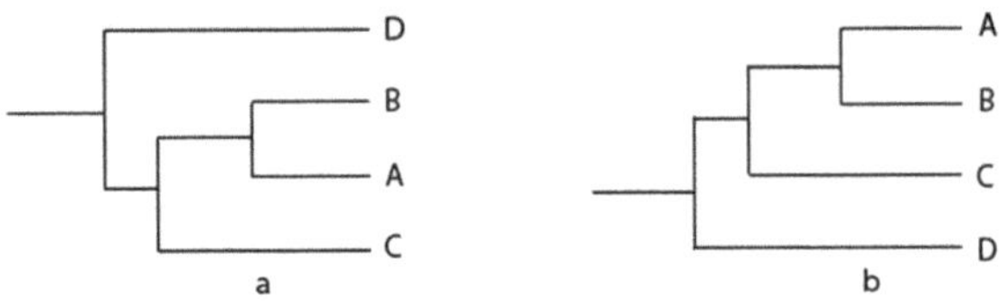

Figure 5.9 Branches of a phylogenetic diagram can be rotated without changing its evolutionary relationship.

promoted through this freedom is an anthropocentric worldview—a worldview that is apparently shared by many scientists, educators, and science communicators. As humans, we very genuinely believe in our own evolutionary superiority (how could it be otherwise for a highly intelligent and self-conscious species!), and wittingly or not, we are guided by that belief when describing the history of lives. As O'Hara (1992) put it, our "common membership in this [human] community blinds us to the effect our perspective has had on our representations of the past"; after all, "there are no coelacanths, no bird's nest fungi, no vestimentiferan worms writing evolutionary history" (p. 144). If these organisms *were* to create phylogenetic diagrams, we could expect to be reminded of our own bias.

If our anthropocentric worldview is understandable, it is certainly not unproblematic. Even if the scientists and science communicators who created these diagrams understand that humans are *not* the height and endpoint of evolution, it is doubtful that public audiences will (see Phillips et al., 2010). Even if the accompanying text correctly explains evolutionary histories, the "graphs appeal to the optical memory" and tend to be remembered as they are (Sandvik, 2009, p. 438). Indeed, textual explanation is a dubious point too. Although this study focuses on the visual representation of phylogenetic diagrams, it is worth noting that the magazine articles found in this study, including recent ones, frequently use words such as "primitive," "advanced," "lower," and "higher" to describe species and taxa, even though terms such as "primitive" should only be used to describe characteristics, not species, while terms such as "lower" and "higher" should be abandoned altogether (O'Hara, 1992).

Labels and Legends

An almost mandatory component in phylogenetic diagrams is taxa labels—labels that explain what organisms or species are being graphed. Most often, the labels take the verbal format: for example, using the word "swordfish" at the tip of a corresponding branch to signal that organism. In addition or alternatively, visual labels are employed, for example, using an iconic image of a swordfish to represent that organism. Forty-four percent of the diagrams found in this study use such visual labels, some of which are specific (as in the swordfish example), and others are exemplary (as in using the image of a dog to represent "vertebrate"). From an accessibility standpoint, these visual labels have several advantages in popular science communication. First, they add visual interest and a sense of familiarity to a diagram that public audiences may find unfamiliar or uninviting. Even though the visual labels cannot, on their own, convey information about evolutionary relationships, they provide some entry points into that information. Second, visual labels help readers recognize, differentiate, or learn about unfamiliar taxa.

Without an image, some readers may not know what a "colugos" is or what "Articulata" are. Even in cases where readers are already familiar with the taxa, visual labels allow at-a-glance recognition. From these perspectives, popular science magazines can benefit from using more visual labels for their phylogenetic diagrams. Certainly, as McDonald and Wiley (2012) cautioned, if readers rely on morphological appearance to infer biological relationships, visual taxa labels may create unintended confusion and allow readers to conflate physical similarity and evolutionary relatedness. But then again, the opposite effect is also possible: By seeing how physically dissimilar organisms are related, readers may be guided to discard misconceptions about evolution and its mechanism.

Another label sometimes seen in phylogenetic diagrams pertains to synapomorphies. Placed on or beside appropriate branches, these labels highlight the morphological, molecular, or behavioral characteristics (the synapomorphies) inherited from a common ancestor and shared by a group of taxa, for example, the character of having an integrated nervous system (Novick, Catley, & Funk, 2010). By distinguishing certain organisms from other potentially related organisms, synapomorphies are the basis to establish clades and nested hierarchical relationships. In this study, 18 diagrams (8%) were found to use such labels, which is even lower than McDonald and Wiley's (2012) finding that only 20% of museum phylogenetic diagrams included synapomorphies. These findings point to a missed opportunity. As Novick, Catley, and Funk (2010) argued, integrating synapomorphies into phylogenetic diagrams can provide readers with relevant evidence for the common ancestry and evolutionary relationships being depicted. In addition, when used in ladder diagrams, these labels led to a significant improvement in students' ability to process the graphed information, because the labels helped to visually break the continuous main line in the ladder and thus offset the influence of the Gestalt principle of continuation (Novick, Catley, & Funk, 2010). As a result, students saw the line not as one single evolutionary event but the multiple events it represents.

A third possible label to add in phylogenetic diagrams is time. Based on previous research, misreading the direction of time is a common mistake in interpreting phylogenetic diagrams (Dodick & Aharonson, 2009; Gregory, 2008; Meir, Perry, Herron, & Kingsolver, 2007). For example, due to the left-to-right reading habit, it is common for English readers to assume that time progresses from the left-most branch tip to the right-most branch tip or from the top-left tip to the root (Gregory, 2008; Meir et al., 2007). Adding a temporal label to explicitly specify that time progresses from the root to branch tips can thus enhance the diagrams' accessibility. 34% of the diagrams found in this study contain time labels. Some of these labels take the form of time measured in "million years ago," others are notations of geologic time divisions such as the Eocene period, and still others are arrows that indicate the direction of time. 34%, however, is comparatively low. In McDonald and Wiley's (2012) study of museum phylogenetic diagrams, 46% included timelines; in Catley and Novick's (2008) study of middle school, high school, and college biology textbooks, 42% phylogenetic diagrams contained some representation of time. These findings suggest another area where popular science magazines can stand to adjust their phylogenetic diagram designs.

Conclusions: Heuristics for Designing Phylogenetic Diagrams for Public Audiences

As Baum et al. (2005) advocated, phylogenetic diagrams are "the most direct representation of the principle of common ancestry—the very core of evolutionary theory—and thus they must find a more prominent place in the general public's understanding of evolution" (p. 980). For phylogenetic diagrams to fulfill this role, their representation in popular media needs to be scientifically correct, cognitively accessible, and socially responsible. This, in turn, requires heuristics that scientists and science communicators can follow so as to make conscious design choices rather than follow their intuition of what looks appealing or seems natural. In Brumberger and Northcut's (2013) words, we need explicit criteria rather than personal preferences. While a single study cannot hope to offer a complete set of such heuristics, the following represents some starting points.

First and foremost, scientists and science communicators must take public audiences' information needs seriously and treat popular science visual representations not (or not merely) as a way to attract reader attention but as meaningful scientific evidence that they are. In the case of phylogenetic diagrams, this means avoiding playful designs (such as using the image of a real tree) and syntactic decorations that result in erroneous or unclear diagrams. Given that not all scientists are necessarily comfortable with phylogenetic diagrams (Baum, 2008a), reducing incorrect designs in popular media also requires scientists and science communicators to adopt a more rigorous approach to these visuals and to fully understand how they are to be constructed and interpreted.

Second, as shown in this chapter, phylogenetic diagrams can assume various alternative designs. This flexibility allows the diagram to accommodate diverse evolutionary data that vary in volume, scope, focus, branching patterns, and supporting evidence. At the same time, this flexibility also results in design variations that make it more difficult for public audiences to learn to process the diagram. It also gives scientists and science communicators a great degree of freedom that probably contributes to erroneous and unclear designs. While I do not suggest that popular science magazines be prescribed only certain phylogenetic designs, it is meaningful to consider which choices facilitate perceptual and cognitive processing and may be more suitable for non-expert readers. As this study shows, circular and radial designs are rare in popular science magazines, probably for good reasons. On the other hand, some designs that have been found difficult to interpret are commonly used. For example, ladder and almost-a-ladder displays account for 32% of the diagrams despite ladders being harder to process than trees due to the influence of the Gestalt principle of continuation (Novick & Catley, 2007). If a ladder is to be used, the down-to-right orientation is recommended (Novick et al., 2012), but the up-to-right orientation is currently prevalent in popular science magazines.

Third, scientists and science communicators should make fewer assumptions about public audiences' prior familiarity with phylogenetic diagrams and instead do more in providing explanations, labels, and general guidance. Descriptions of what a phylogenetic diagram is and what its key visual elements signify should be included whenever possible, preferably in the diagram caption. Assistive labels such as time and synapomorphies should be considered when adding them can facilitate information processing. "Exceptional" designs should be carefully explained, for example, when uneven branch lengths are used to represent an evolutionary scale.

Last, scientists and science communicators need to consider the semiotic impact of otherwise correct phylogenetic diagrams and, in particular, avoid anthropocentric designs. We can do so by purposefully and consciously taking the imaginary point of view of another taxon (O'Hara, 1992). This means, at a minimum, avoiding putting humans at the top or top-right corner of a diagram. More generally, it means using design choices that avoid a visual impression of *any* "superior" organism. For example, rather than "pruning" some branches (as if those organisms do not matter or do not exist) in order to focus on other organisms, an alternative way to highlight certain organisms is to use an inset, as is commonly done in maps (O'Hara, 1992). Or, we may rotate the branches of a diagram (while preserving its pattern) to create "balanced" images where branching events do not appear to visually "progress" in a linear fashion. Figure 5.10 demonstrates this possibility. The diagram in Figure 5.10a has one continuous main branch (terminated in A) from which multiple other branches split. This "one-sided" design creates the impression of A as an evolutionary height and endpoint. By rotating some of the branches, we can depict the same information in Figure 5.10b. In this diagram, splitting events are more evenly distributed, which encourages the visual impression of evolution as "copiously and luxuriantly branching bushes" (Gould, 1994, p. 91).

I hope more researchers will consider studying the use of phylogenetic diagrams in popular communication. Future studies can examine additional media examples to expand and modify current findings, or they can assess current heuristics by testing them with public audiences.

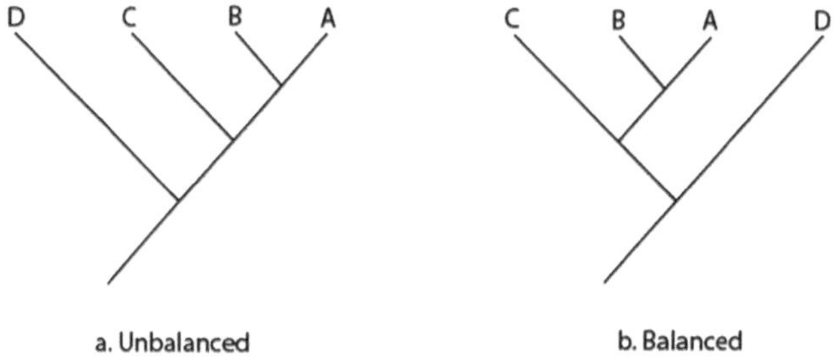

Figure 5.10 Balanced phylogenetic diagrams avoid the visual impression of evolution as a progressive event.

Without these efforts, phylogenetic diagrams will not be able to cultivate the kind of tree-thinking skills scholars and educators believe essential for us to understand, appreciate, and celebrate evolution.

References

Baum, D. (2008a). Reading a phylogenetic tree: The meaning of monophyletic groups. *Nature Education, 1*(1), 190.

Baum, D. (2008b). Trait evolution on a phylogenetic tree: Relatedness, similarity, and the myth of evolutionary advancement. *Nature Education, 1*(1), 191.

Baum, D. A., Smith, S. D., & Donovan, S. S. S. (2005). The tree-thinking challenge. *Science, 310*(5750), 979–980.

Bluis, J., & Shin, D. (2003). Nodal distance algorithm: Calculating a phylogenetic tree comparison metric. In *Proceedings of the Third IEEE Symposium on BioInformatics and BioEngineering* (pp. 87–94). Piscataway, NJ: IEEE.

Brumberger, E. R., & Northcut, K. (2013). *Designing texts: Teaching visual communication.* Amityville, NY: Baywood.

Catley, K. M. (2006). Darwin's missing link—A novel paradigm for evolution education. *Science Education, 90*(5), 767–783.

Catley, K. M., & Novick, L. R. (2008). Seeing the wood for the trees: An analysis of evolutionary diagrams in biology textbooks. *Bioscience, 58*(10), 976–987.

Catley, K. M., Novick, L. R., & Shade, C. K. (2010). Interpreting evolutionary diagrams: When topology and process conflict. *Journal of Research in Science Teaching, 47*(7), 861–882.

Darwin, C. (1859). *On the origin of species*, chapter VI foldout. Retrieved February 29, 2016, from http://darwin-online.org.uk/graphics/Origin_Illustrations.html.

Dodick, J., & Aharonson, M. (2009). Phylogeny exhibits and understanding geological time. *Understanding the tree of life Harvard conference*, Cambridge, MA. Retrieved February 29, 2016, from http://evolution.berkeley.edu/UToL/dodick_aharonson_harvard09.pdf.

Falk, J. H., Storksdieck, M., & Dierking, L. D. (2007). Investigating public science interest and understanding: Evidence for the importance of free-choice learning. *Public Understanding of Science, 16*(4), 455–469.

Gould, S. J. (1994). The evolution of life on the Earth. *Scientific American, 271*(4), 84–91.

Gregory, T. (2008). Understanding evolutionary trees. *Evolution: Education and Outreach, 1*(2), 121–137.

Halverson, K. L. (2010). Using pipe cleaners to bring the tree of life to life. *American Biology Teacher, 72*(4), 223–224.

Kabnick, K., & Peattie, D. (1911). Giardia: A missing link between prokaryotes and eukaryotes. *American Scientist, 79*(1), 34–43.

Kjærgaard, R. S. (2011). Things to see and do: How scientific images work. In D. J. Bennett & R. C. Jennings (Eds), *Successful science communication: Telling it like it is* (pp. 332–354). New York: Cambridge University Press.

Lakoff, G. & Johnson, M. (2003). *Metaphors we live by*. Chicago, IL: The University of Chicago Press.

MacDonald, T., & Wiley, E. (2012). Communicating phylogeny: Evolutionary tree diagrams in museums. *Evolution: Education and Outreach, 5*(1), 14–28.

Meir, E., Perry, J., Herron, J. C., & Kingsolver, J. (2007). College students' misconceptions about evolutionary trees. *American Biology Teacher, 69*(7), 71–76.

Meisel, R. (2010). Teaching tree-thinking to undergraduate biology students. *Evolution: Education and Outreach, 3*(4), 621–628.

National Center for Education Statistics. (2016). *Fast facts: Educational attainment*. Retrieved February 11, 2016, from https://nces.ed.gov/fastfacts/display.asp?id=27.

National Science Board. (2014). *Science and engineering indicators 2014*. (No. NSB 14–01). Arlington, VA: National Science Foundation.

Novick, L., & Catley, K. (2007). Understanding phylogenies in biology: The influence of a Gestalt perceptual principle. *Journal of Experimental Psychology: Applied, 13*(4), 197–223.

Novick, L., Catley, K., & Funk, D. (2010). Characters are key: The effect of synapomorphies on cladogram comprehension. *Evolution: Education and Outreach, 3*(4), 539–547.

Novick, L., Shade, C., & Catley, K. (2010). Linear versus branching depictions of evolutionary history: Implications for diagram design. *Topics in Cognitive Science, 3*(3), 536–559.

Novick, L., Stull, A., & Catley, K. (2012). Reading phylogenetic trees: The effects of tree orientation and text processing on comprehension. *Bioscience, 62*(8), 757–764.

O'Hara, R. J. (1992). Telling the tree: Narrative representation and the study of evolutionary history. *Biology and Philosophy, 7*(2), 135–160.

O'Hara, R. J. (1988). Homage to Clio, or, toward an historical philosophy for evolutionary biology. *Systematic Zoology, 37*(2), 142–155.

Phillips, B., Novick, L., Catley, K., & Funk, D. (2010). Interactive effects of diagrammatic format and teleological beliefs on tree thinking. In S. Ohlsson & R. Catrambone (Eds.), *Proceedings of the 32nd Annual Meeting of the Cognitive Science Society* (pp. 2656–2661). Austin, TX: Cognitive Science Society.

Reading trees: A quick review. (2016). Retrieved February 9, 2016, from http://evolution.berkeley.edu/evolibrary/article/phylogenetics_02.

Sandvik, H. (2009). Anthropocentricisms in cladograms. *Biology & Philosophy, 24*(4), 425–440.

Scientific American media kit. (2016). Retrieved February, 24, 2017, from www.scientificamerican.com/mediakit/.
Scott, M. (2010). The pleasures and pitfalls of teaching human evolution in the museum. *Evolution: Education and Outreach, 3*(3), 403–409.
Scott, M., & Giusti, E. (2006). Designing human evolution exhibitions. Insights from exhibitions and audiences. *Museums & Social Issues, 1*(1), 49–68.
Segata, N., & Huttenhower, C. (2011). Toward an efficient method of identifying core genes for evolutionary and functional microbial phylogenies. *Plos One, 6*(9), 1–10.
Shtulman, A. (2006). Qualitative differences between naive and scientific theories of evolution. *Cognitive Psychology, 52*(2), 170–194.
Society for Science & the Public. (2016). *Science News*. Retrieved February, 24, 2017, from www.societyforscience.org/science-news.
Soltis, D. E., & Soltis, P. S. (2003). The role of phylogenetics in comparative genetics. *Plant Physiology, 132*(4), 1790–1800.
Tversky, B. (1997). Cognitive principles of graphic displays. *Technical Report FS-97-03* (pp. 116–124). Menlo Park, CA: AAAI Press.
Tversky, B., Kugelmass, S., & Winter, A. (1991). Cross-cultural and developmental trends in graphic productions. *Cognitive Psychology, 23*(4), 515–557.
Zimmer, C. (2016). Scientists unveil new "tree of life." *The New York Times*. Retrieved February 25, 2017, from www.nytimes.com/2016/04/12/science/scientists-unveil-new-tree-of-life.html?_r=1.

6 Tweeting the Anthropocene

#400ppm as Networked Event

Lauren E. Cagle and Denise Tillery

On Wednesday, May 6, 2015, the National Oceanic and Atmospheric Administration published a press release reporting that, for the first time since atmospheric carbon dioxide has been tracked, "The monthly global average concentration of this greenhouse gas surpassed 400 parts per million in March 2015" (NOAA, 2015). The press release quotes Pieter Tans, lead scientist for NOAA's Global Greenhouse Gas Reference Network:

> We first reported 400 ppm when all of our Arctic sites reached that value in the spring of 2012. In 2013 the record at NOAA's Mauna Loa Observatory first crossed the 400 ppm threshold. Reaching 400 parts per million as a global average is a significant milestone.
>
> (NOAA, 2015)

The initial crossing of the threshold of 400 ppm in May 2013 prompted Bruno Latour (2014) to argue that the new problem in science studies was "how to understand the active role of human agency not only in the construction of facts, but also in the very existence of the phenomena those facts are trying to document" (p. 2). While the specific measurement of 400 ppm is not new, the monthly global average concentration exceeding that number represents a significant threshold.

Like much contemporary science news, the crossing of the 400 ppm threshold was shared widely across social media platforms, particularly Twitter, where links to the press release were often accompanied by the hashtag #400ppm. To understand how social media's affordances, such as hyperlinks and hashtags, relate to and shape scientific news, we conducted a study of tweets that referred to this news of carbon dioxide's 400 ppm concentration. After building a dataset of all tweets from May 2015 that used the #400ppm hashtag, we applied Latour's actor-network theory to understand how hashtags and hyperlinks can function as both text and technology, thereby accelerating the spread of scientific news. We then analyzed how this scientific news coalesced around common topics and frames apparent in the dataset. Often, those broader ideas draw on elements of the original press release that kickstarted the sharing of this scientific news.

DOI: 10.4324/9781315160191-6

For example, several common frames among the tweets reveal a sense of urgency about climate change that mirrors the urgent, activist tone of the NOAA press release, relayed primarily through quotes attributed to Tans and other experts. Tans refers to "the fact that humans burning fossil fuels have caused global carbon dioxide concentrations to rise more than 120 parts per million since pre-industrial times" (NOAA, 2015). The press release concludes by quoting the director of NOAA's Global Monitoring Division, saying,

> Elimination of about 80 percent of fossil fuel emissions would essentially stop the rise in carbon dioxide in the atmosphere, but concentrations of carbon dioxide would not start decreasing until even further reductions are made and then it would only do so slowly.
>
> (NOAA, 2015)

Typical of science reporting, the story frames the human aspect—the implications and significance for people—and includes a few details about the methods the scientists used (air samples are collected in 40 remote sites across the globe). The press release also refers to the steady recent climb in atmospheric CO_2 concentration and includes a data display as well as photos of the process of sample collection. The urgent tone helps to shape the measurement's significance, and the usual buttons urging users to like, share, and tweet make this web-based article an event that can be shared across social media. Figure 6.1 shows a screenshot of the NOAA website with the sharing buttons prominently displayed between the press release headline and body.

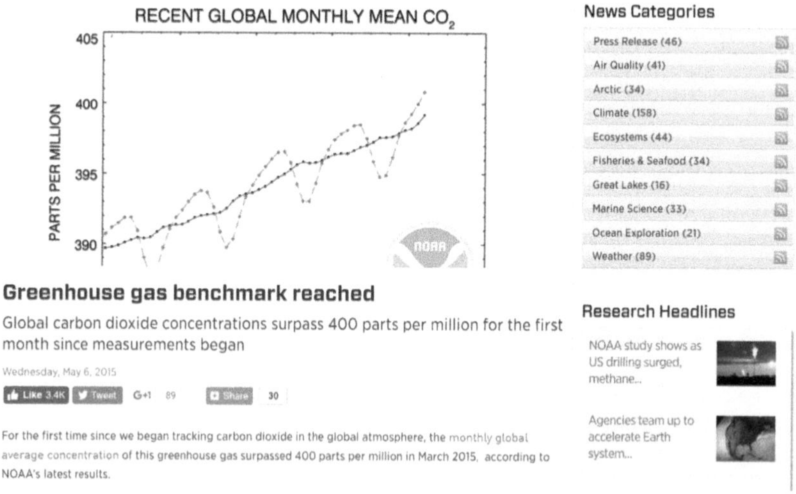

Figure 6.1 Screenshot of May 6, 2015 NOAA press release on official NOAA website (National Oceanic and Atmospheric Association, 2015). Color figure in plate section.

To be clear, of course, the dramatic increase of atmospheric CO_2 is significant whether shared on social media or not. But global warming, as a chronic, slow-moving catastrophe difficult to perceive on an individual scale, requires experts and communicators to find ways to invest particular numbers with news-worthy significance, in order to gain media attention. In essence, they need to create and argue for *kairos*, an appeal to timeliness, which "calls attention to the nature of discourse as event rather than object" (Miller, 1992). And certainly, this event, and NOAA's press release, did inspire attention, including articles in the *Guardian* and *Mother Jones*, both of which were also published on May 6, 2015. The *Guardian* article connected the CO_2 readings to the Paris climate talks, and also mentions that this 400ppm average "comes nearly three decades after what is considered the 'safe' level of 350ppm was passed" (Vaughan, 2015). Similarly, the *Mother Jones* (2015) article refers to 350 ppm as a goal set by activists including Bill McKibben, who argued that 350 ppm represented a "tipping point," after which various risks, including ocean acidification and unreliable monsoons, will increase. That article ends pessimistically with the simple statement, "We're now at 400" (Oh, 2015). These two articles and the press release they are based on, all shared across social media, form part of the network that exemplifies how scientific information is picked up, circulated, disseminated, and discussed in our new media landscape.

In this study, we consider the circulation of the hashtag "#400ppm" on Twitter, which was used to mark the initial eventful moment in May 2015, when NOAA's press release kicked off a swell of discussion across social media among both scientists and non-scientists. Across Twitter in particular, climate scientists, science communicators, and members of the public discussed the news. Often, their tweeted reactions included hyperlinks to the original press release, mainstream news stories about it, and other online media they deemed relevant. Our interest was sparked by the various ways that scientific information is accommodated, distributed, and changed as users hyperlinked to it and shared their responses.

By observing and theorizing the movement and transformation of a single scientific news event through these networks, we hope to suggest some ways that scientists and science communicators can use the networks' affordances. As science communication continues to spread across new media platforms, science communicators should rise to the challenge that Latour identifies of explicating humans' role in both the construction of phenomena and the construction of facts. In the case of the hashtag #400ppm, people communicating about science, both formally and informally, transformed a scientific measurement into an event. The event is initiated by the initial press release and continued by further stories based on that press release. On Twitter, the work of these texts is largely represented by and delegated to hyperlinks.

We borrow Latour's use of the term "delegate" here, which he defines as the act of shifting work from humans onto non-human technologies and objects, allowing humans to rely on technologies to do work, rather than having to do the work themselves. In the case of hyperlinks, they take on the work of transmitting the content of the texts they link to; the human Twitter user's work is limited to clicking a share button or copy and pasting, saving them the effort of retyping or paraphrasing the texts their hyperlinks point to. Hyperlinks are thus just one element within a broader network of humans and non-humans, all working together to make this communication of scientific news on Twitter possible. Within Latour's broader actor-network theory (ANT), each human and non-human connected to this network is an actant, a term that refers to both "individual human actors" and "non-human, non-individual entities" (Latour, 1996). This approach to hyperlinks as technological delegates suggests that hyperlinks and brief descriptions of them function as pivotal actors propelling the transmission of news about science events. In addition to examining the more traditional modes of the press release and news story, science communication researchers should attend to the technologies that allow their circulation through social media networks and actively work to shape those technologies.

Social Media Communication Research on Public Discourse and Environmental Communication

Social media platforms have been widely available for over 10 years, and communication research has responded in part by exploring how Facebook and other social media sites are used for advocacy and public discourse. Earlier studies (Bortree & Seltzer, 2009; Woolley, Limperos, & Oliver, 2010) considered how users and advocacy groups used Facebook, while more recent work considers how activists use Twitter (Bowdon, 2014; Potts, 2014). Poell (2014) traces how social media are "shaped by interacting techno-cultural and political economic relations" (p. 717) through a hyperlink analysis. Langlois, Elmer, McKelvey, and Devereaux (2009) consider how social media platforms, Facebook in particular, "transform public discussion and regulate the coming into being of a public" by imposing specific conditions, possibilities, and limitations of online use (p. 417). Many scholars of social media, including Potts (2014), Poell (2014), and Langlois et al. (2009), approach social media platforms as assemblages, arguing that content cannot be separated from platforms, protocols, and networks.

Environmental issues, including climate change, have also been the focus of recent work on social media, including *Environmental Communication's* 2015 special issue on Climate Change Communication and the Internet (Koteyko, Nerlich, & Hellsten, 2015). Segerberg and Bennett (2011) focused on climate change protests, arguing that social media can play a role as organizing mechanisms or agents, and that these technologies may also reflect larger organizational schemes in a protest ecology (p. 212). Bennet and Segerberg (2012) also analyzed digitally networked actions such as the "Putting People First" movement, which aimed to mobilize the public against the social harms of unrestrained capitalism of the type that led to the 2008 economic meltdown. Bennet and Segerberg (2012) found that this movement succeeded in part by using "personal action frames," such as "we are the 99%," because such frames "are inclusive of different personal reasons for contesting a situation that needs to be changed" (p. 744). Katz-Kimchi and Manosevitch (2015) described a successful campaign undertaken by Greenpeace to use a Facebook campaign to encourage Facebook itself to convert to green energy. These studies show how activists can successfully use the affordances of social media to support their goals.

Although these studies show social media's power to effect positive changes, other researchers note that social media can reveal, or even foster, a sense of passivity or helplessness, or even provide a venue for climate change denialism (Sharman, 2014). Newell and Dale (2015), describing a comprehensive and multi-faceted legislative and policy framework to spur local climate innovation in British Columbia, observed only limited public engagement, which presents a significant barrier to developing inclusive solutions to the problem of climate change. Sharman's (2014) study of the climate change denial blogosphere concluded that climate skeptic blogs offer an alternative site for audiences to access what they consider expert knowledge on the topic. Matthews (2015), in reviewing skeptical comments on climate change blogs, found that a significant motive for skepticism was the belief that warnings about climate change have been overstated; his findings corroborate other research indicating that dire messaging about climate change may encourage skepticism (Feinberg & Willer, 2011). Thus, social media platforms can work against the goals of climate scientists, activists, and communicators, particularly if they fail to engage publics or are interpreted as being "alarmist."

For many of these studies, researchers developed methods that incorporated content analysis and quantitative approaches. Veltri and Atanasova (2015) analyzed a data set of 60,000 tweets relating to climate change and identified four common themes: (1) calls for action and increasing awareness of climate change, (2) discussions about the consequences of climate change such as extreme weather and representations of a risk discourse, (3) policy debate about climate change and energy, and (4) local events associated with climate change (Veltri & Atanasova, 2015). While our study is not near the scale of Veltri and Atanasova's, we used a type of content-based thematic analysis to sort the tweets into thematic categories, finding some overlap between our categories and those of Veltri and Atanasova (2015). These tweets, whatever their category, are often legitimized by their inclusion of hyperlinks, which we describe as key technologies to which some of the tasks of science communication have been delegated.

Methodology

We analyzed a dataset of 122 tweets published in May 2015 that used the hashtag "#400ppm" to explore ways that users engage with eventful science news on Twitter. We begin by theorizing the work done by hyperlinks, drawing on Latour's (1992) conceptualization of nonhuman technology as "delegates" taking on the work of human and other nonhuman actors. This analysis allows us to approach hyperlinks as both discourse and technology, making meaning through the text and images they place in tweets but also enabling users to do connective work more easily through their affordances as technology. Other social media researchers have drawn on Latour's sociological theories. Potts (2009), for example, uses Latour's articulation of actor network theory (ANT) as an approach to studying dynamic online communicative systems. ANT is an ontological approach to studying social and material relations that treats both humans and non-humans as agents within networks of people and things, all helping to collaboratively build reality and create opportunities for action. Latour's insistence on viewing non-human technologies as making vital contributions to communication networks of humans and texts provides a more comprehensive view of how communication works in the spaces of social media. For traditional rhetorical analysts, the temptation when studying text-based data is to stop with a text, focus on the details, and analyze each rhetorical move. But ANT reminds us to consider discursive *movement*—what happens as texts connect to each other and discourse is shared between actants, assembled ultimately into a network (Latour, 2005).

Latour's actor-network theory allows us to account for the roles played by users, links, hashtags, and other actants in the effort to move information through a larger network. However, we also want to account for the content of the language itself, to see how the various rhetorical strategies played a role in shaping and moving a message. In other words, we want to account for hashtags, tweets, and links as pieces of discourse as well as actants. For this purpose, we turned to a concept from classical rhetorical theory: topoi, often translated as "commonplace," defined as recurring lines of argument that circulate within particular exigencies (Ross, 2013). Rhetorical scholars such as Leff (1996) and Dyck (2002) show the role that topoi play in invention. Walsh (2010) argues that topoi are not static and context-free, instead asserting that "topoi are pervasive and dynamic cognitive strategies—linking people, texts, and experiences—that engage particular rhetorical situations" (p. 122). We've adapted the strategy from Walsh (2013) to identify recurrent topoi in the 122 tweets.

Denise and another coder unfamiliar with our theoretical approach reviewed the full dataset of tweets and, based on those initial impressions, individually identified prominent themes in the data (including the text of the tweets, the links, and the hashtags). Following comparison of their identified themes, the two coders agreed to categorize the tweets according to four topoi: "just the facts," alarmism, "we must act now," and human impacts. The coders performed separate content analysis to verify the identified topoi, categorizing each tweet into one or two topoi. (For one tweet, our second coder applied a third topos.) The first coding pass yielded interrater agreement of 74% for at least one of the topoi applied by each coder, with 90 of 122 tweets receiving at least one shared code. The most consistent agreement occurred on applying the topos of "just the facts," while "human impacts" yielded the least agreement. Following this, we discussed the remaining 32 tweets with our independent coder to arrive at consensus for these tweets; this resulted in interrater agreement of 85%, with 18 tweets still coded into different categories. In the results section below, we describe some of the differing interpretations of topoi that led to these remaining differences. We hope that transparency about these differences will offer deeper insight into the topoi under discussion, as well as highlight some of the specific challenges of coding tweets, whose brevity and referential nature can yield multiple interpretations. By layering this topical analysis on the Latourian analysis of the work done by the hyperlinks, we show how various recurring lines of argument circulate within the more complex assemblage comprised of tweeters, news stories, hashtags, hyperlinks, and facts about climate change.

Hyperlinks as Technological Delegates

One of the most striking features of the communication circulating within the May 2015 "#400ppm" dataset was the frequency with which tweets included links to news stories, many of which cited the NOAA press release. Table 6.1 shows the raw number and a percentage breakdown for total tweets in the dataset, the tweets that included hyperlinks, and the tweets with hyperlinks that originated from .gov URLs, suggesting they linked to official reports about the atmospheric CO_2 findings. The non-governmental hyperlinks included 32 other news sources ranging from traditional news outlets such as the *Guardian* (Vaughan, 2015), *National Geographic* (Howard, 2014), and the *New York Post* (Associated Press, 2015) to online only news publishers including *Mother Jones* (Oh, 2015), *Politico* (Restuccia, 2015), and *Democracy Now* (Democracy Now, 2015). Yet others link to sites with a focus on environmental, and often specifically climate-related, news and science, such as *Carbon Brief* (Pidcock, 2015) and *Climate Central* (Kahn, 2015).

The high percentage of tweets in the dataset that included hyperlinks shows that these links clearly functioned as key elements for consumers

Table 6.1 Total Count of Tweets and Tweets with Hyperlinks in the May 2015 "#400ppm" Dataset

	May 2015	
	Raw Count	*% of Total Dataset*
Total tweets	122	100.0
Tweets w/hyperlinks	94	77.0
Tweets w/non-.gov hyperlinks	78	63.9
Tweets w/gov hyperlinks	16	13.1

and sharers of the news event in question. The relatively low percentage of tweets with .gov hyperlinks compared to the total hyperlinks count shows the important role played by other outlets picking up the news first released by a government agency. It is perhaps unsurprising that Twitter users would choose to share hyperlinks, as these sources may have been where they first encountered the news themselves, prompting them to propagate it. Additionally, we suggest that, in cases where users are sharing what they perceive as fact to be learned or acted on, the hyperlinks serve to validate users' promotion of the #400ppm hashtag by lending it credibility as a reference to fact and, moreover, doing so by acting as a delegate for the work of science communication. In other words, the hyperlinks take on the work of providing access to scientific facts and producing news events, replacing the need for a science communicator, whether lay or professional, to do this work in the tweets themselves.

To explain this delegation function played by hyperlinks within the network, we propose that they be understood as examples of Latour's (1992) theorization of technology's imbrication in "programs of action, sections of which are endowed to *parts* of humans, while other sections are entrusted to parts of nonhumans" (p. 254). Latour uses the example of a door-hinge taking on the work of enabling people to pass through walls while maintaining the wall's ability to continue differentiating inside from outside. Without a door, passage through walls would require first breaking a hole in a wall and then rebuilding it. By contrast, the technology of the door-hinge creates the possibility of a passageway through the wall that can repeatedly be opened or closed as necessary. Thus, the work done by the door-hinge replaces the work that would be required to repeatedly break through and rebuild walls. Latour (1992) describes this shift of work as "delegation" or "translation," which names the process by which "we have delegated (or translated or displaced or shifted down) to the hinge the work of reversibly solving the wall-hole dilemma" (p. 229). In this way, we make nonhumans responsible for handling work for us, often in ways that allow us to no longer even think about that work.

For Twitter users sharing the news about #400ppm, the hyperlinks function as a technology to which users delegate the work of explaining the scientific fact and its eventfulness. To understand the work of delegation, Latour (1992) enjoins us to "simply imagine what other humans or other nonhumans would have to do were this character not present" (p. 229). If not for the hyperlink, the Twitter user would have to laboriously copy out or paraphrase the text of the press release or news story to share its content. In other words, the technology of the hyperlink allows communicators to delegate the work of repeating individual texts. Instead, these texts can circulate, not through actual physical motion—the 1's and 0's encoding them never have to leave their original servers—but rather as a metaphorical circulation that is more accurately described as a process of growing their connections to an expanding network of tweets, users, other texts, and so on.

The hyperlink's centrality to the internet's ability to function as a network, rather than a series of siloed texts, is, of course, well-documented. Often, this networking is conceived of as a way to connect related texts, as in internet pioneer Doug Engelbart's (1962) theorization of "'associative linking' possibilities, a notion that was to serve as the forerunner of hypertext and led three decades later to the World Wide Web" (Markoff, 2005 p. 48). Similarly, media scholars Schneider and Foot (2004) conceive of the internet as made up of web spheres, which they describe "as not simply a collection of Web sites, but as a hyperlinked set of dynamically defined digital resources spanning multiple Web sites deemed relevant or related to a central theme or 'object'" (p. 118). The hyperlink brings together otherwise scattered parts of the web. While this use as a textual marker of digital texts' relevance to each other is certainly a key function of the hyperlink, the hyperlink also functions as a critical technological delegate, allowing web users to do much more work with ease than they would without it. Thus, while Latour (1992) distinguishes between text and technology, the hyperlink troubles this distinction by being both discourse and technology. This insight recalls theorizations of computer code that complicate treatments of it as mere text, despite being made up of orthographic marks; these include Mackenzie's (2003) and Berry's (2016) careful distinction between code as text and code as process. Twitter further muddies this distinction between text and hyperlink by affording users the ability to share story headlines and visuals within the tweet when they tweet out a hyperlink, as many of the users in our dataset did.

In a previous ANT study of social media, Potts (2009) highlights hyperlinks as just such key actants in the sharing and validation of information: "Links allowed information to travel throughout these networks, thus providing a sense of consistency when more than one actor would refer to these links in communicating information to each other" (p. 294). Potts (2009) also notes that this centrality of hyperlinks illustrates participant behavior that escapes specific social media networks to draw connections beyond it:

> Rarely sticking to one Web site, participants are actively moving among sites, gathering information and turning that information into knowledge as they share it with others. Presenting this kind of literate activity is something new, requiring a different lens through which to study these experiences.
>
> (p. 284)

The technology of the hyperlink replaces human labor, but also makes methodological demands on us to take its work seriously.

Topoi Analysis of Tweet Dataset

We turned next to the specific topoi that users relied on when sharing hyperlinks and how these topoi relate to the texts that the hyperlinks, as delegated science communicators, work to relay. We identified four major topoi in our dataset. These four topoi can largely be classified according to Aristotle's three types of discourse: political (or deliberative), which focuses on questions of policy and proposed future actions; legal (or forensic), which focuses on questions of guilt and innocence; and ceremonial (or epideictic), which focuses on questions of praise and blame (Aristotle & Kennedy, 2006). Gross (1994) suggests that traditional one-way science communication is epideictic in nature, in that it is designed to increase the audience's appreciation for science. Gross (1994) also argues for the connection between ethical and political considerations: "As Aristotle saw, rhetorical activity is also ethical and political activity: nothing significant can be advocated in the public forum that does not entail judgments of right and wrong" (p. 5). In particular, epideictic rhetoric, with its focus on ethics and shared values (Golden, Berquist, & Coleman, 1992), and deliberative rhetoric, with its focus on public action, at times overlap (Tillery, 2003). As Chapter 7 notes, it is possible to take up questions of policy without making authoritative arguments, but by engaging topoi, data, and moments of identification. In relation to the four topoi we describe below, the alarmist topos seems most often to be epideictic, characterized by self-blame, and the "we must act now" topos is most often deliberative, characterized by calls to action. But even in the coding, we found that these two topoi and functions often overlap. The following section is a description of the topoi, with connections to some of the texts the hyperlinks functioned as summaries of and connections to.

Topos #1: "Just the Facts"

The first major topos is a claim of factual reporting, summed up in a "just the facts" attitude and exemplified by a neutral statement that global monthly averages of CO_2 have surpassed 400 ppm. Typically, these tweets rely on the linked stories to relate alarm or connect to other topoi. This category of neutrally sharing a fact was the largest (with both coders applying the topos to 63 out of 122 tweets), although many of those were categorized as using more than one topos. Several users presented a neutral fact in the tweet but embedded a more alarmist or activist claim within a hashtag, such as #oceanacidification, or more commonly (3 occurrences) #keepitintheground. For example, "Global #CO_2 concentrations surpass #400 ppm for first time #oceanacidification" (Morgan, 2015) offers a neutral statement (a "fact" in Latour's sense) in the text of the tweet, but the hashtag #oceanacidification implies a causal link to concrete and serious environmental impacts. The tweets using the hashtag #keepitintheground allow that hashtag to work as an enthymeme, leaving readers to work out the implications regarding the harmful impacts of fossil fuels.

Several of these more "neutral" tweets also use understatement as a rhetorical device, including references to CO_2 reaching "prehistoric" levels, without comment on any implications. The inclusion of such adjectival qualifications led to some disagreement between coders, raising the question of how to distinguish between tweets sharing facts which happen to be alarming to environmentally-concerned readers from tweets sharing facts already framed as alarming and tweets actively engaging the topos of alarmism. For example, one tweet sharing a story from salon.com reads, "For the first time in recorded history, CO_2 levels averaged above #400ppm for the entire month of March" (Earth Island Journal, 2015). While this could be read as a simple statement of fact, both coders categorized it as both "just the facts" and "alarmism," as the phrase "for the first time in recorded history" highlights the unusual and alarming nature of this fact. This example demonstrates the difficulty of determining precisely what can be interpreted as "just the facts," particularly when our rhetorical training prompts us to read for these kinds of rhetorical moves which frame facts in particular ways for particular purposes. For other tweets, though, the alarmism was clearly layered onto the facts, as when one tweet editorialized with a one-word interjection: "Yikes! The world's carbon dioxide levels just hit a new milestone of #400ppm" (Amazon Aid, 2015). The category of alarmism is discussed in greater detail below.

Most of the tweets in the "just the facts" category, even when also fitting rhetorically into other categories, seem to participate in the goal seemingly established by the NOAA's original press release: to transform this global average measurement from a fact into an event. Movement towards alarmist, activist, or human-centric topoi often happened within the hashtags, suggesting that hashtags can be viewed as affordances, actants, and forms of argument.

Topos #2: Alarmism

The second most prominent category, with 24 out of 122 tweets categorized as such by both coders, consists of expressions of alarm, the apocalyptic, doomsday scenario that has recurred throughout environmental discourses for decades and is often criticized for fostering a defeatist response (Foust & O'Shannon Murphy, 2009). In many cases, that doomsday topos is enabled or invoked by the tweeted link's headline or pull quote; five of the tweets in this category tweeted out a link to the *Mother Jones* article, with the headline referring to a "staggering new milestone." That key phrase was repeated in the tweets. Other alarmist tweets included frequent references to the first time in one million years that atmospheric CO_2 averaged 400 ppm, including one tweeted image of a Tyrannosaurus Rex dinosaur (4USolution, 2015). Alarmist hashtags included prompts to be alarmed because there is #notmuchtime and we are #screwed, as well as highly emotionally charged and apocalyptic phrases like #massextinction, #ecocide, and #fuckingsuicide. The level of alarm called for by these hashtags varies, with the extreme end of the spectrum suggesting that we are past the point of irretrievable damage.

Such a suggestion would seem to foreclose the possibilities of deliberative rhetoric. It may be that these tweets are acting as epideictic rhetoric, seeking to form a network of contacts with similar values or assumptions. In this sense, alarmist environmental discourse may serve to establish values and beliefs, rather than to persuade, and such expressions play a role in fostering a community of actors. More pessimistically, we can see these expressions as symptomatic of the problem Latour (2014) describes:

> I think that it is easy for us to agree that, in modernism, people are not equipped with the mental and emotional repertoire to deal with such a vast scale of events; that they have difficulty submitting to such a rapid acceleration for which, in addition, they are supposed to feel responsible while, in the meantime, this call for action has none of the traits of their older revolutionary dreams.
>
> (p. 1)

Topos #3: "We Must Act Now"

The topic of activism as a response to the message of 400 ppm was also a common response, with 20 of the 122 tweets placed by both coders in this category (many cross-listed with "just the facts"). Unlike the alarmist or just-the-facts tweets, this topos was more often expressed only within hashtags. Along with #keepitintheground, other activist hashtags include #carbontax, #ParisClimate, #emissions or #ZeroEmissions, #divestment, or the less specific #StopThisRise. Like the activist hashtags described by Bennet and Segerberg (2012) such as "#wearethe99%," these climate-related hashtags offer broad frames that can encompass a variety of messages and welcome broad perspectives. In conjunction with the #400ppm hashtag that provides the frame of this study, #keepitintheground, #divestment, and #ZeroEmissions offer concrete goals that groups and individuals can both promote through social media and work towards through deliberative, activist efforts.

By calling for specific actions, these hashtags also function as deliberative rhetoric, making the political dimensions of the #400ppm event explicit. As such, hashtags invoking the "we must act now" topos connect this single climate change milestone to broader policy discourses that prompt us to treat such scientific events not just as isolated, apolitical facts, but as kairotic opportunities to advocate for individual and structural actions (also see Chapter 7). In their attention to specific actions, these hashtags distinguish themselves from alarmist ones which either suggest the time for action is past (such as #screwed) or allow that there may still be time for actions, but do not offer suggestions for what those might be (such as #notmuchtime). This distinction may matter especially to scientists and science communicators invested in encouraging support for pro-environmental action, given research showing that directing people's attention to problem-solving can encourage environmentally beneficial behaviors (Homburg, Stolberg, & Wegner, 2007). Moreover, while social science research has shown mixed results to the fear appeals embedded in the alarmist topos, research suggests they can support critical decision making when used to show causal relationships between actions and environmental outcomes (Meijinders, Midden, & Wilke, 2001). The "we must act now" topos could thus enable hashtags to both open deliberative debates about responding to climate change and to encourage efficacy and thoughtful responses to climate change events.

Topos #4: Human Impacts

Human Impacts was by far the smallest and most contentious category of topoi, with one coder applying the topos to 18 tweets and one coder applying the topos to nine tweets during the first coding pass. Upon discussion, the reason for the disparity became apparent: while one interpreted human impacts to mean "human impacts on the climate causing climate change," the other interpreted human impacts to mean "impacts climate change will have on humans." These are very different topoi, with the former typically applying to epideictic tweets intending to lay blame on humans for the news of #400ppm, and the latter looking ahead to the future in a variety of ways.

The tweets looking to future impacts on humans usually referred to extreme weather events such as hurricanes and cyclones, rather than to longer-term (and perhaps higher-impact) problems such as ocean level rising (one tweet referred to population displacement) or drought (no references). This lack of emphasis on the human impacts of climate change in relation to the #400ppm hashtag was surprising, as the effects of climate change are likely to be severe on human health and infrastructure. However, this lack of emphasis in tweets and hashtags on human impacts was in keeping with most of the press releases and news stories whose links were tweeted, which generally did not emphasize the possible human impacts. The human impacts that were referred to included human populations (e.g., @NO_ACP's tweet including the phrase "500,000 population displacement") or commonly known weather cycles (e.g., @ knollster's tweet pointing out unusual weather: "Massive storm system in the Midwest, tropical cyclone developing in the Atlantic a month before hurricane season… #400ppm"). Certainly, though, human impacts could be inferred from some of the other hashtags coded as alarmist or "we must act now." While the hashtag #screwed, for example, does not specify *who* is screwed, the suggestion that there will be dire consequences to the #400ppm event raises the possibility that those dire consequences will include dire impacts on humans. Again, the topoi in these tweets are open to interpretation depending on the views, values, and knowledge readers bring to them. This is not to suggest a porous topos such as "human impacts" cannot do valuable rhetorical work; rather, as with all communication, even topos-driven tweets are deeply contextual.

Conclusions

As we consider what these results might mean for science communicators, it seems clear that press releases and news stories can be structured in ways to foster certain patterns of sharing on social media. Invoking certain topoi in their press releases and news stories—even less political topoi such as human impacts—would give scientists and science communicators a grounding from which to develop usefully deliberative and epideictic hashtags with which to publicize and circulate their hyperlinked writing, with the acknowledgement that these hashtags and grounding topoi are highly susceptible to variable interpretation in the compressed space of a tweet. Attention to the text and technology of the hashtag suggests it has the potential to encourage the circulation of information in ways that promote activism rather than despair. This robust circulation depends on hyperlinks, the technology underlying both hashtags' and news stories' ability to be shared. Despite being nonhuman technologies, these hyperlinks are themselves important actors in the network of science communication, in both their facets, of text and technology. Against efforts to cleanly distinguish among humans and nonhumans, Latour (1992) counters, "I see only actors—some human, some nonhuman, some skilled, some unskilled—that exchange their properties" (p. 235). Attending to all these actors in social media networks can help both scientists and science communicators leverage news stories and social media platforms to spread science facts as they become important events.

Acknowledgements

The authors would like to thank Stephanie Phillips for her generous help creating coding categories and coding tweets for topoi.

References

4USolution [4USolution]. (2015, May 11). *Global atmospheric #CO_2 is now above #400ppm. It's the first time since...when exactly? http://on.nrdc.org/1FbyTt7 by @PalmerBrian v.@onEarthMag.* Retrieved April 16, 2016, from https://twitter.com/4Usolution/status/597797547407728640.

Amazon Aid [AmazonAidF]. (2015, May 7). *Yikes! The world's carbon dioxide levels just hit a new milestone of #400ppm http://bit.ly/1IlPw4B #ActOnClimate.* Retrieved April 16, 2016, from https://twitter.com/AmazonAidF/status/596408748165455873.

Aristotle, & Kennedy, G. A. (2006). *On rhetoric: A theory of civic discourse.* New York: Oxford University Press.

Associated Press. (2015, May 6). *Carbon dioxide levels reach highs not seen in 2 million years.* Retrieved August 12, 2016, from http://nypost.com/2015/05/06/carbon-dioxide-levels-reach-highs-not-seen-in-2-million-years/.

Bennett, W. L, & Segerberg, A. (2012). The logic of connective action. *Information, Communication & Society, 15*(5), 739–768. doi:10.1080/1369118X.2012.670661.

Berry, D. (2016). *The Philosophy of software: Code and mediation in the digital age.* New York: Springer.

Bortree, D. S., & Seltzer, T. (2009). Dialogic strategies and outcomes: An analysis of environmental advocacy groups' Facebook profiles. *Public Relations Review, 35*(3), 317–319. doi:10.1016/j.pubrev.2009.05.002.

Bowdon, M. (2014). Tweeting an ethos: Emergency messaging, social media, and teaching technical communication. *Technical Communication Quarterly, 31*(1), 35–54. doi:10.1080/10572252.2014.850853.

Democracy Now. (2015, May 8). *Atmospheric concentration of CO_2 tops 400 ppm for longest period on record.* Retrieved April 25, 2016, from www.democracynow.org/2015/5/8/headlines/atmospheric_concentration_of_co2_tops_400_ppm_for_longest_period_on_record.

Dyck, E. (2002). Topos and enthymeme. *Rhetorica: A Journal of the History of Rhetoric, 20*(2), 105–117. doi:10.1525/rh.2002.20.2.105.

Earth Island Journal. [earthislandjrnl]. (2015, May 6). *For the first time in recorded history, CO_2 levels averaged above #400ppm for the entire month of March. #climate http://www.salon.com/2015/05/06/the_world_reached_a_scary_emissions_milestone_in_march_and_its_only_going_to_get_worse/.* Retrieved June 19, 2017, from https://twitter.com/earthislandjrnl/status/596026950080077824.

Engelbart, D. C. (1962). *Augmenting human intellect: A conceptual framework* (Summary Report No. AFOSR-3223). Menlo Park, CA: Stanford Research Institute. Retrieved August 7, 2017, from http://www.dougengelbart.org/pubs/papers/scanned/Doug_Engelbart-AugmentingHumanIntellect.pdf

Feinberg, M., & Willer, R. (2011). Apocalypse soon? Dire messages reduce belief in global warming by contradicting just-world beliefs. *Psychological Science*, *22*(1), 34–38. doi:10.1177/0956761039l911.

Foust, C. R., & O'Shannon Murphy, W. (2009). Revealing and reframing apocalyptic tragedy in global warming discourse. *Environmental Communication*, *3*(2), 151–167. doi:10.1080/17524030902916624.

Golden, J. L., Berquist, G. F., & Coleman, W. E. (2000). *The rhetoric of western thought.* Dubuque, IA: Kendall/Hunt Publishing Company.

Gross, A. G. (1994). The roles of rhetoric in the public understanding of science. *Public Understanding of Science*, *3*(1), 3–23.

Homburg, A., Stolberg, A., & Wagner, U. (2007). Coping with global environmental problems: Development and first validation of scales. *Environment & Behavior, 39*(6), 754–778.

Howard, B. C. (2014, May 27). *Northern hemisphere cracks 400 ppm CO_2 for whole month for first time.* Retrieved April 25, 2016, from http://news.nationalgeographic.com/news/2014/05/140527-400-ppm-carbon-dioxide-global-warming-climate-science/.

Kahn, B. (2015, May 6). *A global milestone: CO_2 passes 400 PPM.* Retrieved April 25, 2016, from www.climatecentral.org/news/co$_2$-400-ppm-global-record-18965.

Katz-Kimchi, M., & Manosevitch, I. (2015). Mobilizing Facebook users against Facebook's energy policy: The case of Greenpeace Unfriend Coal Campaign. *Environmental Communication*, *9*(2), 248–267. doi:10.1080/17524032.2014.993413.

Knoll, A. [Knollster]. (2015, May 7). *Massive storm system in the Midwest, tropical cyclone developing in the Atlantic a month before hurricane season... #400ppm.* Retrieved April 16, 2016, from https://twitter.com/knollster/status/596410106524250113.

Koteyko, N., Nerlich, B., & Hellsten, I. (2015). Climate change communication and the internet: Challenges and opportunities for research. *Environmental Communication*, *9*(2), 149–152. doi:10.1080/17524032.2015.1029297.

Langlois, G., Elmer, G., McKelvey, F., & Devereaux, Z. (2009). Networked publics: The double articulation of code and politics on Facebook. *Canadian Journal of Communication, 34*(3), 415–434. doi:10.22230/cjc.2009v34n3a2114.

Latour, B. (1992). Where are the missing masses? The sociology of a few mundane artifacts. In W. E. Bijker & J. Law (Eds.), *Shaping technology / building society: Studies in sociotechnical change* (pp. 225–258). Cambridge, MA: MIT Press.

Latour, B. (1996). On actor-network theory: A few clarifications plus more than a few complications. *Soziale Welt, 47,* 369–381. Retrieved February 15, 2017, from www.bruno-latour.fr/sites/default/files/P-67%20ACTOR-NETWORK.pdf.

Latour, B. (2005). *Reassembling the social: An introduction to actor-network-theory*. Oxford and New York: Oxford University Press.

Latour, B. (2014). Agency at the time of the Anthropocene. *New Literary History, 45*, 1–18.

Leff, M. (1996). Commonplaces and argumentation in Cicero and Quintilian. *Argumentation, 10*, 445–452. doi 10.1007/BF00142977.

Mackenzie, A. (2003). *The problem of computer: Leviathan or common power?* Lancaster: Lancaster University: Institute for Cultural Research.

Markoff, J. (2005). *What the dormouse said: How the sixties counterculture shaped the personal computer industry*. New York: Penguin.

Matthews, P. (2015). Why are people skeptical about climate change? Some insights from blog comments. *Environmental Communication, 9*(2), 153–168. doi:10.1080/17524032.2014.999694.

Meijnders, A. L., Midden, C. J. H., & Wilke, H. A. M. (2001). Role of negative emotion in communication about CO_2 risks. *Risk Analysis, 21*(5), 955–965.

Miller, C. (1992). Kairos in the rhetoric of science. In S. P. Witte, N. Nakadate, & R. D. Cherry (Eds.), *A rhetoric of doing: Essays on written discourse in honor of James L. Kinneavy*. Carbondale, IL: Southern Illinois University Press.

Morgan, K [kamorgan91]. (2015, May 6). *Global #CO_2 concentrations surpass #400ppm for the first time #oceanacidification #climatechange www.politico.com/story/2015/05/global-co$_2$-concentrations-surpass-400-ppm-for-first-time-117689*. Retrieved April 16, 2016, from https://twitter.com/kamorgan91/status/595986217403047937.

National Oceanic and Atmospheric Administration. (2015, May 6). *Greenhouse gas benchmark reached*. Retrieved March 29, 2016, from http://research.noaa.gov/News/NewsArchive/LatestNews/TabId/684/ArtMID/1768/ArticleID/11153/Greenhouse-gas-benchmark-reached-.aspx.

Newell, R., & Dale, A. (2015) Meeting the climate change challenge (MC3): The role of the internet in climate change research dissemination and knowledge mobilization. *Environmental Communication, 9*(2), 208–227. doi:10.1080/17524032.2014.993412.

NoACP [NO_ACP]. (2015, May 8). *#400ppm will boil the oceans and create dead zones. 500,000+ population displacement #Virginia coast in ten years*. Retrieved April 16, 2016, from https://twitter.com/NO_ACP/status/596761525454565376.

Oh, I. (2015, May 6). *The world's carbon dioxide levels just hit a staggering new milestone*. Retrieved April 25, 2016, from www.motherjones.com/blue-marble/2015/05/carbon-dioxide-global-concentrations.

Pidcock, R. (2015, May 6). *Monthly global carbon dioxide tops 400ppm for first time*. Retrieved April 25, 2016, from www.carbonbrief.org/monthly-global-carbon-dioxide-tops-400ppm-for-first-time.

Poell, T. (2014). Social media and the transformation of activist communication: Exploring the social media ecology of the 2010 Toronto G20 protests. *Information, Communication, and Society, 17*(6). doi:10.1080/1369118X.2013.812674.

Potts, L. (2009). Using actor network theory to trace and improve multimodal communication design. *Technical Communication Quarterly, 18*(3), 281–301. doi:10.1080/10572250902941812.

Potts, L. (2014). *Social media in disaster response: How experience architects can build for participation*. New York: Routledge, Taylor & Francis Group.

Restuccia, A. (2015, May 6). *Global CO_2 concentrations surpass 400 ppm for first time.* Retrieved April 25, 2016, from www.politico.com/story/2015/05/global-co2-concentrations-surpass-400-ppm-for-first-time-117689.html.
Ross, D. G. (2013). Common topics and commonplaces of environmental rhetoric. *Written Communication, 30*(1), 91–131.
Schneider, S. M., & Foot, K. A. (2004). The Web as an object of study. *New Media & Society*, 6(1), 114–122. doi:10.1177/1461444804039912.
Segerberg, A., & Bennett, W.L. (2011). Social media and the organization of collective action: Using Twitter to explore the ecologies of two climate change protests. *The Communication Review*, *14*(3), 197–215.
Sharman, A. (2014). Mapping the climate sceptical blogosphere. *Global Environmental Change, 26*, 159–170. doi:10.1016/j.gloenvcha.2014.03.003.
Tillery, D. (2003). Radioactive waste and technical doubts: Genre and environmental opposition to nuclear waste sites. *Technical Communication Quarterly*, *12*(4), 405–421.
Vaughan, A. (2015, May 6). Global carbon dioxide levels break 400ppm milestone. *The Guardian*. Retrieved August 16, 2016, from www.theguardian.com/environment/2015/may/06/global-carbon-dioxide-levels-break-400ppm-milestone.
Veltri, G. A., & Atanasova, D. (2015). Climate change on Twitter: Content, media ecology and information sharing behavior. *Public Understanding of Science, 26*(6), 1–17. doi:10.1177/0963662515613702.
Walsh, L. (2010). The common topoi of STEM discourse: An apologia and methodological proposal, with pilot survey. *Written Communication 27*(1), 120–156. doi:10.1177/0741088309353501.
Walsh, L. (2013). Resistance and common ground as functions of mis/aligned attitudes: A filter-theory analysis of ranchers' writings about the Mexican Wolf Blue Range Reintroduction Project. *Written Communication*, *30*(4), 458–487.
Woolley, J. K., Limperos, A. M., & Oliver, M. B. (2010). The 2008 presidential election, 2.0: A content analysis of user-generated political Facebook groups. *Mass Communication and Society*, *13*(5), 631–652. doi:10.1080/15205436.2010.516864.

7 From Questions of Fact to Questions of Policy and Beyond

Science Museum Communication and the Possibilities of a Rhetorical Education

Gregory Schneider-Bateman

This chapter examines the public side of science communication as it occurs in science museum exhibits, which remain some of the most popular and wide-reaching sites for scientific ideas to reach the public—95 million visits were made to science museums worldwide in 2013 alone (Association of Science-Technology Centers, 2014). While exhibits in these institutions are still committed to the broad educational goals of science literacy and the public understanding of science, criticism from scholars within Museum Studies, Rhetorical Studies, Cultural Studies, Science Studies, and Political Science have led exhibit designers to become in a sense self-aware. More sensitive to their power to constitute kinds of publics, they are also more willing to embrace and assert their often implicit political dimension. For example, in her recent work on postmodern science exhibits, Fiona Cameron (2010b) argues for exhibits where "audiences are conceived as assemblages, as unique, historical and historically contingent, rather than as a one-dimensional citizen and as an object of discipline" (p. 126). Consequently, for some exhibits it no longer suffices to expect that mere understanding will generate political change (e.g., that a public that understands how evolution works will champion it at their hometown school board meetings). This trend reflects the modern science museum's concern with rhetorical agency, both its own and its visitors'.

The result is an ongoing dramatic refashioning of the museum as a form of public address. Visitors to museums now encounter exhibits quite different from those curated just a few decades ago. Gone are heavy-handed, authoritative exhibits. In their place museums have created exhibits that are more inclusive of other voices and that do not advance a single message or unified narrative structure. Many of these are designed as so-called free choice exhibits, where an open plan and lack of a guided route relaxes the curator's authorial control even further by allowing visitors

DOI: 10.4324/9781315160191-7

the freedom to wander as they please. The ubiquity of inquiry-based, interactive exhibits in these spaces grants visitors even more agency and autonomy over their own experiences. Not all modern science museums offer exhibits curated in these ways, but most do, if not in exhibits of their own design then through the traveling exhibits they rent.

This chapter examines the shifting landscape of informal science museum rhetoric through a pair of climate change exhibits I visited in London and Paris in 2012 while on a two-month tour through European science museums. These two large public exhibits—*Atmosphere* at London's Science Museum and *Ocean, Climate, and Us* at Paris's Cité des Sciences—stood out as significant because they represented major efforts in each city (perhaps each nation) to engage its citizens in climate science. They also stood out because, unlike exhibits on mathematical models or the most recent dinosaur discovery, they directly engaged climate change in a way that introduced its obvious and significant political and policy implications. After discovering each exhibit, I extensively documented it with recordings and photographs, each over multiple days, and then analyzed these documents in an effort to understand how their designs engage visitors as political agents in the world (my analysis was also informed by informal conversations with exhibit designers at each institution). By rhetorically reconstructing each exhibit's organizational design and narrative, I concluded that though they approached climate science from two distinctly different perspectives (atmosphere vs. ocean), they each progressed through the four stases: beginning with science literacy questions of fact and definition and moving to questions of value and policy.

While including questions of value and policy represents an important shift in science museum rhetoric, their mere presence is not sufficient to effectively empower audiences. As such, in the second half of this chapter I argue that it is how these questions are presented—how they allow visitors to constitute themselves—that is most significant. With climate change as a focal topic (also see Chapter 6), I will illustrate how exhibits that take up questions of value and policy might position visitors as voters, deliberators, and cosmopolitical citizens. Exhibits do this not by making specific authoritative arguments themselves; instead, they present arguments, topoi, authorities, data, and moments of identification that prepare visitors to meaningfully engage questions of value and policy outside the museum's walls. In demonstrating this, I describe how modern science exhibits have begun to offer a distinct form of rhetorical education, one that empowers visitors to invent arguments (and indeed themselves) in ways that might transform their political agency.

Stasis Theory and Science Museum Communication

First established in Hermagoras' *Ad Herennium,* stasis theory names an ancient rhetorical framework for inventing arguments in legal contexts. In that classic formation, stasis theory describes four questions, each pinpointing where issues remain unsettled in a controversy. For Hermagoras, those four questions are fact (did something happen and what was it?), definition (what was the nature of the act?), quality or value (what are the aggravating circumstances?), and jurisdiction/action (is this the correct venue to try this act and what should be done?) (Kennedy, 1963). Take the case of a stolen car: following its original formation, prosecutors must first determine if the car is missing (did something happen) and who took it. Second, if the car is indeed missing, they must next establish if it was stolen or was simply *borrowed*, a question of definition. Next, if the car was indeed stolen without permission, the defendant might debate the quality of that definition by agreeing that she stole the car but that it was needed to rush the president to the hospital (a fact which might exonerate the act). Finally, the ultimate verdict might hinge on whether the defendant had been read her Miranda rights.

Widely adapted and developed over its long history by rhetoricians and scholars from Cicero and Quintilian to Toulmin (1958), Turner (1991), and others, stasis theory in its modern form has become a broader, more flexible, and more useful framework for inventing and analyzing texts (including visual texts) across any rhetorical situation. Particularly relevant here are the ways in which stasis theory has been theorized within scientific contexts by Prelli (1989), Gross (2004), Fahnestock and Secor (1988), and Northcut (2007). However, my focus here on how science gets accommodated to the public will retain a conventional application of Hermagoras' basic theory, because as I will show, it is that earlier theory that better explains my observations. Still, two important amendments must be added. First, because jurisdictional questions are irrelevant in many public scientific controversial contexts, Fahnestock and Secor (1983) revise the fourth stasis into a question of policy, a generalized concern with what should be done. This move enlarges stasis theory's range of application without radically revising its basic structure. Second, Fahnestock and Secor (1985) have added a fifth stasis to recognize that in extra-legal contexts, the question of cause is another sticking point (indeed, insanity defenses also raise causal issues). This additional stasis question—the question of cause (why did the act happen?)—is particularly relevant in scientific contexts where facts are explained, not simply established. Thus, in public deliberations over climate change, for example, the modern stasis questions might look like this:

FACT: Is the climate changing and how do scientists know?
DEFINITION: What kind of change is it? By what measure?
CAUSE: Why is the climate changing? Is it a natural cycle or because of human causes?
QUALITY: How do we evaluate it? Is it the necessary result of progress or detrimental to our future security?
PROCEDURE/POLICY: What, if anything, should we do and who should do it?

As a tool for invention, the full set of stasis questions allows rhetors to gain a more abstract sense of where arguments can be sourced, how they might be developed, and how they should be interpreted. As Fahnestock and Secor (1988) state: "The stases tell a writer 'where to think,' not 'what to think'" (429).

The value of stasis theory as a tool for invention comes from two characteristics identified by Fahnestock and Secor (1985). First, the questions are recursive: raising questions at one stasis can reintroduce or reshape questions at other stases. For example, how we settle the causal question has important ramifications for how the definitional question shows up, and which values are established inevitably shape the range of policy options. This feature makes stasis theory flexible, responsive, and adaptable to the complexity of modern argument situations where information and values shift over time. Second, stasis theory is hierarchical without being necessarily linear:

> Questions of fact or conjecture in the first stasis are prior to those of definition in the second; definitions in turn must be established before quality [or value] is debated; and finally all three must be answered or assumed before an action can be recommended or taken in a specific case.
>
> (Fahnestock & Secor, 1983, p. 139)

While nothing requires the questions to be taken up in order or indeed to be taken up together, this hierarchy often acts as a "natural pull" for both those inventing arguments and those encountering them. Because of this "natural pull," rhetors who work through the stasis questions often generate a matching organizational strategy—texts, speeches, even exhibits that move stepwise up the stasis ladder. A similar natural pull exists for audiences—learning about facts often leads to questions of evaluation and action. As Fahnestock (1986) writes,

> Even if the scientific report were translated from insiders' to outsiders' language with minimum amount of distortion and no attempt to provide an epideictic exigence for the report, the public as readers would move the information themselves into the higher stases and ask, 'Why is this happening? Is this good news or bad news? What should we do about it?'
>
> (p. 292)

For example, the natural pull shows up when the fact of sea level rise immediately leads to negative evaluations and questions about what to do to prevent it. Thus, stasis theory provides a structured framework for thoughtfully and recursively exploring how topics and arguments might be presented to best clarify or convince.

But stasis theory is not just a useful tool for inventing political and scientific arguments. As Fahnestock and Secor (1988) demonstrate, it is also a powerful analytic tool for understanding how rhetors constitute audiences within a specific rhetorical situation. The initial clue to this potential lies in stasis theory's hierarchical character: identifying which stasis questions exhibits deployed illuminates how they position audiences as pedagogical, political, and/or civic agents. For example, if the conversation begins with questions of policy (e.g., What should we do about climate change?), then one has assumed the audience already agrees that climate change is real, that it's human caused, and that it's bad. Contrarily, to never raise the value or policy questions positions visitors pedagogically. Alternatively, to leap from facts to action implies that values and definitions can be taken for granted. In these cases, the speaker, writer, or designer assumes certain characteristics about its audience—that they all share certain values, that they already understand certain facts, or that they are ready and willing to change behavior.

This insight into the relationship between which stasis questions are deployed and how audiences are constituted allows us to understand how the museum has altered the way it positions the public. Prior to becoming deeply concerned with the public, science and natural history museums engaged their audiences in stases aimed at developing scientific knowledge. Thus, questions of fact, definition, and cause were central. This kind of environment aligns with a classicist museum: "object-drenched spaces whose installations have intentionally omitted explanatory labels that might help most mere mortals" (Gurian, 2010).

When the modern science and natural history museum embraced its public educational role, the emphasis on questions of facts, definition, and cause remained the same, but these were now connected to a sense of general civic engagement where the knowledge of those facts, definitions, and causes would by some route generate a civic public. The museum displayed authoritative scientific information to visitors who might then make more informed personal and political decisions, enthusiastically support scientific research, or become scientists themselves. But in their modernist form, museums rarely made this link explicit; rather, an informed, scientifically literate public was to simply act differently. In this sense, the absence of the upper stases of value and policy indicates that visitors' political and civic agency was not fully engaged *inside* the museum. Essentially, visitors were expected to move from authoritatively established starting points to questions of value and policy when they encountered them in their own individual contexts.

The failures of this approach in modern science museums have been well documented, and they might best be summarized in two ways. The first criticism comes from Tony Bennett's "exhibitionary complex"—the now canonical application of Foucault to the museum environment that demonstrated how public institutions disciplined populations through an insidious pedagogical ideology of pedagogy (Bennett, 1995; see also Bennett, 2015; Hetherington, 2015; Greenhill, 1992). Thus, remaining within the lower stases of fact, definition, and cause does not eliminate the political; it simply hides it behind the guise of scientific objectivity. Chakrabarty (2002) provides the second criticism by outlining the deep problems with pedagogic models of citizenship that require audiences to first be educated before they can be taken seriously as citizens. For Chakrabarty, the public arrives already politicized, and to ignore this fact means museums fail to engage a meaningful sense of visitor agency. Where Bennett's application of Foucault demonstrates that the museum was inherently politicized and disciplinary, Chakrabarty does the same for visitors.

One of the museum's responses to these criticisms was to reinvigorate the visitor's position in the museum. Viewed from within stasis theory, the general approach was to revise which stases show up by first rethinking the lower stases and second by adding in the upper stases. Following John Durant (1994), museums aimed to help visitors understand how science *really* worked, which went beyond facts, definitions, and some inaccurate plastic version of the scientific method. In this way, questions of fact, definition, and cause became in a sense reflexive, and the civic-minded visitor wasn't meant to just understand scientific knowledge but to understand how that knowledge was constructed in the first place. While an improvement on the old science literacy model, this Public Understanding of Research (PUR) model nevertheless still focused on the oversimplified facts/action stasis structure, where facts now were about science itself and the political/policy step again remained implicit.

The second way museums respond to criticisms has been to explicitly introduce questions of value and policy. Thus today museum exhibits on controversial topics like climate change, evolution, and race directly take up the political. As Mike Hulme (2015) argues, "museums need to be more political and less scientific" (p. 14). Because facts about the climate or claims about scientific consensus alone won't solve climate change, museums can't limit themselves to simply displaying climate science (p. 12). Topics relevant to public issues must be displayed for disagreement. And yet as Rutherford's (2011) analysis of green governmentality at the American Museum of Natural History convincingly documents, the problem of power persists even in cases where visitors are openly and explicitly engaged in value and policy questions. For Rutherford, a shift from questions of fact to questions of value and policy may not be enough if the museum retains its authorial control over those questions—if, essentially, they use them to tell a single story. Thus, it is not simply which stasis questions show up, it's *how* they show up that matters.

Museum Agency and Shifting the Power of Invention

Because the simple addition of questions of value and policy does not inherently resolve the critique of museum power, the science museum's rhetorical form needs to be reconsidered. Recent work by Fiona Cameron (2010c) and other museum scholars provides a route to articulate the ways in which stasis theory might be used to reframe science museum communication. Relying on a broad array of social theorists, including Latour, Beck, Urry, Baumann, and Deleuze and Guattari, Cameron has worked to establish a foundation for postmodern exhibits, which she terms liquid museums. Responding to the Cartesian, classicist museum characterized by certain knowledge, subject/object dichotomies, the absence of controversy, politically passive visitors, and centralized authority, Cameron's postmodern liquid museum offers another model: one that embraces uncertainty, presents controversy, offers multiple perspectives, serves as a site for peer review, functions as a node within larger digital, cultural and transnational networks, transparently expresses ideological and political commitments, and promotes individuals' self-understanding. In short, the postmodern museum does not discipline populations, it presents information in such a way to allow them to actively construct themselves and their beliefs and actions. In this model, visitors aren't constructed as simple rational citizens. Instead, "audiences are conceived as assemblages, as unique, [and] historical and historically contingent" (Cameron, 2010b, p. 126).

By embracing the uncertainty and complexity of questions of value and policy as well as the political controversies that attend them, Cameron's liquid museum re-situates the visitor as a subject able to act creatively. In other words, the museum can no longer be viewed simply as disciplining the visitor in one single way or from one particular angle. And it can't even be seen as disciplining them in multiple ways or from multiple angles. The fact that the museum is (or can become) a liquid institution means that discourses come from anywhere, and most importantly from museum visitors. It means, essentially, that the museum gives up control in order to move audiences from "being objects of intervention to being subjects for action" (Cameron, 2011, p. 91). Instead of the kinds of disciplinary power identified by Tony Bennett, Cameron's "institutions need to assist the 'self-interested' visitor in forming, planning, and designing themselves as individuals on their own terms" (2010a, p. 69).

The argument for the liquid, postmodern museum does not just have a theoretical foundation, however. Cameron also grounds it in empirical audience surveys that show how audiences now bring a completely different set of expectations to the museum. While visitors still want to learn something, they are much more likely to demand that the museum treat them seriously as self-inventing human beings. Visitors want to be treated as intelligent individuals who can make sense of information on their own. As such, visitors request things like more points of view, less certainty, and a wider array of cultural voices. And this doesn't simply mean they want to encounter the more human questions of value and policy; they also want exhibits to represent the deeper complexity in the questions of fact and definition that are too often treated purely educationally, as certain knowledge to be transferred.

The result of Cameron's theoretical and empirical arguments is nothing less than the recreation of the museum as a new ontological assemblage, a self-designing and self-organizing entity that continuously refashions itself by displaying controversial topics; engaging with risk conflicts; framing topics far into the future; embracing deep uncertainty, complexity, and nonlinearity; incorporating transnational and cosmopolitical perspectives; and integrating post-humanism. This refashioning gains clarity through the lens of stasis theory by illustrating how the museum might deploy its inventional power. In other words, the route through the political critique of the science museum cannot simply mean museums invent exhibits that present answers to questions of value and policy. As Hulme (2015) argues, "The purpose of museums is not to get everyone thinking the same thing about climate change" (14). Instead those questions should be used to encourage people to "disagree about climate change" and to "allow people for whom the idea of climate change provokes different stories of meaning—different visions of a good and desirable future—to listen and to learn from each other" (14). Truly embracing stasis as a tool for invention in controversial exhibits would help to accomplish this, for the visitor would be invited to invent (and invent themselves) alongside and within the science museum (Cameron, 2011, p. 100). Doing so means the museum can truly "assist the 'self-interested' visitor in forming, planning, and designing themselves as individuals on their own terms" (Cameron, 2010a, p. 69). In the end, the burden must be shared. After all, exhibits must be invented by institutions. But how those exhibits open up and deploy their stasis questions can dramatically alter the kinds of agencies visitors can embrace.

The Possibilities of Rhetorical Agency in Climate Change Exhibits

In an effort to illustrate how this might be achieved through science museum exhibit communication, in the next section I analyze how two climate change exhibits deploy stasis questions and how that deployment might point a way towards renewing visitor agency.

Deliberating the Atmosphere at London's Science Museum

The first climate change exhibit takes us across the Atlantic to London's Science Museum and its large, immersive experience titled

Atmosphere: Exploring Climate Science. The visual shift is dramatic when moving from traditional text-heavy, highly organized exhibits to *Atmosphere's* dark, open, ethereal, immersive, and highly interactive exhibit. Curved glass is lit blue across the ceiling, grey panels and objects are lit brightly from below, digital kiosks glow throughout the space, the floor is a mosaic of projected digital images that change and shift, and people are everywhere staring at screens, pushing buttons, and running around. Upon entering it seems as if no immediately obvious pathway presents itself, though one becomes clear if we attend to the exhibit's broader thematic structure. Emphasizing the lack of a specific path is the first installation, which is a large multi-person interactive experience where visitors attempt to allocate resources to save different London neighborhoods from the Thames River that is flooding due to climate change. Instead of introducing visitors to scientific principles of the greenhouse effect or even introducing them to the atmosphere, visitors are thrust into an engaging experience where they must respond to the hypothetical results of climate change. In interactive terms, *Atmosphere* somewhat inverts the read-text-then-interact structure found in more traditional exhibits.

A rationale for beginning with this interactive element rather than starting with, say, a more natural choice like the exhibit's section on the carbon cycle can be found in the exhibit team's planning documents. One of the most striking planning documents the exhibitors shared with me was the audience research that led them to choose *Atmosphere* as their title for the climate change gallery. TWResearch, the team that conducted the study, evaluated multiple titles: Climate Change, A Climate of Change; Our Changing Climate; Climate Science; Our Planet, Our Challenge; Our Changing World; Atmosphere; and Changing Our World. These titles were assessed based on how well they appealed to the target audiences and whether they were informative. Each was also evaluated for gallery expectations, tone of voice, and fit with Science Museum. What the TWResearch (2010) found was that "in terms of titles, the phrase 'climate change' drives a fatigued response" (p. 6). "Atmosphere," on the other hand, they concluded "ambiguous but with an element of intrigue" that seemed "appealing and motivating" (p. 18). It allowed for multiple interpretations (with the worry that it could be misleading) and implied "a sense of scale, wonder, marvel [and is] interactive, engaging, interesting" (p. 18). In their conclusions, TWResearch suggested Atmosphere as the first choice but noted that it would need a "more informative strapline," which the museum took to heart by adding "Exploring Climate Science" to the exhibit's final title that embraced other title elements the research team suggested (p. 21).

Just like the decision to lead with the Thames River interactive, the title *Atmosphere* works, partly, as an appealing enticement, less staid and authoritative and more vague and exciting. Even though "climate science" is retained in the exhibit's subtitle, the gallery is often referred to as the "Atmosphere Gallery," a choice that seems consistent with the aim of enticing visitors and moving away from aspects that might engender a fatigued response. It is also a conclusion that museums in the United States might take to heart. In the past few years, the most prominent exhibits on climate change have all offered titles that would, if TWResearch's work applied to US audiences, engender a tired response. For example, *Climate Change: The Threat to Life and a New Energy Future* at the American Museum of Natural History (on display 2008–2009); *Climate Change in Our World* at the Smithsonian (on display 2009–2010); *Climate Change* at The Field Museum (on display summer 2012); and *Climate Change: Our Global Experiment* at the Harvard Museum of Natural History (ongoing). *Atmosphere*, then, is not *simply* or *only* or at least *not at first glance* primarily about climate science.

Before arriving at the final culminating experience, visitors to *Atmosphere* must first traverse exhibit elements that cover the traditional ground of facts and definitions. As discussed above, while there is no clear order to the structure of the gallery, the exhibit's five major thematic sections can be linked into a kind of logical order. These sections are as follows, with the order of the first three largely interchangeable, and the last two clearly building from the historical and current knowledge gained in the first three.

- Exploring the carbon cycle: Science can show us carbon's global pathways and how we're causing them to change.
- Exploring Earth's energy balance: Science can show us how greenhouse gases work and why they really matter.
- Exploring the climate system: Science can show us how the climate works and what causes it to change.
- Exploring what might happen: Science can track what's already changing and help us imagine the future.
- Exploring our future choices: Science and technology are already helping…what are our options for tomorrow?

In practice, the order is not set in the *Atmosphere* gallery, and visitors can wander from installation to installation as they please, especially since the always popular interactive elements aren't free at exactly the moment they want to play with them. This results in much moving back and forth. This sense of visitor freedom (what the museum world terms free-choice) allows visitors a much richer sense of agency and is a good example of the ways in which museums are trying to reduce the strong authoritative feel and structure for which museums have often been criticized.

While each of the five main thematic sections covers different content, each area includes similar elements: interactive games, objects, historical anecdotes, and detailed explanations. For example, in the fifth section "Exploring our future choices," visitors play with an interactive game that asks them to play act as a reporter gathering information for news stories about alternative energies like solar, wind, and hydroelectric power. They also click and choose their way through a digital interactive element that allows them to follow threads of information about geo-engineering the climate that ask "big questions" about strategies for CO_2 reduction and solar reflection. These interactive features are accompanied by real objects and their stories. Again, visitors are encouraged to click through a digital component to learn more about each thing. The objects here include energy harvesting concrete slabs, carbon collectors, artificial trees, plastic solar cells, hybrid cars, and high efficiency solar cells. The interactive, object-based, and digital information development of this theme is symmetrical with the exhibit's other themes. *Atmosphere's* five thematic sections aim to establish the fact and definition stasis upon which an informed public can engage questions of policy.

The larger policy questions are hinted at in the thematic section on "Exploring our future choices...," but they are taken up in earnest after visitors have made their way through the immersive, interactive exhibit space. At this point, visitors are offered seats in front of computer terminals where they can tap through four questions about what should be done about climate change:

- Should we engineer the climate?
- Is money better spent cutting emissions now or adapting to impacts later?
- Is it an individual's responsibility to curb greenhouse gasses?
- Should rich countries give away low-carbon technologies?

As should be evident, questions prompting yes/no answers seem appropriate in museum contexts. However, rather than positioning visitors to vote up or down on them, the exhibit recognizes that these questions do not have easy answers and instead presents five positions on each question. For example, to the question that asks, "Is money better spent cutting emissions now or adapting to impacts later?" the exhibit offers three "now" arguments:

- It's unfair to leave the problem to future generations. Let's take responsibility for the problem we created.
- It'll be cheaper to tackle the problem now than to wait and deal with it in the future.
- If we don't tackle emissions now, climate change could be so devastating that we're unable to respond in the future.

The side that answers "later" offers two arguments:

- Human ingenuity always manages to cope. Let's adapt to the impacts in the future when technology is more advanced.
- The economy is continually growing. Let's tackle emissions when we have more money.

While these positions might seem a bit facile at first glance, they do present some basic frameworks for answering the question: appeals to fairness, economics, uncertainty about the future, and faith in technology. Distilling these responses down to their essential frameworks captures the common stasis points around which most policy questions are contested. In short, these answers provide a set of topoi—they are what we would hope science communication students might invent if they were to give an essay topic like "Should we engineer the climate?" And, what's more, if the visitor taps on one of the arguments, they get a short video interview of a London citizen from the street making a more personal, developed case.

Together the quick answers and the brief videos offer a range of responses around each major policy question, wherein both sides of an argument are, in some measure of balance, presented for the visitor to understand. The goal here can't be to come away with a specific, knock down argument for why we should act now or wait. So instead of understanding a policy choice and asking visitors to choose, the museum is presenting a problematic, and it presents it in such a way that visitors are offered competing commonplace arguments but not actively convinced of one or the other. It asks them, essentially, to understand and appreciate the contingency and rhetoricity of each viewpoint rather than pick a side or check a box. Thus, visitors are encouraged to become familiar with a range of arguments, an approach that addresses them rhetorically and provides the topoi as resources for visitors to understand and perhaps employ when arguing these questions outside the museum.

As a result, *Atmosphere* positions visitors not as voters but as deliberators and prepares them to participate in public discussions. The difference between treating visitors as deciders on policy and visitors empowered to deliberate on policy might seem slight, but there are some important differences. As deliberators, visitors are expected to participate and engage in ongoing debates, not to simply decide one way or another. This embraces a vision of citizenship that accords real agency in the real world. Positioning visitors as deliberators rather than voters also implies that climate change policies are complex and that a simple "yes" or "no" vote oversimplifies and would curtail discussions that require much public debate. This choice to leave open the large policy questions and not ask visitors to employ or exhaust their civic mindedness in the exhibit itself orients them to the future, to contexts and conversations outside the space and time of the exhibit. This future-orientation is hinted at in the exhibit's opening panel, which seeks to "inspire hundreds of thousands of visitors" and states, "Working together, we can push ideas further and find solutions faster." Whether this happens remains an open question and one that exhibit evaluators will continue to explore. Nevertheless, *Atmosphere* imagines a deliberative (not simply a voting) public and clearly frames future visitor action on climate change policy in a way that shows them where to think, not what to think, thereby empowering and encouraging them to find a voice.

Identifying with Oceanic Peoples at Paris's Cité des Sciences

Ocean, The Climate, and Us, the French counterpart to *Atmosphere*, shares many attributes with its British cousin. Like *Atmosphere*, *Ocean* is titled and sub-titled in such a way that its focus on climate change does not hit the visitor square in the face. Visitors can be enticed to visit the gallery because they might hope to learn something about the ocean (and indeed they will). Also like *Atmosphere*, *Ocean* is an immersive experience, dark and blue, watery, and filled with sound. On the other hand, *Ocean* is a more organized and directive space, with a set of three progressive units that visitors are directed through. Across its first two units *Ocean* is a highly interactive exhibit, but it falls short of the game-filled, digital interactive environment found in *Atmosphere*. The exhibit's culminating experience, however, offers an intriguingly emotional experience that addresses visitors as if they were someone else, as if they occupied a different subject position. As I will show, positioning the public as someone other than who they are has important implications for how they respond in the exhibit and how they think about future action.

Ocean is organized in three units, which, unsurprisingly, can be seen to reflect the stasis order displayed in *Atmosphere*. The first two sections, titled "How does the ocean influence the climate?" and "How is the ocean changing today?" respectively, set out to address how questions that establish the fact and definition stasis questions. In the first of these, visitors learn about the role played by things like ocean currents, salinity, wind, temperature, plankton, and El Nino/Nina. This section also includes contents that deal with the historical display of climate change metrics to show that it is human caused and not a result of increased solar energy or volcanoes. A similar exhibit element exists in *Atmosphere* as well.

The exhibit's second section shifts from processes and mechanisms to effects, ongoing changes, and the methods by which scientists study the ocean today. This section includes objects for measuring the ocean (e.g., ocean buoys and probes), computer kiosks that display up-to-the-minute data on ocean temperature, and sections on ocean diversity and the food chain, ocean acidification, and overfishing. Together, these two sections build on and document the claim that opens the exhibit's introductory panel: "It's a fact, climate change due to human activities is under way." This strong assertion also frames the question asked by the exhibit's final thematic component: "How can we adapt to change?" These two sentences reflect the inevitability of *Ocean*: the effects of climate change cannot be mitigated. While potentially discouraging, during my interview curator Marie-Christine Hergault indicated that the final section was meant to end the exhibit on a positive note: the ocean will change in ways we might regret, but we will be able to adapt. The final section prepares the visitor for this eventuality.

Through six videos that represent six different geographical regions affected by the changing ocean, *Ocean's* final section asserts that, "Nation states will have to adapt and the world must work together to implement two complementary strategies: cut greenhouse emissions and adapt to inevitable changes." The six videos focus more on adaptation than prevention in the Maldives, the Mediterranean, the Arctic, the Netherlands, Bangladesh, and Senegal. After reading a brief introductory panel that outlines what's at stake for the region, visitors enter a small room where they sit for a five to six minute video where a representative from the country/geographical region discusses the problem and their adaptive strategies. These videos are well produced, and the speakers are professional actors. After watching the video, visitors move on to the next geographical region, and after the final installation they exit the exhibit.

Two things are striking about this description of *Ocean's* culminating experience. First, the civic dimension is not as clear as it was in the *Atmosphere* galleries. We might ask: Aren't these videos merely informative? How do these videos position visitors for the future? Second, the level of physical interaction is low compared to *Atmosphere*. Instead of voting by pressing buttons (highly active) or tapping through questions/answers that interest them (choice, self-directed learning), visitors here simply sit and passively watch a video. Nevertheless, the ways in which these videos interpellate or position visitors as subjects has important ramifications for them as world citizens.

How are they positioned? Ironically, visitors in this exhibit are positioned other than they are. The mode of address changes. The "you" or "we" of other exhibits is no longer inclusive of the visitor *qua* visitor. Instead of a public visitor who needs to understand climate science so that she can make choices as a citizen, here she is instead addressed as an interested, engaged expert. A couple of examples will illuminate the significance of this shift. In the Mediterranean example, an introductory panel describes the challenges of the region and ends with a question "Will we be able to take the necessary sustainable measures?" Here, the "we" is inclusive of the visitor qua visitor and positions her with the responsibility and agency to make future choices. This positioning shifts dramatically when the visitor sits down to watch the video and is immediately greeted as "Dear sir" by the "deputy mayor in charge of town planning." As you listen, it becomes clear that you are being addressed as an architect or landowner who has applied to build a housing complex on the coast. The video diary is the deputy mayor's opportunity to reject your application and provide a detailed rationale. In making his case, the mayor describes the low lying geography, points to the way climate change exacerbates this problem, explains that the sea is already a problem, and provides historical examples and cites experts. Besides building a strong case against your application, the mayor also addresses you honestly and respectfully. He says, "If the coast was deserted, [then the cyclical ten year storms] wouldn't raise any major problems and we could just adopt a 'laisser-faire' [sic] approach. But that isn't the case. And you are well placed to know it." Here the "you" is both you and not you. This mode of address occurs again at the end of the video: "I hope you have found my explanation convincing. In any case, I suggest we meet up to discuss a solution for your housing estate. In fact, wouldn't it be better to build it inland rather than by the sea?" In the role of the housing builder, you are addressed respectfully, as a rational, knowledgeable actor who understands the climate and empathizes with the geographical context.

In the case of Bangladesh, the visitor is addressed as "Dear professor," an opening that is used in other videos as well. Here the position of the speaker is unclear, but she begins with a rhetorical question that seems to repeat a question that the professor must have asked: "So, how could my country, Bangladesh, adapt to the transformations due to climate change?" Other scenarios might explain this introduction, but it is quickly clear that you—the professor—have some knowledge of Bangladesh and seek the narrator's expert opinion. Not only is she responding to you as an expert, but she also recognizes that "the professor" has a degree of knowledge about the context of which she speaks. For example, before telling an historical anecdote she says, "Maybe you remember. …" and later she says, "As you know" with regard to the fact that climate change has reduced the number of seasons from six to three. Along the way, the speaker provides examples of the challenges and opportunities for Bangladesh. Near the end she again seems to use an asked question as a transition, "What will my country look like in 40 years' time?" Her answer—the culture will adapt its lifestyle and international solidarity will allow for infrastructural developments—is "the fruit of her imagination." These are her hopes for the future, and the burden is on her people and the international community.

Both of these examples help to capture what occurs when the visitor sits down and attends to each of the six videos. In each case, the visitor is addressed by the person on the screen not as a visitor but as someone who she is not. Importantly, that person is intimately aware of the local context discussed on screen (coastal Mediterranean, Bangladesh villages, etc.), and so the visitor is accordingly addressed *as if* they also know these things. Beyond knowledge is something much more important: interest, empathy, and engagement. The videos address visitors as subjects who identify with and care about the people and cultures on the screen. Even if the visitor does not know where the Maldives is, the video assumes she remains deeply concerned. In short, through this contrived shift in audience positions, the exhibit assumes not only that visitors want to hear this information but that they are already on board with the climate science and that they empathize with the population. The appeals and examples the exhibit selected have been deliberately, purposefully, and powerfully tailored. In other words, instead of addressing a general, hard-to-pin-down visitor, the culminating experience places visitors within a specific conversation where they are addressed by a person they "know" and respect and who can address their individual concerns.

Because the videos address a singular, individual audience with a set of predefined and specified interests and responses, the videos build in values, knowledge, and emotions that would not be guaranteed to exist in every visitor who attends the video. This interpellation cultivates a different kind of rhetorical response in visitors, who, after all, are not primarily professors or building moguls trying to get approval to construct a housing complex. The videos ask the visitor to identify not only with the speaker on the screen but also to identify as the expert the speaker addresses. As such, the visitor is encouraged to emotionally respond, to act and think *as if* she were the person being addressed. The question of whether or not this works remains to be seen. What is evident, however, is that the visuals in this portion of the exhibit employ what Charles Hill (2004) terms "vivid information" to increase the videos' rhetorical presence. The degree to which visitors embrace or resist this rhetoric will mean a different kind of emotional response to the video on display. Even though that emotional response is part of a particularly passive experience (simply watching a video), it emerges through a high level of emotional identification, a deeper level of interaction than is typically meant by the museum community's use of the term. In the end, the effect of emotionally embracing the interested, empathic, engaged, curious subject position is to cultivate and experience those values as one's own. Thus, the civic potential of these videos is not located in specific interactive exhibit elements asking visitors to vote or deliberate. Instead, *Ocean's* culminating experience, if embraced by the visitor, plays the traditional epideictic role of establishing and reinforcing cultural values that serve as the basis of future action. By letting visitors identify as someone who believes in climate change, appreciates its effects on the ocean, empathizes with a wide range of peoples and cultures, and, finally, actively seeks adaptation strategies, *Ocean* cultivates a public that not only understands climate change but also knows what it feels like to choose to adapt to its inevitable and already ongoing effects.

Conclusions

The two exhibits I've discussed here follow the same stasis order: establish knowledge on questions of fact and definition about climate change, demonstrate that those facts indicate human caused warming, and engage the policy question. In short, they all move from fact to policy, which simply points up the utility and power of stasis as a means of explaining the organization of certain argument types. Where these exhibits differ is in the ways they position visitors and the kinds of publics they call into being. The question is, now that visitors know, what will they do, and in what capacity will they do it? As voters? As deliberators? As world citizens empathizing with the plight of others? Questions about the kind of public we want to create or imagine (and how we want the public to imagine itself) are critical for shifting the public debate on climate change.

Neither of these exhibits fails. They both generate public understanding, establish the authority of science (sometimes in rather sophisticated ways), and encourage visitors to think about the future. Yet a common rhetorical vision for moving someone from understanding facts to taking action starts with identification, moves to deliberation, and culminates in decision. To emphasize the final step without respect for the first two, as many climate change exhibits do, is partly a result of the urgency of climate change; we must make choices now. However, it's also a result of a narrow rhetorical application of stasis theory. What is so powerful about *Atmosphere's* open-ended policy questions and *Ocean's* epideictic appeal for identification is that they call into being publics with a broader sense of agency. In *Atmosphere*, visitor agency is deliberative: by establishing facts and definitions the prepares visitors to meaningfully voice their opinions. In *Ocean*, visitor agency emerges through cultural identification: it cultivates the foundational values by which a public becomes receptive to arguments about facts, definition, value, and policy. This analysis illuminates the utility of stasis theory as a powerful and popular exhibit organizing structure that builds from facts to civic engagement. What *Atmosphere* and *Ocean* indicate is that if we want to call into being an empathic, deliberative public, then it is profitable to position visitors in ways that ask them to do more than simply understand and vote.

In the end, some modern science museums, specifically those that exhibit topics relevant to public policy in the public sphere, have begun to move visitors up the chain of stases. When done well, these exhibits include moments where the uncertainty inherent in policy questions is left wide open and where visitors are granted a renewed political agency to truly engage with those questions both inside and outside the exhibit. As a result, these exhibits provide a kind of rhetorical education. This rhetorical education takes two forms. First, it highlights the rhetorical nature of questions of value and policy. By opening up and not closing down these questions, the exhibit reinforces for the visitor the contingency of future decisions. Second, by presenting policy questions in an honest way, the exhibit provides not one position but many positions on how to think about what humans might do. In presenting these positions, the museum addresses visitors as creative, politically vibrant individuals who might encounter these policy questions outside the museum walls. Thus they cultivate not an understanding of facts or definitions or even specific correct policy choices, but an appreciation of the variety of arguments and positions surrounding a policy, many of which might serve as topoi and talking points to be presented or countered. In this sense, by embracing the rhetorical core of policy questions and offering up a few of the rhetorical resources for engaging those questions as available means of persuasion, the museum has begun to embrace and reallocate the inventional burden in a way that truly contributes to the visitor's rhetorical education.

References

Association of Science-Technology Centers. (2014). *2013 Science center and museum statistics*. Retrieved March 13, 2017, from www.astc.org/wp-content/uploads/2014/10/2013-Science-Center-Statistics.pdf.

Bennett. T. (1995). *The birth of the museum: History, theory, politics*. New York: Routledge.

Bennett. T. (2015). Thinking (with) museums: From exhibitionary complex to governmental assemblage. In K. Message & A. Witcomb (Eds.), *Museum theory* (pp. 3–20). London: Wiley Blackwell.

Cameron, F. (2010a). Risk society, controversial topics and museum interventions: (Re) reading controversy and the museum through a risk optic. In F. R. Cameron & L. Kelly (Eds.), *Hot topics, public culture, museums* (pp. 53–75). Cambridge: Cambridge Scholars Publishing.

Cameron, F. (2010b). Liquid governmentalities, liquid museums, and the climate crisis. In F. R. Cameron & L. Kelly (Eds.), *Hot topics, public culture, museums* (pp. 112–128). Cambridge: Cambridge Scholars Publishing.

Cameron, F. (2010c). Introduction. In F. R. Cameron & L. Kelly (Eds.), *Hot topics, public culture, museums* (pp. 1–17). Cambridge: Cambridge Scholars Publishing.

Cameron, F. (2011). From mitigation to creativity: The agency of museums and science centres and the means to govern climate change. *Museum and Society*, *9*(2), 90–106.

Chakrabarty, D. (2002). Museums in late democracies. *Humanities Research*, *9*(1), 5–12.

Durant, J. (1994). What is scientific literacy? *European Review*, *2*(01), 83–89.

Fahnestock, J. (1986). Accommodating science: The rhetorical life of scientific facts. *Written Communication, 15*(3), 330–350.

Fahnestock, J. R., & Secor, M. J. (1983). Grounds for argument: Stasis theory and the topoi. In D. Zarefsky (Ed), *Argument in transition: Proceedings of the third summer conference on argumentation* (pp. 135–146). Annandale, VA: Speech Communication Association.

Fahnestock, J., & Secor, M. (1985). Toward a modern version of stasis theory. In C. W. Kneupper (Ed.), *Oldspeak/newspeak: Rhetorical transformations* (pp. 217–226). Arlington, TX: Rhetoric Society of America.

Fahnestock, J., & Secor, M. (1988). The stases in scientific and literary argument. *Written Communication, 5*(4), 427–443.

Greenhill, E. H. (1992). *Museums and the shaping of knowledge.* London: Routledge.

Gross, A. G. (2004). Why Hermagoras still matters: The fourth stasis and interdisciplinarity. *Rhetoric Review, 23*(2), 141–55.

Gurian, E. H. (2010). Forward: Celebrating those who create change. In F. R. Cameron & L. Kelly (Eds.), *Hot topics, public culture, museums* (pp. xi–xv). Cambridge: Cambridge Scholars Publishing.

Hetherington, K. (2015). Foucault and the museum. In K. Message & A. Witcomb (Eds.), *The international handbooks of museum studies volume 1: Museum theory* (pp. 21–40). London: Wiley Blackwell.

Hill, C. A. (2004). The psychology of rhetorical images. In C. A. Hill & M. Helmers (Eds.), *Defining visual rhetorics* (pp. 1–40). London: Lawrence Erlbaum Associates.

Hulme, M. (2015). Why we *should* disagree about climate change. In F. R. Cameron & B. Neilson (Eds.), *Climate change and museum futures* (pp. 9–130). New York: Routledge.

Kennedy, G. (1963). *The art of persuasion in Greece.* Princeton, NJ: Princeton University Press.

Northcut, K. M. (2007). Stasis theory and paleontology discourse. *The Writing Instructor,* 9, 1–22.

Prelli, L. J. (1989). *A rhetoric of science: Inventing scientific discourse.* Columbia: University of South Carolina Press.

Rutherford, S. (2011). *Governing the wild: Ecotours of power.* Minneapolis: University of Minnesota Press.

Toulmin, S. (1958). *The uses of argument.* Cambridge: Cambridge University Press.

Turner, M. (1991). *Reading minds.* Princeton, NJ: Princeton University Press.

TWResearch. (2010). *Naming the climate change gallery: Qualitative research findings.* London: Science Museum.

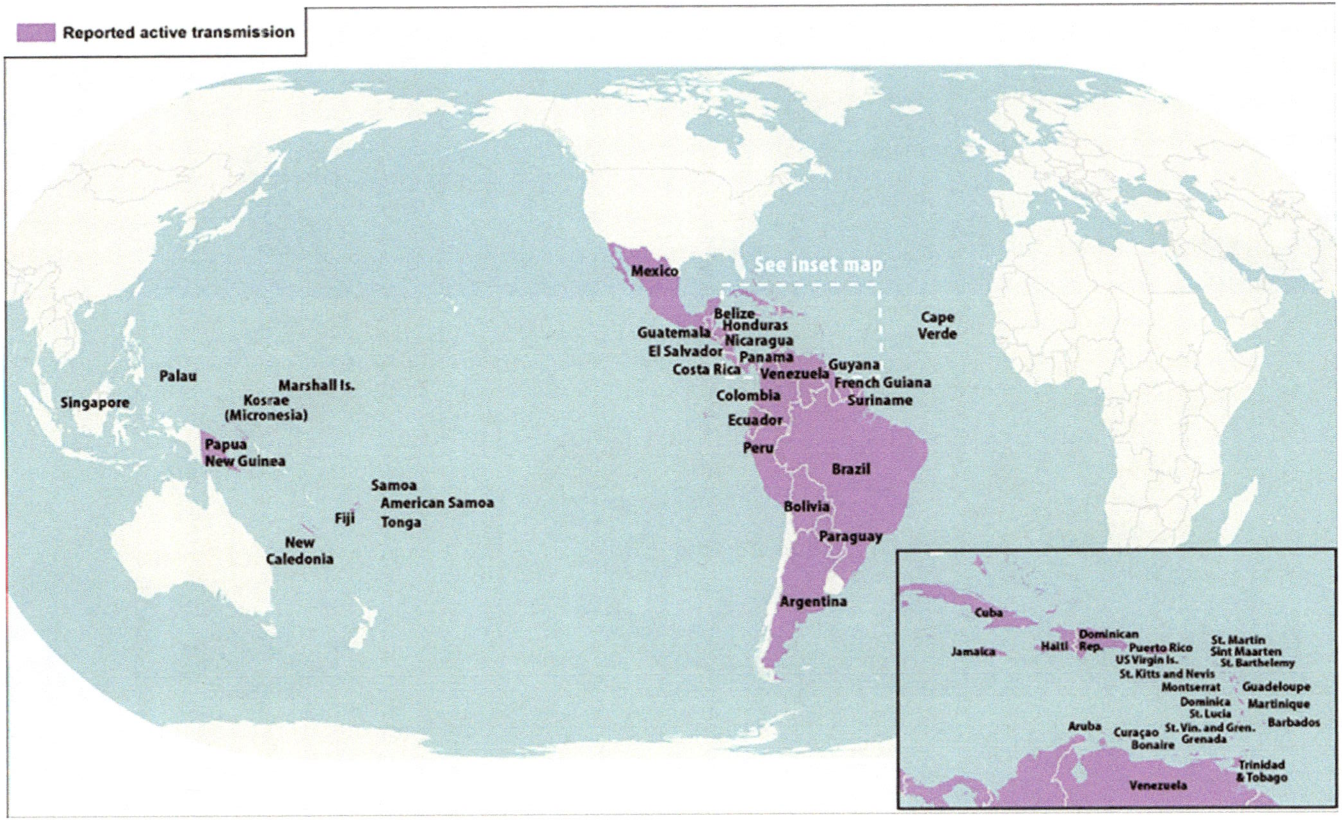
Reported active transmission
See inset map
Mexico
Belize
Honduras
Guatemala
Nicaragua
El Salvador
Panama
Costa Rica
Venezuela
Guyana
French Guiana
Suriname
Cape Verde
Colombia
Ecuador
Peru
Brazil
Bolivia
Paraguay
Argentina
Palau
Singapore
Marshall Is.
Kosrae (Micronesia)
Papua New Guinea
Samoa
American Samoa
Fiji
Tonga
New Caledonia
Cuba
Jamaica
Haiti
Dominican Rep.
Puerto Rico
US Virgin Is.
St. Kitts and Nevis
St. Martin
Sint Maarten
St. Barthelemy
Montserrat
Guadeloupe
Dominica
Martinique
St. Lucia
Aruba
Curaçao
Bonaire
St. Vin. and Gren.
Grenada
Barbados
Trinidad & Tobago
Venezuela

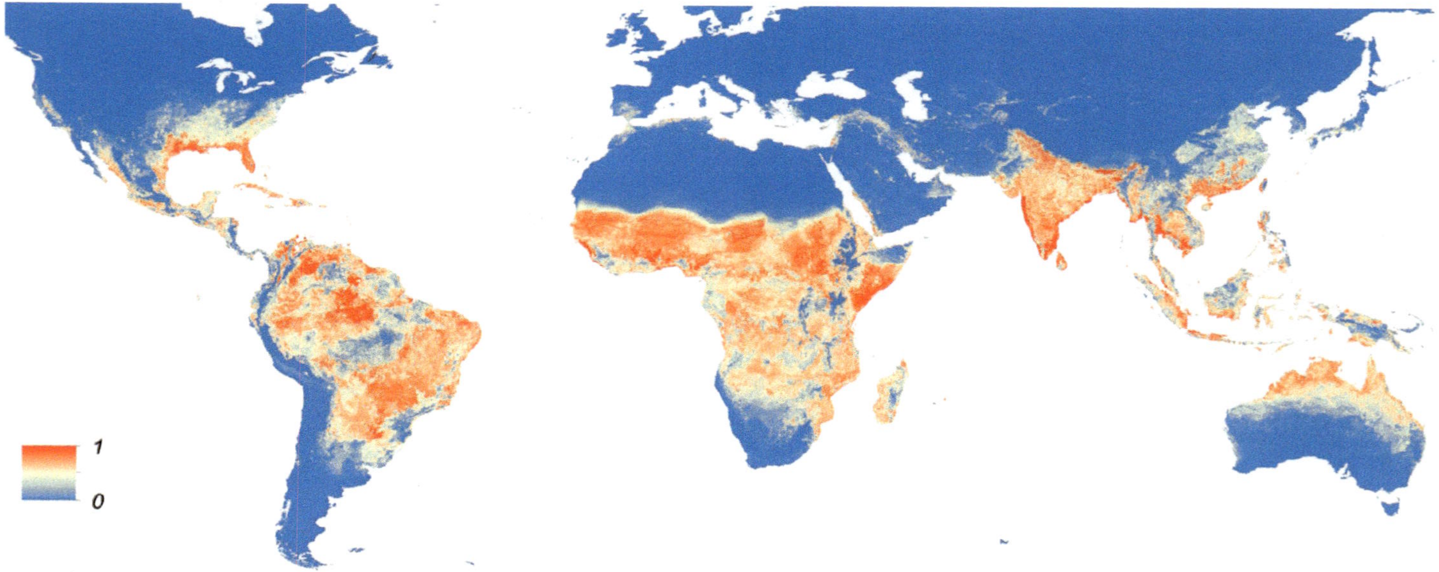
1
0

Mexico
Dominican Republic
United States Virgin Islands
Jamaica
Haiti
Puerto Rico
Saint Martin
Honduras
Guadeloupe
Martinique
Guatemala
El Salvador
Curaçao
Barbados
Venezuela (Bolivarian Republic of)
Guyana
Suriname
Nicaragua
Costa Rica
Panama
Colombia
French Guiana
Cabo Verde
Gabon
Ecuador
Brazil
Bolivia (Plurinational State of)
Paraguay
World Health Organization

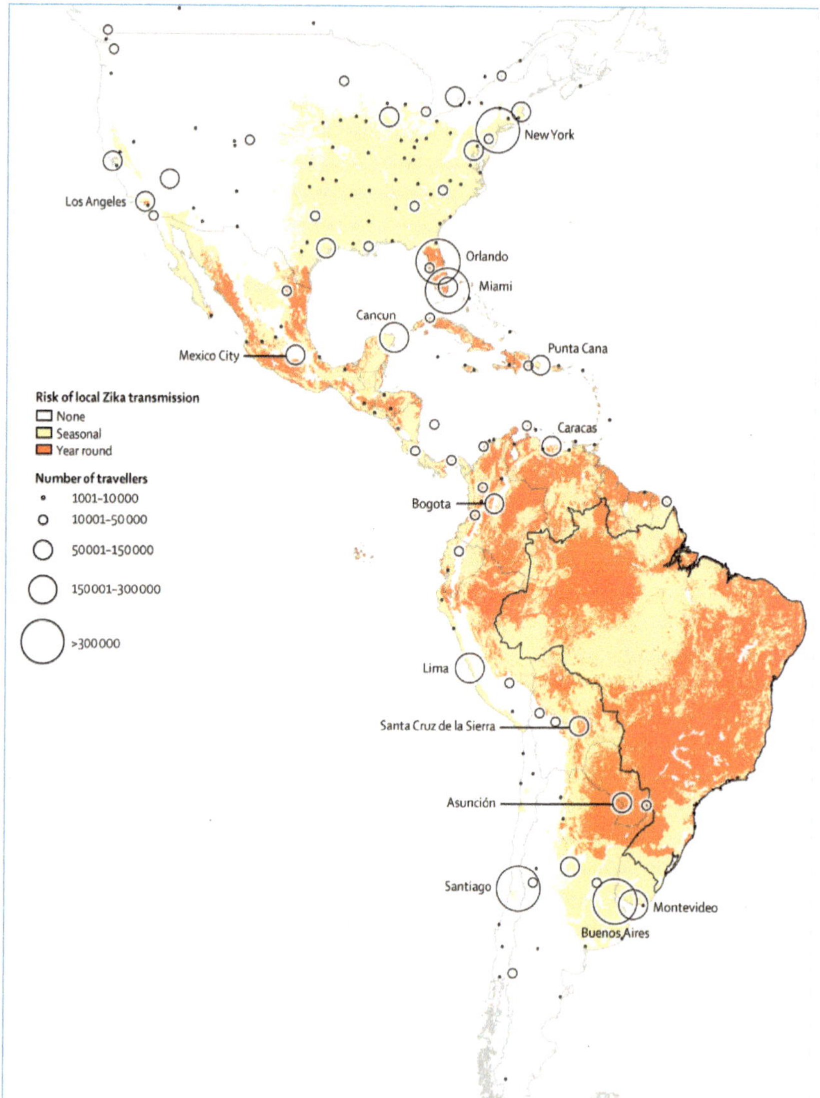

New York
Los Angeles
Orlando
Miami
Cancun
Mexico City
Punta Cana
Risk of local Zika transmission
None
Seasonal
Year round
Caracas
Number of travellers
1001–10 000
10 001–50 000
50 001–150 000
150 001–300 000
>300 000
Bogota
Lima
Santa Cruz de la Sierra
Asunción
Santiago
Montevideo
Buenos Aires

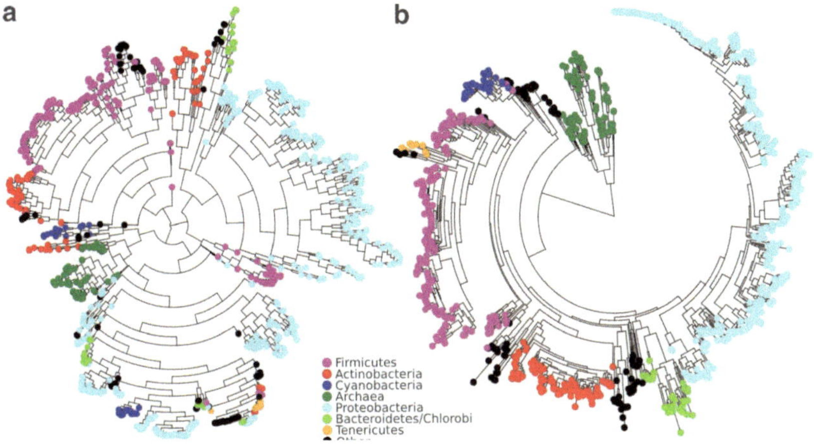

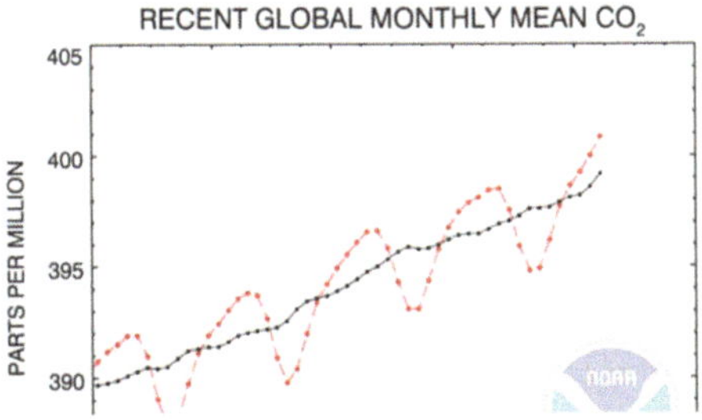

Greenhouse gas benchmark reached

Global carbon dioxide concentrations surpass 400 parts per million for the first month since measurements began

Wednesday, May 6, 2015

Like 3.4K Tweet G+1 89 Share 30

For the first time since we began tracking carbon dioxide in the global atmosphere, the monthly global average concentration of this greenhouse gas surpassed 400 parts per million in March 2015, according to NOAA's latest results.

News Categories

- Press Release (46)
- Air Quality (41)
- Arctic (34)
- Climate (158)
- Ecosystems (44)
- Fisheries & Seafood (34)
- Great Lakes (16)
- Marine Science (33)
- Ocean Exploration (21)
- Weather (89)

Research Headlines

NOAA study shows as US drilling surged, methane...

Agencies team up to accelerate Earth system...

Seasonal fishing in Wilmington, North Carolina

[redacted]

ENG 315
UNC WILMINGTON

INTRODUCTION

Wilmington, North Carolina is a fairly developed port city on the southeast coast. With a growing population and beautiful beaches, Wilmington attracts fishermen from all parts of North Carolina and surrounding areas. The social structure of the fisherman is sometimes based on the area of the fishery. Wilmington fisheries are great with solid offshore artificial reefs holding many different species of fish. Due to the artificial reefs placed in the waters off of the coast of Wrightsville Beach, various species of fish corrugate--causing an ideal fishing environment. Wilmington also provides several types of fishing environments for fishermen from all walks of life. For example, saltwater fishing on the piers of Wrightsville Beach, Carolina Beach, Kure Beach, and freshwater fishing in the Masonboro Inlet. When fishing around the Wilmington area, there are a variety of species of fish that can be found. The relationship between the fisherman or woman's preference of fish and the weather both seem to correlate to the species of fish that are sought after. Overall fishing is a common practice within all types of communities and functions to provide an activity of leisure as well as satisfy the biological need of hunger.

METHOD

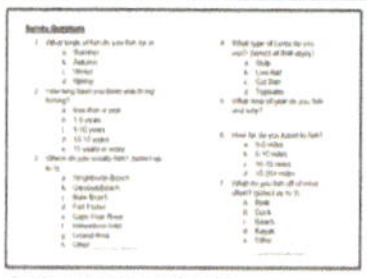

The population for this study included fishermen of Wilmington, North Carolina as well as students of UNCW. The purpose of including residents of the Wilmington area was to broaden our sample size. The reason we permitted students of UNCW to participate in this study was because there are many students on campus and they can be easily reached. The study was conducted using a seven question survey containing multiple types of questions that was physically given to our sample (*see above for surveying instrument*). The questions in the survey included short answer, multiple choice, and ratio-scale questions. The survey helps to answer the research questions by getting at the detailed habits of fishermen and women. It attempts to understand which types of fish are most fished in Wilmington as well as when and where they are fished. A question about what kind of structure fishermen/women fish off of is also asked to better determine fishing habits. A question about lures was included to determine if certain lures matter in the types of fish caught. All of the survey questions are important to narrowing the different reasons that fishermen/women fish where, when, and how they do in order to determine what they catch.

DATA AND RESULTS

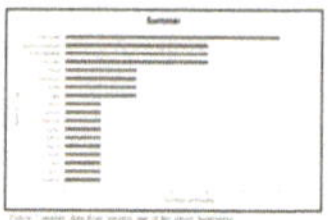

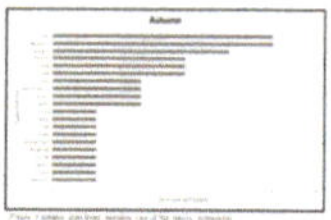

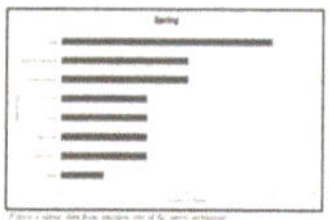

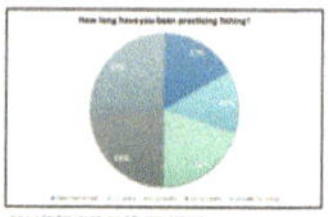

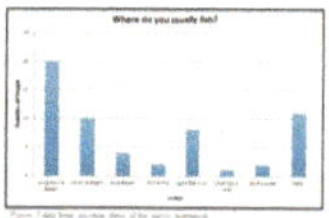

CONCLUSIONS

According to the fishermen and women of the local area that we surveyed, autumn is the most popular fishing season. This can be concluded because we had the most amount of data for the autumn season. King mackerel and Spot were the top two listed species of fish in autumn. The next most popular fishing season is summer because summer had the second most amount of data. From the survey results, Mahi Mahi is the most fished for species in the summer season. The third highest amount of data that was collected, was for the spring astronomical season. The most popular species of fish according to our results for the spring season is the Bass from the Genus, *Morone*, which includes all species of temperate basses (ITIS, 2015). Moving forward, the winter astronomical season had the least amount of survey data, most likely because winter is known for having harsh weather conditions unsuitable for human beings. To get a better idea of how experienced the participants were, we asked the question: "How long have you been practicing fishing?" Answer choices for this question included: "less than a year, 1-5 years, 5-10 years, 10-15 years, or 15 years or more". The most popular answer choice, answered by 25% of those surveyed, was "15 years or more". The least popular answer choice, chosen by 13% of those surveyed, was "1-5 years". The question "Where do you usually fish?" was posed, with the answer choices being: Wrightsville Beach, Carolina Beach, Kure Beach, Fort Fisher, Cape Fear River, Masonboro Inlet, Leland Area, and a fill in the blank "Other" section. Participants were allowed to choose up to three of these areas. We found that Wrightsville beach was chosen the most out of all of these choices, with "Other" surprisingly was the second most chosen answer. Our "Other" section received answers like Cape Fear River, Sutton Lake, Hatteras, Topsail, and a few more. If another group of scientists were to conduct a similar study in the future, they could ask a more broad set of questions that apply to different states in in the U.S., for example, questioning which states are their favorite to fish in, as well as what cities do you like to fish in North Carolina. Asking those types of research questions creates vast choices for the participants.

REFERENCES

- ITIS Standard Report. (n.d.). Retrieved November 1, 2015, from http://www.itis.gov
- New Hanover County QuickFacts from the US Census Bureau. (n.d.). Retrieved October 14, 2015, from http://quickfacts.census.gov/qfd/states/37/37129.html

Living shorelines: Protecting our coasts

Kelsey Potlock, Department of Environmental Studies, UNCW

UNCW Environmental Studies

Brunswick Town/Fort Anderson History

Brunswick Town/Fort Anderson (BTFA) is an important location in Wilmington's history. Founded in 1726 by Maurice Moore, the town served as a port city for domestic and foreign trade. The most common exports were tar, pitch, and turpentine, all products of the longleaf pine that grew abundantly in the area. During the American Revolution in 1776 the British Navy burned the town down where it remained rubble until 1861 where it was rebuilt as Fort Anderson, a garrison stronghold for the Confederacy during the American Civil War. Fort Anderson served as a part of the Cape Fear River defense for Wilmington. On February 19, 1865, Confederate forces were ordered to retreat leaving the Fort to fall to the Union. Three days later, the port city of Wilmington fell to the Union. Within weeks, the American Civil War would end.

Why Brunswick Town?

The BTFA site provides a perfect introduction for this pilot living shoreline system as it is a historical site. As the waves crash against the marsh shoreline, artifacts and wooden structures that make up the historical compone C:\Users\Potlock\Desktop\shoreline.jpg ed, or washed away. This project aims to protect these.

Erosion Basics

It is important to understand that erosion is a naturally occurring process. Simply put, erosion is the process of wind or water removing sediments from the shore and transporting it elsewhere. Shorelines are typically areas of huge movement because of the waves that continuously impact the area. As waves collide against the shoreline, areas become severally undercut, gradually weaken, and eventually, fall into the water. Erosion, while natural and gradual, can be accelerated through human activities. Building along coastlines, dredging, and shipping traffic all increase wave energy leading to increased erosion.

Shoreline Protection Methods

There are several types of shoreline protection methods ranging from "hard" and "soft" structures. Hard methods are designed to halt or reflect wave energy and soft methods are designed to more naturally slow wave energy.

Hard

Groins

Hard

Riprap Wall

Soft

Oyster Reef

Soft

Vegetation

Living Shorelines

Living shorelines are a more natural method of shoreline protection. They can consist of oyster beds or grass plantings but are designed to mimic the natural state of the environment while preserving the integrity of the shoreline. Living shorelines tend to be cheaper and be easier to maintain and repair over time.

Works Cited

Coastal shore protection structures and techniques. (n.d.). Retrieved August 21, 2015.

North Carolina historic sites. (2011, August 31). Retrieved April 20, 2015, from http://civilwarexperience.ncdcr.gov/brunswic/narrative-anderson2.htm

North Carolina historic sites. (2015, February 28). Retrieved April 20, 2015, from http://www.nchistoricsites.org/brunswic/brunswic.htm

Reefmaker marine ecosystems. (n.d.). Retrieved August 21, 2015

ReefMaker Ecosystem Units

A "hybrid" living shoreline method pioneered by ReefMaker. These are trays with pieces of concrete or granite attached that will encourage oyster spat to attach. Over time, the oysters will grow allowing the water to irregularly reflect off the shells, slowing the wave energy and allowing the shoreline to slowly rebuild.

Main page
Contents
Featured content
Current events
Random article
Donate to Wikipedia
Wikipedia store

Interaction
Help
About Wikipedia
Community portal
Recent changes
Contact page

Tools
What links here
Related changes
Upload file
Special pages
Permanent link
Page information
Wikidata item
Cite this page
Expand citations

Print/export
Create a book
Download as PDF
Printable version

Languages
Add links

Morris J. Berman oil spill

From Wikipedia, the free encyclopedia

This is an old revision of this page, as edited by Mercy11 (talk | contribs) at 02:17, 17 February 2014 *(wikilinks, rewording)*. The present address (URL) is a permanent link to this revision, which may differ significantly from the current revision.

(diff) ← Previous revision | Latest revision (diff) | Newer revision → (diff)

The barge **Morris J. Berman** grounded off Punta Escambron, San Juan, Puerto Rico, on January 7, 1994. The barge spilled approximately 800,000 gallons of #6 oil on the reef.[1]

Response

The responsible party initially assumed responsibility for the spill, but very quickly expended the ten million dollar limit of their insurance policy. Full federal funding of the spill occurred at 0600 on January 14 and it became a United States Coast Guard (USCG) directed response. The USCG Gulf Strike Team (GST) was brought on scene and immediately began lightering operations for the barge.[2]

The vessel was cleaned out and taken to deep water where it was intentionally sunk under government supervision.[1] The cost of the cleanup amounted to approximately $130 million.

The Government of the Commonwealth of Puerto Rico and the Government of the United States, along with several other parties, sued the owners of the vessels for recovery of clean up costs, natural resource damages, and third party costs. Also involved were a number of clean up contractors, spill response managers, and representatives of the U.S. Coast Guard and other governmental regulatory bodies.

Responsibility

Criminal prosecutions resulted from issues of crew negligence and the management knowingly sending a vessel to sea in an unseaworthy condition. Caribbean Petroleum owned the cargo that spilled from the *Morris J. Berman*. The Berman Case was settled on January 19th, 2001. The Government of the Commonwealth of Puerto Rico and Government of the United States settled their claims against Metlife Capital Corp, Water Quality Insurance Syndicate and Caribbean Petroleum Corporation, among others, in the amount $83.5 million. Water Quality Insurance Syndicate paid $5 million, Metlife Capital Corp paid $62 million and Caribbean Petroleum Corporation paid $16.5 million.[1]

References

1. ^ a b c "Fact Sheet: Morris J. Berman Oil Spill" (PDF). Retrieved 5 January 2010.
2. ^ "BARGE MORRIS J. BERMAN". Retrieved 5 January 2010.

This Environmental disaster-related article is a stub. You can help Wikipedia by expanding it.

Categories: Environmental disaster stubs | Oil spills in the United States | 1994 industrial disasters | Disasters in Puerto Rico | 1994 in the environment | 1994 in Puerto Rico | 1994 disasters in the United States

Morris J. Berman oil spill

From Wikipedia, the free encyclopedia

(diff) ← Previous revision | Latest revision (diff) | Newer revision → (diff)

The **Morris J. Berman oil spill** occurred on January 7, 1994, when the *Morris J. Berman*, a single-hull 302-foot-long barge, with the capacity to carry over 3 million gallons of oil, collided with a coral reef near San Juan, Puerto Rico,[1] causing the release of 750,000 gallons of heavy grade oil.[1] The spill affected the tourism and fishing industries as well as wildlife along the shores of Puerto Rico, Isla Culebrita, and Isla de Viques. The spill had major long-lasting impacts on the biological and natural resources of the entire Puerto Rican area.[2] This spill was also the first to occur in U.S. waters after the passing of the Oil Pollution Act of 1990.[1]

Contents [hide]

The incident

An example of an oil barge being towed by a towing cable

The *Morris J. Berman* left the Port of San Juan, Puerto Rico in the early morning hours of January 7th, 1994 in tow behind the tug boat *Emily S.*, carrying a total of 1.5 million gallons of oil.[1] The barge was nearly two hours into its trip to Antigua when the towing cable connecting the *Berman* to the *Emily* snapped for the first time. Repairs were made to the cable, and the voyage continued. At 3:50 AM, the cable snapped a second time, and the barge broke free striking a coral reef off of Escambron Beach, near San Juan. The impact caused extensive damage to two of the barge's nine holding tanks, releasing 750,000 gallons of oil.[1] An intense northerly wind added to the damage done to the vessel by pounding the stranded barge with heavy surf.[1] Oil washed on shore in Puerto Rico between Punta Boca Juana and Punta Vacia Talega. Oil also affected the shores of Isla Cuberta, an island to the east of Puerto Rico.[2] Most of the areas threatened and impacted by the spill were sheltered lagoons, biota-rich intertidal zones, and beach dunes, which are all habitats for many species. Several historic locations were also threatened and impacted by the oil, some of them being pre-Columbian ruins, historic walls and bridges, and Spanish era forts. The vast majority of all of these affected areas were located on the north shore of Puerto Rico.[1]

Environmental effects

The spill affected most of the environmental resources present in the shore and offshore areas near San Juan. 1,100 square miles of surface waters along the north coast of Puerto Rico were projected to have felt effects from the oil.[3] It was also projected that 105 miles of ocean shore line and 66 miles of bay shoreline were covered in oil.[3] Seagrasses and sediments were also contaminated in the spill area. It is projected that 40,000 square meters of seagrasses in lagoons near the spill area were impacted by sunken oil.[3] This caused most of the grasses to die. All of the affected areas were covered in varying amounts oil which resulted in large amounts of pollution, loss of tourism, loss of habitat, and the loss of life for living organisms.[3]

The spill made many important habitat areas unusable by many species for a period of time. One example of lost habitat was the loss of intertidal bay and shallow waters that were used for fish as nursery areas. This affected the population levels of certain species of fish in the years following the spill. Another example of lost habitat was the reef area lost near the grounding site of the *Berman*. Over 152 species were estimated to have resided in these reefs.[2] Some of these species include fish, shellfish, algae, and sponges. The reef was the main source of food and shelter for all of these species. The reef also provided a natural breakwater that helped to diminish storm surge from hurricanes. The effectiveness of the reef as a breakwater and habitat area were greatly decreased after it was damaged by the barge.[3] Large pieces of the reef were lost after the barge ran into it.

Effects on tourism

Condado Beach, San Juan, Puerto Rico

Oil was present at many beach areas surrounding the spill site. Beaches near the immediate site of the spill were closed to all visitors, either because of the presence of heavy oil or cleanup efforts.[1] While other beaches in the surrounding areas did have some oil, they were left open to visitors. People were strongly discouraged from visiting these open beaches, though. Tourists and residents that continued to visit oil affected beaches were not able to use the beach normally due to the presence of oil. Beach goers reported damaged swimming gear as well as headaches that were caused by the oil fumes.[4]

The San Juan National Historic Site was also affected after the spill. Due to its close proximity to the water, intense oil vapors were present inside of the fort located at the site for up to three weeks following the spill. National Park Service workers reported that the fumes were strong enough to induce headaches.[5]

The fishing industry was also affected by the spill. Both recreational and commercial fishing are vital parts of Puerto Rico's economy. Fishing was not able to be done in oil covered waters. This resulted in the loss of potential charter fishing trips and seafood sales. Normal fishing operations were able to eventually resume once the cleanup was complete.[6][1]

Due to the lack of appropriate data, it is impossible to determine the exact amount of tourism dollars that were lost because of the spill.[3] It is reasonable to infer though that the negative results of the spill felt in these tourist areas did indeed deter some tourists form visiting Puerto Rico.[3] The negative societal impacts described in this section were all caused by negative ecological impacts.

Effects on wildlife

In the days following the incident 5,268 organisms, the majority of which were dead, were estimated to have washed ashore.[2] Many live organisms were also found both on and offshore covered in oil. Of the 5,268 organisms 152 species were picked up after the spill by scientists and volunteers from the Caribbean Stranding Network (CSN) and various other organizations.[2] These organisms were both living and dead. Organisms found were sponges, anemones, sea worms, crustaceans, mollusks, sea stars, sea urchins, fish, birds, and sea turtles.[2]

Biological resources in the spill area

Several different biological resources were present in the initial areas affected by the spill, some of these include the sandy-intertidal invertebrate communities, rocky-intertidal invertebrate communities, and coastal and offshore communities of fish (many of which hold much importance to commercial and recreational fishing). Many endangered and protected marine mammals also used the affected areas during their times of migration.[2] Some of these mammals include the West Indian manatee, several species of dolphins, humpback whales, and sperm whales.[2] Even though these species of whales frequented the spill area, no whales were observed to have been affected by the oil.[2] Three species of endangered sea turtles, the green turtle, the hawksbill turtle, and the leatherback turtle also frequently nested in the affected areas.[2] The situation for these turtles was critical after the spill as the turtle nesting season was to begin in the months following the spill.[1] Several endangered and threatened species of birds also used areas surrounding the spill to rest and feed. These species of birds include the royal tern, sandwich tern, common tern, roseate tern, least tern, brown pelican, magnificent frigatebird, Audubon shearater, Caribbean coot, white-cheeked pintail, osprey, and the peregrine falcon.[2] All of the endangered and threatened species listed, plus other non-threatened species, were affected in some way by the spill. Most of the species affected were in the immediate landing site of the barge. Many species at the sinking site of the barge were also affected. Many of those species resided in areas of high ecological value, such as the shores of the Piñones State Forest.[2]

Most affected species

The Brown Boobly, a bird that was present at the spill site

The sally lightfoot crab, the periwinkle, the common West Indian Chiton, the rock boring urchin, and the brown booby were the most affected (least resilient)species after the spill. The sally lightfoot crab was the most affected crustacean after the spill.[2] This crustacean was the most affected because it was the most abundant of all crustaceans in the area. The sally lightfoot crab is prominent in the Caribbean as well as the West coast of the United States.[4] The periwinkle and the common West Indian chiton were the most affected of the mollusks observed after the spill.[2] Of all sea urchins observed after the spill, the rock boring urchin fared the worst. Rock-boring urchins were the most affected organism compared to all observed species after the spill, as they accounted for 29% of all recorded affected species.[2] Of all birds observed after the spill the brown booby was the most affected.[2] The main reason that most of these species were the most affected after the spill was simply because they were the most abundant species located in the spill area. Only 63% of the sea turtles and birds that were treated after the spill survived.[2]

Legal response

The owners of the *Morris J. Berman* initially assumed responsibility for the spill, but the ten million dollars that was provided from their insurance policy for oil spill cleanup was quickly spent. The federal government provided funding for the spill on January 14th and it became a United States Coast Guard directed response.[1]

The Governments of Puerto Rico, the United States, and other groups, sued the owners of the two vessels for clean up costs and natural resource damages. Criminal prosecutions were brought against the owners of the two vessels due to issues of crew negligence and the act of knowingly sending a vessel to sea in an unseaworthy condition.[1] The three owners of the barge were charged with criminal negligence based on laws from the Oil Pollution Act of 1990. One of the barge managers and the captain and first mate of the 'Emily S' were also charged with felonies for violations of the Clean Water Act.[1] The court case was officially settled on January 19th. The Governments of Puerto Rico and the United States were paid a total of $83.5 million from Metlife Capital Corp, Water Quality Insurance Syndicate and Caribbean Petroleum Corporation.[6] The cargo that was spilled from the *Berman* was owned by the Caribbean Petroleum Corporation.[6]

Environmental response

Treatment of wildlife

After the spill, as a part of the Local Spill Response Plan for Puerto Rico and the US Virgin Islands, the CSN was sent out to document damages to biological resources and to prevent further damages to the environment and living organisms.[2] The CSN collected many organisms, both live and dead, after the spill in order to document the damage done to them and to treat as many of the living organisms as possible. Most of the organisms collected by this group were off of the Northshore of Puerto Rico, but some were recovered from the south, east, and west shores of Puerto Rico, as well as near the shores of Isla Culebrita, and Isla de Viaques.[2]

Collected living wildlife were brought to a non permanent triage facility where crabs, birds, and sea turtles were treated for oil contamination and other injuries. Of all animals collected, these were the only ones that were able to be treated. Animals that were not able to be treated were either too small or in too bad of a condition. Nearly 400 crabs that were brought to the facility were successfully treated and released.[2] 28 birds and two sea turtles were also brought to the facility where 18 were successfully treated and released and 11 died at certain points during treatment.[2] All affected organisms were held until they were deemed completely healthy by a veterinarian. The oiled birds and turtles required an average 30 days of treatment before being released.[2] These organisms were treated with doses of olive oil and non-steroid antibiotics before they were cleaned in order to provide relief from the affects of the toxic oil. Solutions of olive oil, Simple Green, and BioSolve were then used to clean the organisms in tubs filled with luke-warm water. The organisms were then rinsed and dried.[2]

Cleanup of oil

Over 1,000 workers from 15 agencies and groups came together after the spill to clean up affected areas.[1] In total the workers put in over 1.5 million man hours of work.[4] This complete cleanup and assessment of the spill took 114 days.[2]

Eight days after the spill on January 15th, the barge was towed to an area 37 km Northeast of San Juan and intentionally sunk under government supervision to a depth of 2 km in an underwater canyon, where it is still located today.[2][3] This decision was ultimately made after it was determined that transporting the remaining oil off of the barge would not be feasible.[1] The barge was sunk at an isolated site that does not experience much ship traffic to avoid further damage. The cleaning of the majority of the oil was completed by April of 1994. The cost of the cleanup amounted to approximately $130 million.[1]

The general consensus among several newspaper articles and reports is that the overall response to the spill was a success, as based on the parameters set forth by the Oil Pollution Act of 1990.[1] Despite the overall success of the cleanup it was found that there were several areas for improvement though. One glaring example of possible improvement was the actions of the crew of the *Emily S.* The crew did not have proper supplies and were not properly trained to deal with the situation that they were faced with. The cable that broke during the incident was in a severe case of disrepair and had broken once five months prior to the spill.[1] A new cable was supposed to be installed, and if one had been, the spill could have been completely avoided. The crew also did not have sufficient supplies on board their vessel to properly repair the broken line. It was also found in federal investigations following the spill that negligent behavior was encouraged aboard the *Emily S.* This made it obvious that the crews of oil carrying vessels needed to be properly trained and equipped.[1]

Unlike the crew aboard the *Emily S.*, the governments of the United States and Puerto Rico were well prepared to deal with the spill. The two countries also met most of the requirements that were set forth by the Oil Pollution Act of 1990. For example, a spill response plan for the area had been developed and it was executed to near perfection. Millions of dollars of oil cleanup equipment was already stationed in Puerto Rico and was deployed into action within a matter of hours after the spill.[1] As mentioned earlier, despite the overall success of Puerto Rico's oil spill response plans there were still some areas for improvement. For example, there were some mechanical failures with some skimmer boats, inappropriate use of oil soaking materials, and an insufficient number of properly trained personnel.[1] There were also problems with communication between agencies as well as some organizational issues during the cleanup process. The Berman spill served as a field test of how well the plans set forth by the Oil Pollution Act of 1990 worked, and provided many examples on how future spill response efforts could be improved.[1] Several, though not all, of the things learned during this spill affected response efforts for future spills such as the BP Oil Spill in the Gulf of Mexico in 2010.[7]

References

1. ^ a b c d e f g h i j k l m n o p q Kurtz, Rick (September 2008). "Coastal Oil Spill Preparedness and Response: The Morris J. Berman Incident". *Review of Policy Research*. **25** (5): 473–486.
2. ^ a b c d e f g h i j k l m n o p q r s t u v w x y z aa ab Mignucci-Giannoni, A.A (1 February 1999). "Assessment and rehabilitation of wildlife affected by an oil spill in Puerto Rico". *Environmental Pollution*. **104**: 323–333.
3. ^ a b c d e f g h i j k "RESTORATION PLAN / ENVIRONMENTAL ASSESSMENT FOR MORRIS J. BERMAN OIL SPILL SAN JUAN, PUERTO RICO" (PDF). TETRA TECH, INC. April 2007. Retrieved 7 March 2014.
4. ^ "Sally Lightfoot Crabs". Quasar Expeditions. Retrieved 1 April 2014.
5. ^ "BARGE MORRIS J. BERMAN". Retrieved 5 January 2010.
6. ^ a b c d "Fact Sheet: Morris J. Berman Oil Spill" (PDF). Retrieved 5 January 2010.
7. ^ Alpen, Ghassen; Mancuso, Louis; Omar, Adnan; Dedogne, Natalie (2010). "A GUMBO OF CATASTROPHIC EFFECTS – THE EFFECTS OF THE BP OIL SPILL MEASURED DURING THE OIL SPILL AND SIX MONTHS AFTER THE OIL SPILL ON THREE TOWNS IN SOUTH LOUISIANA". *Entrepreneurial Executive*. **17**: 1–11. Cite uses deprecated parameter |coauthors= (help)

External links

- Collection of News Articles regarding the spill
- The US Coast Guard's Contingency Plan for oil spills near Puerto Rico

Categories: Oil spills in the United States | 1994 industrial disasters | Disasters in Puerto Rico | 1994 in the environment | 1994 in Puerto Rico | 1994 disasters in the United States

Part II
Pedagogy and Curriculum

8 Science and Writing

A Transectional Account of Pedagogical Species

Jonathan Buehl and William T. FitzGerald

Increasingly, scientists are turning to writing workshops to develop their communication skills as a complement to their training as researchers. They recognize the need to communicate with fellow scientists in precise technical language and with non-scientists in clear, jargon-free prose. Yet as one writing consultant observes, "educators don't have a real consensus about how—or even what—scientists should be learning in order to become better communicators" (Ossola, 2014). The teaching of scientific writing remains underdeveloped, which presents opportunities for teachers and scholars who work with scientific discourse.

Successful pedagogical interventions in science and writing look beyond a single course to issues of sustainability in the face of growing demand and the need to think vertically in a developmental arc from science novice to science professional. Such programmatic concerns must be situated within institutional and disciplinary spaces that are best described through an ecological framework. To articulate such a framework here, we adapt the concept of a "transect" to characterize the various intersections of science and writing in university and professional curricula and thereby estimate the health of pedagogical "species" occupying specific curricular niches in diverse programmatic terrain. By transecting, or hiking through, habitats from first-year composition to postdoctoral education in which science and writing instruction are collocated, we approach writing programs as ecosystems, identify opportunities for developing sustainable curricula, and describe strategies for cultivating cultures of writing that engage scientific discourse.

Transecting the Terrain of Science and Writing: Metaphor, Method, and Motives

Transects are data collection techniques used to sample populations. Scientists plot a course through a specific terrain, walk through that terrain, and collect data about species along the way. The collected data can serve as an index of the overall population. Typically, transects are executed on humble scales. For example, a researcher might count the number of trees within ten meters along a five-kilometer line extending through a

DOI: 10.4324/9781315160191-8

forest. When transects are scaled up, they become "megatransects." For example, a 2005 study reported on a series of megatransects through which scientists counted baobab trees in all the major geological regions of Benin (Assogbadjo, Sinsin, Codjia, & Van Damme, 2005).

Scientists conduct transects and megatransects to advance new knowledge claims. The baobab study revealed new information about how pulp, seed, and kernel production vary depending on climate, soil chemistry, and soil type. However, ecologists also conduct transects for other purposes. In 1999, conservation biologist Michael Fay embarked on a thousand-mile hike across central Africa. The purpose of "The Megatransect" was to collect information about the environment. Science writer David Quammen described the process for *National Geographic*:

> [Fay's] immediate goal is to collect a huge body of diverse but intermeshed information about the biological richness of the ecosystems he'll walk through and about the degree of human presence and human impact. He'll gather field notes on the abundance of elephant dung, leopard tracks, chimpanzee nests, and magisterial old-growth trees. He'll make recordings of birdsong for later identification by experts. He'll store away precise longitude-latitude readings (automatically, every 20 seconds throughout the walking day) with his Garmin GPS unit and the antenna duct-taped into his hat. He'll detect gorillas by smell and by the stems of freshly chewed *Haumania danckelmaniana*, a tangly monocot plant they munch like celery. Eventually he will systematize those data into an informational resource unlike any ever before assembled on such a scale—with the ultimate goal of seeing that resource used wisely by the managers and the politicians who will decide the fate of African landscapes.
>
> (Quammen, 2000)

Ultimately, Fay's work had programmatic implications. As a direct result of the data Fay collected on his Megatransect (and the publicity it garnered), the nation of Gabon created thirteen new national parks to preserve precious habitat.

In this chapter, we mimic Fay's project by collecting course descriptions, grant awards, textbooks, and scholarship related to science education and composition studies. Like the trees, chewed leaves, and dung tracked by Fay, these signs of curricular life provide a snapshot of science and writing from first-year composition to professional training. By understanding the factors that allow particular interventions to succeed or fail, we are better equipped to administer at a programmatic level the intersection of science and writing. Indeed, understanding the threads connecting science and writing requires attending to both global conceptual considerations, such as the theories that guide our teaching, and local material concerns, in particular, what is possible in a given context.

The need to map a theoretical construct for science-and-writing pedagogy is evident when we consider practices in the disciplinary formation of science professionals and writing pedagogies. Lerner (2009) analyzed a century of writing-to-learn approaches in science laboratory curricula and argues that these "school-based" (p. 52) approaches to writing are reductive and discourage students from developing the argumentation skills required for professional work. Lerner sees these failures as opportunities for science and writing educators to rethink their shared objectives and pedagogical investments. For Lerner, "crises" (p. 49) in science and writing pedagogy involve challenges to achieve "authenticity" in learning experiences (p. 51). In ecological terms, Lerner's concern for authentic writing in science is one of sustainability in the resources required to provide "meaningful tasks" (p. 49). Lerner's critique recognizes that the ecologies of school and professional domains are distinct; however, they must be brought into reasonable alignment if students are to navigate successfully from one domain to the other.

In this context, what does it mean to transect? As teachers and program administrators engaged with scientific discourse, we were familiar with the terrain in which science and writing interact. To better know that terrain, however, we needed to transect it and collect additional data. Of course, our transect was not a GPS-supported march through uncharted territory. Rather, we observed and coded a large sample of evidence of science-and-writing instruction collected over years developing courses and programs.

Various disciplines generate data about science-and-writing instruction. Teacher-scholars of rhetoric, composition, and technical communication are interested in the challenges of teaching scientific discourses, but so are applied linguists, science education researchers, and researchers in scientific fields. Educators across these disciplines develop courses tailored to specific institutional demands (see Table 8.1). Courses in "science writing," "scientific writing," "writing in the sciences," and "writing about science" appear in the undergraduate and graduate catalogs of English and writing-studies departments as well as the catalogs for science departments (also see Chapter 11, this volume). Yet even courses that share the same name will have different missions depending on their institutional mandates and student populations. For example, a course called "scientific writing" looks very different when designed as an intermediate-level writing course, a professional writing requirement, an elective for a graduate-level certificate, a requirement for a medical school curriculum, or a Massive Open Online Course (MOOC).

Traditional and online courses are not the only signs of diverse scholarly and pedagogical activity surrounding science and writing. Scholarship on science and writing pedagogy is found in journals from rhetoric, composition, technical communication, science education, and scientific fields (see Table 8.2). Moreover, grant funding from the

Table 8.1 Selected Science-and-Writing Courses Sorted by Program Type

Course Title	*Institution*	*Program Type*	*Curricular Level*	*URL*
Writing in biology	University of Massachusetts – Amherst	Biology	Upper-level Undergraduate	www.bio.umass.edu/biology/undergraduate/biology-courses/course-catalog/312-writing-biology
	Excerpted description: "Students ... will be able: to communicate biological ideas: in writing, orally, and visually; to write effectively, in paragraphs, appropriate to a scientific audience; to adopt a scientific style of writing and use scientific conventions correctly; to use a scientific figure correctly in a document, presentation, or poster; to assess the significance and reliability of diverse information sources; to navigate the primary literature; to format a manuscript for submission for publication."			
Scientific writing	University of Pennsylvania	Biology	Graduate	www.med.upenn.edu/camb/user_docs/CAMB695ShortSyll2013.pdf
	Excerpted description: "This 7-week course is designed to introduce students to basic scientific writing skills and is ideal for second year graduate students preparing for qualifying examinations."			
Writing in the sciences	University of Nebraska – Kearny	Biology	Graduate	www.unk.edu/academics/msbio/_files/BIOL%20877%20-%20Writing%20in%20the%20Sciences.pdf
	Excerpted description: "This course is particularly geared towards students who have been away from academic writing for many years, or perhaps decades. This will be an introduction into this type of writing, complete with an overview of the materials available to you as a distance student. The primary purpose of this course is to improve your written communications skills. We will focus on your ability to prepare and write technical papers in a professional scientific format."			
Scientific writing	Northern Virginia Community College	English	Lower-level Undergraduate	novaonline.nvcc.edu/descriptions/eng114common/eng114ci116.html
	Excerpted description: "Develops rhetorical expertise in the conventions of scientific argumentation and writing through reading scientific literature and composing scientific writings. Introduces plain style and common genres of scientific writing. Develops the ability to communicate scientific knowledge to diverse audiences. Guides the student in achieving typical voice, tone, style, audience, and content in formatting, editing, and graphics."			

Communication for science and research	North Carolina State University	English	Upper-level Undergraduate	http://english.chass.ncsu.edu/undergraduate/prof_writing/pw333.php
	Excerpted description: "Students will become familiar with the purposes, audiences, and conventions of written communication in the contexts they expect to work in after graduation: scientific and research environments."			
Science writing	University of Maryland	English	Upper-level Undergraduate	www.english.umd.edu/academics/professionalwriting/courses/engl390
	Excerpted description: "Students learn the conventions of scientific prose used in research articles and proposals; they also learn to accommodate scientific information to general audiences. In addition, students learn how to use stylistic and visual devices to make information more accessible and how to edit their own work as well as that of their peers."			
Science writing	Rutgers University	Independent writing program	Upper-level Undergraduate	http://wp.rutgers.edu/informationforstudents/152–342
	Excerpted description: "Science Writing offers students an opportunity to refine their skills in presenting technical and scientific issues to various audiences while they critically examine social aspects of scientific information. The course examines new opportunities for covering science (especially on the Internet), the skills required to produce clear and understandable prose about technical subjects, important ethical and practical constraints that govern the reporting of scientific information, and the cultural place of science in our society."			
Scientific and technical writing	Rutgers University	Independent writing program	Upper-level Undergraduate	http://wp.rutgers.edu/informationforstudents/147–302
	Excerpted description: "Scientific and Technical Writing offers students practice in the forms and discourses of scientific and technical writing as they develop, research, and revise an independent project. The purpose of the class is to prepare students for their professional lives in scientific, technical, or public service fields by helping them organize their knowledge while exploring ways of applying it, thus developing their professional expertise."			

(*Continued*)

Course Title	*Institution*	*Program Type*	*Curricular Level*	*URL*
Scientific writing MOOC	Stanford University	Open access	Graduate / Post Graduate	https://lagunita.stanford.edu/courses/Medicine/HRP213/Writing_in_the_Sciences/about
	Excerpted description: "This course teaches scientists to become more effective writers, using practical examples and exercises. Topics include: principles of good writing, tricks for writing faster and with less anxiety, the format of a scientific manuscript, and issues in publication and peer review."			
Scientific writing	San Diego State University	Rhetoric and writing studies	Upper-Level Undergraduate / Graduate	http://rhetoric.sdsu.edu/certificates/professional_writing/classes.htm
	Excerpted description: "RWS 508 develops the writing skills necessary for scientific and medical research and communication. It introduces the scientific documentation process and covers adherence to standards and regulations. It is intended for majors in the sciences and professional scientific writers. Documentation covered includes the research article, research proposal, case report, review, abstract, and promotional material."			
Writing about science	Princeton University	Science and technology general education program	Upper-Level Undergraduate	https://registrar.princeton.edu/course-offerings/course_details.xml?courseid=007848&term=1152
	Excerpted description: "This workshop-style course is designed to teach students in both science and non-science majors how to write about science 'broadly defined to include physical science, biomedical science, environmental science, engineering and technology' in a way that non-scientists can follow. The goal is to instill not only an understanding of scientific results, but also their context, along with the nature of the scientific process itself."			

Table 8.2 Selected Scholarly Articles and Books Describing Curricular Initiatives, Sorted by Curricular Level

Title (Author, Year)	*Field of Publication*	*Curricular Level*
Composition and the rhetoric of science: Engaging the dominant discourse (Zerbe, 2007)	Rhetoric and composition	First-year undergraduate
"Primary science communication in the first-year writing course" (Moscovitz & Kellogg, 2005)	Rhetoric and composition	First-year undergraduate
"Teaching scientific writing: A model for integrating research, writing & critical thinking" (Krest & Carle, 1999)	Science education	First-year undergraduate
"Scientific writing: A humanistic and scientific course for science undergraduates" (Carlisle & Kinsinger, 1977)	Science education	Lower-level undergraduate
"A chemist's view of writing, reading and thinking across the curriculum" (Powell, 1985)	Rhetoric and composition	Lower-level undergraduate
"From concept to application: Student narratives of problem-solving as a basis for writing assignments in science classes" (Rich, Miller & DeTora, 2011)	Rhetoric and composition	Lower-level undergraduate
"Writing-to-teach: A new pedagogical approach to elicit explanative writing from undergraduate chemistry students" (Vaźquez et al., 2012)	Science education	Lower-level undergraduate
"Writing-to-learn in undergraduate science education: A community-based, conceptually driven approach" (Reynolds, Thaiss, Katkin, & Thompson, Jr, 2012)	Science education	Undergraduate science courses
Learning to communicate in science and engineering: Case studies from MIT (Poe, Lerner & Craig, 2010)	Technical communication	Upper-level undergraduate

(*Continued*)

Title (Author, Year)	*Field of Publication*	*Curricular Level*
"More than a picture: Helping undergraduates learn to communicate through scientific images" (Watson & Lom, 2008)	Science education	Upper-level undergraduate
"Writing science" (Hamilton, 1978)	Rhetoric and composition	Upper-level undergraduate
"Engaging biology undergraduates in the scientific process through writing a theoretical research proposal" (Stanford & Duwel, 2013)	Science education	Upper-level undergraduate
"Client-based writing about science: Immersing science students in real writing contexts" (Kiefer & Leff, 2008)	Rhetoric and composition	Upper-Level Undergraduate
"The use of cognitive and social apprenticeship to teach a disciplinary genre: Initiation of graduate students into NIH grant writing" (Ding, 2008)	Technical communication	Graduate
"Camping in the disciplines: assessing the effect of writing camps on graduate student writers" (Busl, Donnelly, & Capdevielle, 2015)	Rhetoric and composition	Graduate
"Developing an English for academic purposes course for l2 graduate students in the sciences" (Douglas, 2015)	Rhetoric and composition	Graduate
"Revitalizing instruction in scientific genres: Connecting knowledge production with writing to learn in science" (Keys, 1999)	Science education	Not specified

National Science Foundation (NSF) has supported a range of science and writing related projects, from the creation of a graduate-level scientific writing program and the development of technologies to support lab-report writing to the creation of case-study collections for teaching about science writing in the public sphere (see Table 8.3). The existence and considerable size of these grants demonstrates opportunities for scholars of rhetoric, composition, and technical communication. The diversity of student needs and pedagogical approaches is further

Table 8.3 Selected Projects on Science-and-Writing Instruction Funded by the National Science Foundation

Project Title	*Institution*	*Award Amount*	*Project Description (Excerpted from Award Abstracts)*
Science writing and rhetorical training: a new model for developing graduate science writers	University of Rhode Island	$499,977	"The primary goal of this program is to work with graduate students and faculty to implement a new cross-disciplinary model of science communication training that integrates diverse types of science writing and communication from the beginning of and throughout graduate students' scientific training. A key to the training model offered is grounding in rhetoric, the academic discipline devoted to the persuasive power of language that includes studies of argument, public discourse, and civic engagement."
LabWrite: A national web-based initiative to use the lab report to improve the way students write, visualize, and understand science	North Carolina State University	$489,159	"Our first goal is to revise and disseminate for national use our online prototype .... LabWrite, a series of instructional and faculty development modules, encourages and enhances use of the lab report so that students and instructors can take advantage of the opportunities it offers to develop and expand students' scientific literacy. Our second goal is to build an instructional infrastructure for improving the teaching and learning experience of the laboratory nationwide."
Implementing the science writing heuristic: An advanced POGIL workshop	Iowa State University	$499,990	"Materials developed in this proposal allow instructors to implement an advanced guided-inquiry teaching technique in their laboratories that is

(*Continued*)

Project Title	*Institution*	*Award Amount*	*Project Description (Excerpted from Award Abstracts)*
			supplemented by reflective writing. The existing infrastructure of POGIL is used as administrative hub for advertising workshops, enrolling participants, production of resources, and recruitment of future workshop leaders from participants who show an ability to implement the SWH (Science Writing Heuristic)."
An intelligent ecosystem for science writing instruction	University of Pittsburgh	$44,009	"Teachers, employers, and college faculty lament the inability of many high school graduates to write clearly. This deficit in writing is due in part because teachers do not have the time to provide appropriate, timely feedback to students on their writing. This project would help teachers help students achieve these skills through automating an effective feedback process, in ways that are customized to particular disciplines and local classroom needs, particularly in high needs districts."
Cases for teaching responsible communication of science	Iowa State University	$254,157	"The project draws on ten important instances of public science communication across five natural science and engineering disciplines, examining the ethical principles at work in communicating scientific knowledge to a non-expert audience, and using these cases to train young scientists to achieve the broader impacts society rightfully expects."

Table 8.4 List of Textbooks and Guides on Scientific Writing and Science Writing

Title (Author, Year)	*Target Readers*
Science and society - A Longman topics reader (Grinnell, 2006)	First-year students
A short guide to writing about biology, 9e (Pechenik, 2015)	Undergraduate biology majors
Write like a chemist: A guide and resource (Robinson, Stoller, Costanza-Robinson, & Jones, 2008)	Undergraduate chemistry majors
The MIT guide to science and engineering communication, 2e (Paradis & Zimmerman, 2002)	Undergraduate science and engineering students
Writing in the sciences, 3e (Penrose & Katz, 2010)	Undergraduate science majors
A field guide for science writers, 2e (Blum, Knudson, & Henig, 2006)	Aspiring science journalists
The science writers' handbook: Everything you need to pitch, publish, and prosper in the digital age (Hayden & Nijhuis, 2013)	Aspiring science journalists
Science research writing for non-native speakers of English (Glasman-Deal, 2009)	Graduate students and research scientists
How to write and publish a scientific paper, 8e (Day & Gastel, 2016)	Graduate students and research scientists
Scientific writing and communication: Papers, proposals, and presentations, 2e (Hofmann, 2013)	Graduate students and research scientists

demonstrated by the range of textbooks and guidebooks on science and writing, from "science and society" readers appropriate for a first-year composition course to specialized textbooks for graduate students who learned English as a second language (see Table 8.4). Diverse engagements with scientific discourse are further reflected in Table 8.5, which lists various graduate and undergraduate programs in science communication. Although the science journalism strain of science writing seems more likely to be taught at the graduate level, new programs, such as Stanford's Science Communication Notation, demonstrate that writing about science and scientific writing can both be formally valued through undergraduate programming.

A variety of workshops and "boot camps" from university research centers, writing centers, and even for-profit companies are offered (see Table 8.6). Some workshops focus on starting or finishing a dissertation or other research project; others help researchers explain their ideas to the public (see also Chapter 9). Regardless of their aims, these intensive professional development opportunities demonstrate the ongoing need for writing instruction for graduate students and working

Table 8.5 List of Undergraduate and Graduate Programs in Scientific Writing and Science Writing, Sorted by Curricular Level

Program	*Institution*	*Level*	*Description Excerpted from Promotional Material*
Notation in science communication	Stanford University	Undergraduate	"The Notation in Science Communication (NSC) provides undergraduates with a new opportunity to develop their ability to communicate technical information to a variety of audiences. Through a combination of coursework, advising and reflection, selected students can earn a special designation on their official transcripts that indicates their advanced work in science communication." (Stanford University, 2017)
Professional science and technology writing certificate	Washington State University	Undergraduate	"WSU's online undergraduate Professional Science and Technology Writing Certificate gives students the ability to bridge the gap between the work of scientists and how the public understands that work." (Washington State University, 2017)
SciWrite	University of Rhode Island	Graduate	"SciWrite (Science Writing and Rhetorical Training) serves cohorts of Graduate Students and Faculty Fellows and Mentors and the wider university community through workshops, courses, and internships that emphasize academic and non-academic writing." (University of Rhode Island, 2017)
Science Writing (MS)	Massachusetts Institute of Technology	Graduate	"A program for English and science majors, freelance writers or journalists seeking a specialty, working scientists, and others in which to learn

			the art and discipline of science writing. An opportunity to contribute to public understanding of science, medicine, engineering, and technology." (Massachusetts Institute of Technology, 2017)
Certificate and MA in science writing	The Johns Hopkins University	Graduate	"The online / low-residency Science Writing Program at Johns Hopkins University strives to guide the next generation of writers and editors who will help us learn how increasingly complex science, medicine and technology affect our lives." (Johns Hopkins University, 2017)

Table 8.6 Professional Development Workshops Sponsored by Universities, Nonprofit Groups, and For-profit Organizations

Workshop Title	*Sponsoring Organization*	*Organization Type*	*Description*
Distilling your message	Alan Alda Center for Communicating Science / SUNY Stony Brook	University center	"This interactive session introduces participants to general principles in how to craft short, clear, conversational statements, intelligible to non-scientists, about what you do and why it matters." (University of Maryland, 2016)
Kenyon institute in biomedical and scientific writing	Kenyon Institute	University center	"Think of it as a three-day boot camp in effective scientific writing, with seasoned scientists as your teachers, your own project as a focus, and an emphasis on results. The goal: a publishable research paper, a fundable grant proposal, compelling technical writing." (Kenyon Institute, 2017)

(*Continued*)

Workshop Title	*Sponsoring Organization*	*Organization Type*	*Description*
Mellon-Wisconsin dissertation writing camp	University of Wisconsin-Madison (Writing Center and Graduate School) / Mellon Foundation	University center	"The dissertation writing camps focus on three core components: 1. They provide participants with intensive, focused time to write in a supportive atmosphere amid other writers, 2. ... participants have multiple opportunities to discuss their work, [and] 3. ... camps offer brief daily writing exercises and workshops on topics such as setting realistic goals, managing one's time, organizing a major project, obtaining useful feedback, and staying motivated." (Universityof Wisconsin-Madison, 2017)
Skills & careers in science writing	Science in Society / Northwestern University	University center	"This free skills development course is designed to help Northwestern PhD trainees write clearly and speak confidently about their own research..." (Northwestern University, 2017)
Communicating science workshops	AAAS Center for Public Engagement with Science & Technology	Nonprofit organization	"AAAS Communicating Science workshops ... are specifically designed to address the needs of scientists and engineers to communicate scientific or technical information in a variety of public and professional interactions, such as media interviews, writing grant proposals, discussing ideas with students, testifying

			before Congress, or participating in public forums." (American Association for the Advancement of Science, 2017)
Scientific writing retreat	Cold Spring Harbor Laboratory	Nonprofit organization	"The CSHL Scientific Writing Retreat is designed for postdoctoral fellows and junior faculty in all areas of biology who are actively working on professional pieces of writing such as manuscripts, grant proposals, job applications, or research/teaching/ personal statements." (Cold Spring Harbor, 2017)
Scientific writing	BioScience Writers	For-profit company	"Our scientific writing workshops will improve your ability to prepare your research manuscripts for publication in top international journals. Workshop participants learn the critical components of a clearly written manuscript and how to structure each manuscript section." (Bioscience Writers, 2017)

scientists. Indeed, such workshops are perhaps the most efficient and cost effective way for a writing program administrator to draw attention to scientific writing at his or her institution.

Finally, a number of technologies have been developed to help students and scientists write or learn to write more effectively (see Table 8.7). These include digital heuristics for writing laboratory reports, platforms for managing classroom peer review, and tools for manuscript preparation and collaboration. Some of these tools, like Manuscript Architect, are no longer available, suggesting that they did not fill their intended niches. However, the efficacy of other tools, such

Table 8.7 List of Technological Tools for Teaching (and Working on) Scientific Genres

Program	*Sponsoring Institution*	*Program Purpose*	*Description*
Calibrated peer review	Department of Chemistry and Biochemistry, UCLA	Help instructors incorporate writing assignments into science classes by shifting assessment to a facilitated peer review system	"Calibrated peer review (CPR) is a web-based, instructional tool that enables frequent writing assignments in any discipline, with any class size, even in large classes with limited instructional resources. In fact, CPR can reduce the time an instructor now spends reading and assessing student writing." (Calibrated Peer Review, 2017)
LabWrite	Department of English, North Carolina State University	Help students manage the process of writing lab reports	"LabWrite is an online resource designed to help students take full advantage of one of the most important activities for learning in the sciences—writing good lab reports. But LabWrite is not just about writing lab reports. It shapes the entire lab experience as a learning experience by structuring each lab into four stages that provide the foundation for the students' website: PreLab, InLab, PostLab, and LabCheck."(Carter et al., 2004)
Manuscript architect (inactive)	Center for Excellence in Surgical Outcomes, Duke University Medical Center	Help scientists manage collaborative writing	"This application had as its main objective the separation of the multiple tasks associated with scientific writing into smaller components. It was also aimed

			at providing a mechanism where sections of the manuscript (text blocks) could be assigned to different specialists." (Pietrobon et al., 2005)
Scientific Writing AssistaNt (SWAN)	School of Computing, University of Eastern Finland	Help scientists evaluate the effectiveness of writing	"Scientific Writing AssistaNt (SWAN) is a rule-based, computer-assisted tool that combines text quality metrics and natural language processing. SWAN provides feedback on the parts of a scientific paper that create the first impressions: the title, abstract, introduction, conclusions, and the structure (headings and subheadings) SWAN does not give overall grading for a paper. Instead, SWAN points out problems at the local level, as well as assesses text fluidity (both automatic and manual options are available) and cohesion. The newest SWAN versions also contain metrics for assessing the relationships between visuals (figures and tables) in a paper." (Turunen, 2013)

as LabWrite and Calibrated Peer Review (CPR), has been documented through scholarship; Google Scholar searches for these programs return 126 and 812 citations, respectively. Each of these tools demonstrates specific points of tension requiring some kind of intervention. That is, the work each technology attempts to do demonstrates a perceived need: LabWrite can be used to teach process, Manuscript Architect can help to teach teamwork, CPR facilitates peer review, and all can be used as part of instruction in scientific argumentation. The development of these tools indicates a desire to find efficiencies and economies of scale in writing instruction.

After collecting the data represented in Tables 2 through 7, we attempted to categorize them within the ecological frame suggested by the megatransect. We cannot claim empirical precision or statistical validity for our findings; the tables list only representatives of each collected sign of science-and-writing instruction. However, this accumulated data can be categorized into a useful framework for thinking about where and how and why teachers and program administrators might engage scientific discourse in their work. The next section presents our categories for *habitats* (curricular niches within distinct programmatic terrain) and *species* (curricular and co-curricular programs that occupy specific niches).

From Transect to Catalog: Pedagogical Species Adapted to Curricular Niches in Programmatic Terrain

Our metaphorical megatransect allowed us to consider diverse signs of pedagogical life by establishing analogs between the physical space traversed in an ecological transect and the conceptual space of science-and-writing pedagogy. Table 8.8 maps these analogs using the terms *terrain, niche*, and *species*, and each term is explained in the sections that follow.

Table 8.8 Transecting Science and Writing: A Map of the Metaphor

Term	*The Natural World*	*Teaching Science and Writing*
Transect	Traveling through a physical terrain in a systematic way and recording data about the terrain and the species that occupy it	Reviewing various sources that describe teaching science and writing in a systematic way and recording data about how and why people teach scientific discourses
Terrain	Physical environments defined by their topography, geology, and climate	Programmatic spaces defined by the experience of their students
Niches	Localized areas within a terrain in which species have adapted (or are adapting) to the features of the terrain and the presence of other species	Curricular areas that shape and are shaped by the goals of programs, students, institutional stakeholders, and external interests
Species	Plants and animals adapted (or adapting) to the niches they occupy	Specific sites of writing instruction that involve some engagement with scientific discourse Specific sites of science instruction that involve some engagement with writing

Programmatic Terrain

After transecting the ecosystem of science-and-writing pedagogy, we decided to segment the pedagogical *terrain* into legible zones according to the experience level of the intended audience. Basing terrain types on the experience level of students (undergraduate, graduate, etc.) might seem obvious, but this is not the only possible scheme. We could define terrain by field (rhetoric, professional writing, etc.), which was the approach used by Zerbe (2007) in *Composition and the Rhetoric of Science*. His exploration of "the *place* of scientific discourse in rhetoric and composition" (Zerbe, p. 50, emphasis added) divides the terrain into seven tracts: composition readers, composition courses, ecocomposition, writing across the curriculum, rhetoric of science, medical rhetoric, and technical communication. Although these divisions are useful for Zerbe's project to connect composition studies with the rhetoric of science, they have limitations for thinking about science-and-writing pedagogy more expansively (see Sullivan, 2008). Our programmatic terrain includes (1) the first-year experience, (2) post-first-year-experience undergraduate education, (3) graduate education, and (4) post-graduate and professional training. Co-curricular interventions (such as formal or informal instruction through a writing center) and independent interventions (such as fee-for-service workshops and tutoring) can apply to any of these terrains. Within each are distinct niches occupied by pedagogical species adapted (or not) to the needs, resources, and constraints of particular programs.

Curricular Niches

Within each programmatic terrain, we identified four curricular niches in which scientific discourse and writing instruction occupy the same space: (1) writing about the discourses of science, (2) science in writing, (3) writing to learn science, and (4) writing science. These categories extend various grammatical arrangements that have been invoked previously to describe courses collocating science and writing. For example, Hamilton (1978) described how he settled on a course titled "Writing Science" after considering and discarding the less forceful "science writing," "scientific writing" (which suggested, for him, a scientific process of writing), "writing *in* the sciences" (which suggested a readings course), "writing *for* the sciences" (English courses in the service of science departments), and "writing *from* the sciences" (science journalism). Hamilton's terms do not quite match current conceptions of the available range of courses and approaches, and part of this has to do with how the various science-and-writing phrases are parsed by different people (also see editors' introduction "Science and Communication: High Stakes, Great Responsibility"). For example, Hamilton objected to "writing in the sciences" because it "implies a reading course that analyzes matter already written" (32). We do not agree, and presumably neither do Penrose and Katz; their *Writing in the Sciences: Exploring the Conventions of Scientific Discourse* (now in its third edition) does not have the limitations Hamilton suggests come with that title. Similarly "science writing" in many contexts is synonymous with science journalism rather than the writing involved in conducting and reporting on scientific work, though some programs use that phrase to describe disciplinary writing. The larger point is that for any label of a science-and-writing collocation, the grammar matters—at least to *some* stakeholders. Thus, we added language to clearly differentiate our categories: (1) writing about the discourses of science, (2) science in writing, (3) writing to learn science, and (4) writing science. The first two categories align with teaching the rhetoric of science and science journalism, respectively; the latter two align with the respective goals of writing across the curriculum (WAC) and writing in the disciplines (WID) models. Table 8.9 demonstrates the utility of this taxonomy by applying it to the programmatic terrain of the first-year experience.

Within this terrain and depending on its purpose, a first-year course could engage scientific discourse through reading, writing, writing about, and/or rhetorical analysis of scientific discourse. These categories apply equally well to other terrains. Table 8.9 further suggests that *any* course or intervention may be considered a species suited (or ill-suited) to its particular niche. In the next section, we discuss specific examples of species that needed to adapt as we reimagined them for new curricular niches in different programmatic spaces.

Table 8.9 An Example of Pedagogical Speciation across the Terrain of First-Year Experience

Terrain	*Niche*	*Sample Species*
First-year experience	Writing about the discourses of science	A first-year seminar on the rhetoric of science (e.g., Fox et al., 2013)
First-year experience	Science in writing	A first-year composition course that trains students to read and write about scientific discourse (e.g., Moskovitz & Kellogg, 2005)
First-year experience	Writing to learn science	An introductory science course requiring writing either about the scientific research of others or about the results of planned labs (e.g., Krest & Carle, 1999)
First-year experience	Writing science	A first-year course (in science or composition) that requires students to design a study and write about the results (e.g., Zerbe, 2007, pp. 169–180)

Pedagogical Species

A curricular niche must be filled with particular species—courses or other interventions. Here, we illustrate the process of pedagogical speciation by tracing how we, as two instructors, engaged scientific discourse in writing classes ranging from first-year composition to workplace training for pharmaceutical scientists. In doing so, we demonstrate how understanding scientific discourse can help an instructor to adapt existing courses for new audiences and to develop new courses at all levels.

One particular species involved a "mutation" of an undergraduate service course for the University of Maryland's (UMD) Professional Writing Program. Since 1980, all UMD students have been required to complete an upper-level English course in professional writing. Courses fulfilling the requirement initially included only Advanced Composition and Technical Writing, but later they were expanded to include Business Writing, Legal Writing, and Medical Writing. In 2004, a new wave of speciation introduced many new courses, including Writing about Economics, Writing for the Arts, and Science Writing. An offshoot of Technical Writing, Science Writing was designed by us to meet the needs of aspiring science professionals. Using Penrose and Katz's *Writing in the Sciences* (2004), the course took a rhetorical approach to scientific genres, with students analyzing research reports and writing review articles in addition to writing personal statements for graduate school and accommodations of scientific research for non-experts. Pilot sections were well received, and the course continues to be taught. However, when we each migrated the course to other settings, it soon became clear how much local ecologies affected course design.

When William took the course to a regional campus of the state university, an institution also strong in the sciences, the course did not easily take root, probably because it did not exist within an established program in which students were required to take a course suitable to their career interests. As a result, the course drew a different set of students across the humanities and the sciences. Over two iterations, the course shifted toward a focus on popularizing scientific information to a range of public audiences as opposed to engaging with the professional discourses of science for undergraduates on a path to graduate school.

Similarly, when Jonathan considered teaching writing about science in a new context, Ohio State University (OSU), he had to reimagine the UMD course. The curricular niche his new course could occupy—a special topics course in professional communication—had few institutional mandates and thus would enroll English majors, Professional Writing minors, and students from all over campus. The resulting new species and its assignment sequence are described in detail in "Style and the Professional Writing Curriculum: Teaching Stylistic Fluency through Science Writing" (Buehl, 2013). That chapter's title alone alludes to a major adaptation. Whereas the UMD course was focused on the professional genres of science, the main learning outcome of the OSU course was tied to a transferable understanding of style. Regardless of their scientific expertise or major, students could develop their stylistic ranges by reading and writing about scientific discourse. A similar rationale supports situating "writing about science" within first-year composition, as Moskovitz and Kellogg (2005) have demonstrated.

Among the points of entry to science and writing in university culture are learning communities that establish cohorts of students formed around particular interests. In the late 1990s, William participated in one such community—Science, Technology, and Society. In this instance, a section of first-year composition was linked to other entry-level courses and was expected to feature reading and writing assignments centered on science and technology. With a background in math education and coursework in the rhetoric of science, William was well suited to this task. However, some composition programs might have few instructors prepared to help students engage the discourses of science and technology. For that matter, first-year students expressing interest in STEM fields are typically unprepared to wrestle with the language of science. Thus, any program-wide initiative to incorporate scientific discourse into first-year writing might require cultivating the expertise of program faculty through training or other professional development.

The concept of pedagogical species extends to activities beyond coursework. Although students engage with the intersection of science and writing in introductory science courses, typically in laboratory-based writing assignments (see Chapter 12), a growing emphasis on experiential learning and undergraduate research marks the development of aspiring science professionals. Students learn to "write science" not in formal classwork but through practices of mentoring that let students conduct collaborative or independent research later presented in conferences and in papers in peer-reviewed journals. William has seen firsthand the crucial role of mentoring in the work of a colleague in computer science who has guided several generations of students through pathways that lead to top-tier graduate programs. In this labor-intensive process, students learn to write in authentic genres while jumpstarting their scientific careers.

Finally, we turn to the challenges and opportunities of pedagogical species at graduate and postgraduate levels. When Jonathan was teaching Science Writing at UMD, he also consulted with a pharmaceutical company. Its managers wanted a course on report writing for their bench scientists who needed to reconsider how they wrote reports for a business development context. The six-day course that Jonathan created looked nothing like his undergraduate course. Each training session filled an entire day with no mandated readings or homework assignments. Nevertheless, the course drew from common principles regarding scientific genres and styles. Most recently, Jonathan planned "boot camp" style trainings for postdocs, graduate students, and junior faculty similar to those represented in Table 8.6. Clearly, scientists at every level need help to write effectively and efficiently, and writing faculty can help to fill that niche through coursework, service, or fee-for-service models.

This brief review of pedagogical species illustrates the diverse terrain in which an instructor can operate, if properly trained. Fortunate in our training in the rhetoric of science and in professional writing, we did not experience the confidence gap that others without that training may face in teaching scientific discourse. Nevertheless, if any of these species discussed were to be scaled up, robust professional development would be needed. The remainder of this chapter outlines types of training useful for such purposes.

From Transect to Training: Fostering Rhetorical Engagement with Scientific Discourse

Although transects are ecological research methods, their results can be used to promote policy change. We thus pivot from description to address matters of sustainability and development. How can an administrator develop robust and respectable programs that can engage and help others to engage scientific discourses? If hiring instructors well-versed in scientific communication is not an option, how can one train people with a firm rhetorical foundation to understand, edit, and comment upon scientific discourses? The remainder of this chapter outlines topics and tools that a program administrator might use to develop training modules for students, instructors, and colleagues to help them better understand the rhetoric of scientific discourse, the particular genres of scientific communication, and the style of scientific writing.

The Rhetorical Nature of Science

Though some might disagree with the claim that science is rhetorical "without remainder" (Gross, 1990, p. 33), few scientists or scholars of science studies would dispute the fact that science relies on persuasion through formal argument. Scientific knowledge depends on the rhetorical conventions of scientific communities as much as it depends on the material machinery of scientific work. Numerous case studies demonstrate this point, including that of Marshall and Warren, whose Nobel-prize-winning research on stomach ulcers was initially rejected because its arguments were not persuasively presented (Penrose & Katz, 2010). This case, described in Penrose and Katz's *Writing in the Sciences*, can be joined with others to demonstrate the rhetorical work of constructing scientific knowledge. Students (or instructors) might read Prelli's "Rhetorical Construction of Scientific Ethos" (1997) and then watch the archived NASA press conference on GFAJ-1, a bacteria species that allegedly synthesized arsenic into its biomolecules. The press conference itself, the controversy over the research, and the formal and informal documents through which the research was challenged provide a robust case for discussing standards of argument, the peer-review process, and the rhetorical construction of scientific credibility. (See Buehl [2016] for additional details about this case.)

The Special Moves of Scientific Genres and Genres Accommodating Scientific Discourse

Scientific genres are shaped by the material and rhetorical needs of scientists (Bazerman, 1988; Gross, Harmon, & Reidy, 2002). The discursive products of scientists are also raw material for science communicators who popularize research for non-expert publics (Fahnestock, 1986). Instructors must therefore understand three scientific genres: research articles, proposals, and accommodations of primary research. Their significant variations require that instructors and students take a rhetorical approach to reading, analyzing, and producing them.

The *research article* is the most studied scientific genre because it is vital for scientific work, but also because its history, structure, and purpose provide fertile ground for developing theories of genre (see Chapter 1). The challenge for an instructor is to communicate that knowledge concisely in ways that move a trainee away from cookie-cutter approaches and toward a rhetorical understanding that connects micro-level elements of scientific discourse to the larger goals of scientific argumentation. At a minimum, one must understand how scientists establish exigency, how and why they qualify the certainty of claims, and how visuals participate in scientific arguments. These topics are summarized well by the chapter on research reports in *Writing in the Sciences* (Penrose & Katz, 2010), and O'Connor and Holmquist's "Algorithm for Writing a Scientific Manuscript" (2009) describes a process for moving from visualized data to rhetorically situated claims.

Like the research article, the *proposal* varies across disciplines and contexts. In "Proposal Writing from Three Perspectives," Northcut, Crow, and Mormile (2009) describe genre differences between proposals for business and nonprofit development and proposals by scientists and engineers. The piece can spark discussion of working within and across disciplinary lines. To further demonstrate how scientific proposals are read and rated, one can show clips from "NIH Peer Review Revealed" of a mock review panel session after introducing trainees to the rhetoric of science. Trainers can pause the video after particular statements to have trainees explain what's happening using their new rhetorical vocabulary. In one scene, for example, a proposal reviewer questions the ethos of an inexperienced investigator, but her concerns are addressed by another reviewer who points to evidence demonstrating how that inexperience is balanced by the involvement of experienced mentors on the project. Such an example can help trainees to imagine how sophisticated readers respond to scientific claims.

A range of genres *accommodate research* to non-specialist audiences. The domain of science journalists and public information officers, these genres include vehicles for describing results to a broad audience of science professionals reading outside their fields and even more accessible texts that accommodate these findings for untrained readers. From the "science news" sections of journals and newspapers to "gee whiz" magazine articles, university publications, and YouTube videos, there is an abundance of examples for students to discuss and imitate. However, they need strategies for adapting scientific discourse into engaging educational and entertaining forms. More than thirty years after publication, Fahnestock (1986) is still relevant and eminently teachable to adults, as argued by Gigante (Chapter 9). The frameworks of Aristotelian genre, stasis theory, and Latour and Woolgar's taxonomy of certainty are clear and useful tools for both analysis and planning. Indeed, one can find any science news article that references a scientific research article, put the two pieces side by side, and demonstrate the genre shift from fact-making to celebratory prose, the shift in stasis from fact to value, and the ways hedged statements in primary research shift to highly certain or radically speculative claims in the accommodation piece. By augmenting Fahnestock's toolkit with discussions of audience-appropriate definitions, metaphor, narrative, and multimodal composing, instructors can prepare science writers and aspiring scientists to communicate complex topics to non-expert audiences (see Chapter 10).

Reading and Writing Scientific Style

Critiques of scientific discourse often identify jargon as the main difficulty for non-expert readers. However, technical terminology is necessary for experts communicating with other experts and is but one of many features that can be used well or poorly in scientific contexts. Like all styles, scientific styles are collections of choices regarding words, grammar, sentence structure, and passage construction. To help others understand these choices, we suggest discussing major concepts from two compatible readings: Halliday's "Some Grammatical Problems with Scientific English" (1993) and Gopen and Swan's "The Science of Scientific Writing" (1990).

The problems that Halliday discusses are stylistic features we can teach people to identify and deploy. Novice readers struggle with word-level choices typical of scientific texts, such as working with interlocking definitions (needing to know one term to understand another), taxonomies (needing to understand implicit categories), and special expressions (needing to know grammatical constructions unique to a specific field). However, these features are only problems when they are poorly managed for a specific audience. Consider this passage, the first paragraph of an article in *Marine Ecology Progress Series*:

> How settlement-stage coral reef fishes navigate from the plankton to suitable reef habitats is poorly understood, although we know that some larvae are capable of locating and returning to their natal reefs (Jones et al. 1999, Swearer et al. 1999). Many potential navigation cues exist (see reviews in Montgomery et al. 2001, Kingsford et al. 2002, and Myrberg & Fuiman 2002), but evidence to date supports only 2: chemicals and sounds. While chemical signals can influence settlement of reef fishes at small (10s to 100s of metres) spatial scales (Sweatman 1988, Danilowicz 1996, Arvedlund et al. 1999), they are only available downstream of reefs and in areas of high current flow, so may only be of value to larvae that are strong swimmers (Armsworth 2000).
>
> (Simpson et al., 2004, p. 263)

This passage is entirely appropriate for the intended reader of marine ecologists who know about the life cycles of reef fish. However, this passage would not be effective for a broader scientific readership. Thus, when writing about similar research for the multidisciplinary journal *Science*, the same authors provided additional description to help readers navigate the interlocking definitions and technical taxonomies of marine ecology:

> Most reef populations are replenished with recruits that settle out from an initially pelagic existence. The larvae of nearly all coral reef fish develop at sea for weeks to months before settling back to reefs as juveniles. Although larvae have the potential to disperse great distances, recent studies show a substantial portion recruit back to their natal reefs (1, 2). Larvae are not passively dispersed but develop a high level of swimming competence (3). How they use these capabilities to influence their dispersal is an open question. We show here that recruits respond actively to reef sounds, potentially providing a valuable management tool for the future.
>
> (Simpson et al, 2005, p. 221)

Still written for scientists, this passage unpacks many assumptions of the first sentence of the more technical piece. Similarly, when the same research was accommodated for non-expert readers of *The New York Times*, technical terms (such as "natal reefs") and special expressions (such as "recruit back") were replaced with more accessible language that nonetheless introduces key concepts:

> After they hatch on a reef, most fish larvae live a peripatetic existence, floating with currents for weeks. Studies show that many

> juvenile fish can and do go home again when they settle down. "So the question has been what they were doing to detect their reef," Dr. Simpson said.
>
> He and his colleagues focused on the possibility of sound as a cue because "coral reefs make an enormous amount of noise," he said. "And fish have good hearing."
>
> (Fountain, 2005)

Comparing such examples of successful audience accommodation with passages that do a poor job managing terminology can emphasize the importance of audience appropriate descriptions, definitions, and explanations.

Like his terms for word-level problems, Halliday's terms for sentence-level issues are equally useful for helping novices understand, write, and edit scientific prose. These terms include "lexical density" (a high ratio of content words to particles), "syntactic ambiguity" (noun and modifier piles create potentially polysemous phrasing), "grammatical metaphor" (for example, actions expressed as nouns instead of verbs), and "semantic discontinuity" (leaps in logic). Each of these features can be managed effectively by applying the advice from the second suggested reading: Gopen and Swan's "The Science of Scientific Writing" (1990). This text operationalizes for scientific contexts the major principles of Joseph Williams' "Little Red Schoolhouse" tradition of cognitive stylistics. The authors convincingly show how principles of emphasis position, given-to-new and topic-to-comment sentences, subject-verb connection, and sentence-level agency can be taught without asking scientists to write in a non-scientific dialect. Once equipped with a vocabulary for characterizing scientific style, instructors and students should be able to identify effective and less effective instances. Ultimately, they should be able to comment on and produce scientific passages effectively.

If a curricular niche requires accommodating technical material for non-technical contexts, one can do so by leveraging expertise of technical styles. Jonathan's "Writing about Science" course transitions from scientific discourse addressed to other science professionals to writing about science in "plain language" for public audiences and less-expert stakeholders (Buehl, 2013). To help students perform this new style well, he contrasts key differences between the ideal features of scientific and plain styles. For example, a statement such as, "Sufficient preparation is required for effective achievement of difficult tasks," can be used to remind students of common features of technical scientific prose: nominalized actions (grammatical metaphor), passive constructions, etc. Such sentences can then be contrasted with plain language revisions emphasizing agency through concrete subjects and active verb constructions; for example, "We must train students to complete difficult tasks well." Although his course moves from the technical to the plain style, one might also move in the other direction, teaching plain style first and building on students' knowledge of style to approach the specific demands of technical styles.

Conclusions: Tending the Ecosystem

The field of technical and scientific communication should obviously continue to pay attention to how scientific discourse is taught and learned, in the same way that rangers and wardens tend to the species and terrain of parks and wildlife sanctuaries. Science is more than a field of study; it is a broad culture, and it changes constantly. That dynamism is reflected within and acts upon the texts its members produce. Recognizing the moves, styles, genres, and evolution of scientific discourse is easier when guided by a knowledgeable instructor who is familiar with scientific methods, including the methods of knowledge generation and legitimation. Sophisticated scientific literacies can be incorporated into curricula—starting with K-12 education, at every level of higher education, and beyond—as we attempt to nurture lifelong learners. By training graduate students, instructors, and fellow faculty members to work with scientific discourse, program administrators can better position their programs to serve their communities and to expand their programmatic boundaries productively and sustainably. By regularly transecting this programmatic terrain, scholars and administrators of technical communication can identify and engage new opportunities to demonstrate the value of our discipline.

References

American Association for the Advancement of Science. (2017). "March for Science Communicating Science Workshop." Retrieved June 24, 2017, from https://www.aaas.org/event/march-science-communicating-science-workshop.

Assogbadjc, A. E., Sinsin, B., Codjia, J. T. C., & Van Damme, P. (2005). Ecological diversity and pulp, seed and kernel production of the baobab (Adansonia digitata) in Benin. *Belgian Journal of Botany*, *138*(1), 47–56.

Bazerman, C. (1988). *Shaping written knowledge: The genre and activity of the experimental article in science*. Madison: University of Wisconsin Press.

BioScience Writers. (2017). "Scientific Writing Workshops." Retrieved June 24, 2017, from http://www.biosciencewriters.com/WritingWorkshops.aspx.

Blum, D., Knudson, M., & Henig, R. M. (2006). *A field guide for science writers*. (2nd ed.). New York: Oxford University Press.

Buehl, J. (2013). Style and the professional writing curriculum: Teaching stylistic fluency through science writing. In M. Duncan & S. M. Vanguri (Eds.), *The centrality of style* (pp. 279–308). Anderson, SC: Parlor Press.

Buehl, J. (2016). Evolution or revolution? Casing the impact of digital media on the rhetoric of science. In A. G. Gross & J. Buehl (Eds.) *Science and the Internet: Communicating knowledge in a digital age* (pp. 1–10). Amityville, NY: Baywood Publishing Company, Inc.

Busl, G., Donnelly, K. L., & Capdevielle, M. (2015, August 25). Camping in the disciplines: Assessing the effect of writing camps on graduate student writing. *Across the Disciplines, 12*(3). Retrieved June 6, 2016, from http://wac.colostate.edu/atd/graduate_wac/busletal2015.cfm.

Calibrated Peer Review: Home. (2017). "Calibrated Peer Review." Retrieved June 24, 2017, from http://cpr.molsci.ucla.edu/Home.aspx.

Carlisle, E. F., & Kinsinger, J. B. (1977). Scientific writing: A humanistic and scientific course for science undergraduates. *Journal of Chemical Education, 54*(10), 632–634.

Carter, M., Wiebe, E. N., Ferzli, M., & Lambert, J. (2004). "What is LabWrite?" Retrieved June 6, 2016, from https://www.ncsu.edu/labwrite/Professors/whatis_lwr_prof.htm.

Cold Spring Harbor Laboratory. (2017). "Scientific Writing Retreat." Retrieved June 24, 2017, from https://meetings.cshl.edu/courses.aspx?course=C-WRITE&year=16.

Day, R., & Gastel, B. (2016). *How to write and publish a scientific paper* (9th ed.). Cambridge: Cambridge University Press.

Ding, H. (2008). The use of cognitive and social apprenticeship to teach a disciplinary genre: Initiation of graduate students into NIH grant writing. *Written Communication, 25*(1), 3–52.

Douglas, J. (2015, August 25). Developing an English for academic purposes course for L2 graduate students in the sciences. *Across the Disciplines, 12*(3). Retrieved June 6, 2016, from http://wac.colostate.edu/atd/graduate_wac/douglas2015.cfm.

Fahnestock, J. (1986). Accommodating science: The rhetorical life of scientific facts. *Written Communication, 3*(3), 275–296.

Fountain, H. (2005, April 12). For young fish, it seems, the call of the reef is music. *The New York Times*. Retrieved June 6, 2016, from www.nytimes.com.

Fox, J., Birol, G., Han, A., Cassidy, A., Welsh, A., Berger, J. D., ... Samuels, A. L. (2013). Enriching educational experiences through UBC's first-year seminar in science (SCIE113). *Collected Essays in Learning and Teaching, 7*(1), 1–18.

Glasman-Deal, H. (2009). *Science research writing for non-native speakers of English*. London: Imperial College Press.

Gopen, G. D., & Swan, J. A. (1990). The science of scientific writing. *American Scientist, 78*(6), 550–558.

Grinnell, R. M. (2006). *Science and society (a Longman topics reader)*. New York: Pearson.

Gross, A. G. (1990). *The rhetoric of science*. Cambridge, MA: Harvard University Press.

Gross, A. G., Harmon, J. E., & Reidy, M. S. (2002). *Communicating science: The scientific article from the 17th century to the present*. New York: Oxford University Press.

Halliday, M. A. K. (1993). Some grammatical problems in scientific English. In M. A. K. Halliday & J. R. Martin (Eds.), *Writing science: Literacy and discursive power* (pp. 69–85). Pittsburgh, PA: University of Pittsburgh Press.

Hamilton, D. (1978). Writing science. *College English, 40*(1), 32–40.

Hayden, T., & Nijhuis, M. (Eds.). (2013). *The science writers' handbook: Everything you need to know to pitch, publish, and prosper in the digital age*. Boston: Da Capo Press.

Hofmann, A. H. (2013). *Scientific writing and communication: Papers, proposals, and presentations* (2nd ed.). New York: Oxford University Press.

Johns Hopkins University, Advanced Academic Programs. (2017). "Science Writing." Retrieved June 24, 2017, from http://advanced.jhu.edu/academics/graduate-degree-programs/science-writing/.

Kenyon Institute. (2017). "Kenyon Institute in Biomedical and Scientific Writing." Retrieved June 24, 2017, from http://www.kenyoninstitute.org/programs/biomedical-scientific-writing/.

Keys, C. W. (1999). Revitalizing instruction in scientific genres: Connecting knowledge production with writing to learn in science. *Science Education*, *83*(2), 115–130.

Kiefer, K. & Leff, A. (2008, November 22). Client-based writing about science: Immersing students in real writing contexts. *Across the Disciplines, 5*. Retrieved June 4, 2016, from http://wac.colostate.edu/atd/articles/kiefer_leff2008.cfm.

Krest, M. & Carle, D. (1999). Teaching scientific writing: A model for integrating research, writing & critical thinking. *The American Biology Teacher*, *61*(3), 223–227.

Lerner, N. (2009). *The idea of a writing laboratory*. Carbondale, IL: Southern Illinois University Press.

Massachusetts Institute of Technology. (2017). "MIT Graduate Program in Science Writing." Retrieved June 24, 2017, from http://sciwrite.mit.edu.

Moskovitz, C., & Kellogg, D. (2005). Primary science communication in the first-year writing course. *College Composition and Communication*, *57*(2), 307–334.

Northcut, K. M., Crow, M. L., & Mormile, M. (2009, July). Proposal writing from three perspectives: Technical communication, engineering, and science. In *2009 IEEE International Professional Communication Conference(IPCC 2009)*. Waikiki, HI. (pp. 1–8).

Northwestern University. Science in Society. (2017). "Skills and Careers in Science Writing Course." Retrieved June 24, 2017, from http://scienceinsociety.northwestern.edu/audiences/nu-graduate-students.

O'Connor, T. R. & Holmquist, G. P. (2009), Algorithm for writing a scientific manuscript. *Biochemistry and Molecular Biology Education, 37*, 344–348.

Ossola, A. (2014). How scientists are learning to write. *The Atlantic*. Retrieved June 6, 2016, from www.theatlantic.com.

Paradis, J. G., & Zimmerman, M. L. (2002). *The MIT guide to science and engineering communication* (2nd ed.). Cambridge, MA: MIT Press.

Pechenik, J. A. (2015). *A short guide to writing about biology* (9th ed.). New York: Pearson.

Penrose, A. M., & Katz, S. B. (2004). *Writing in the sciences: Exploring conventions of scientific discourse* (2nd ed.). New York: Longman.

Penrose, A. M., & Katz, S. B. (2010). *Writing in the sciences: Exploring conventions of scientific discourse* (3rd ed.). New York: Longman.

Pietrobon, R., Nielsen, K. C., Steele, S. M., Menezes, A. P., Martins, H., & Jacobs, D. O. (2005). Manuscript Architect: a Web application for scientific writing in virtual interdisciplinary groups. *BMC Medical Informatics and Decision Making*, *5*(15), 1–11.

Poe, M., Lerner, N., & Craig, J. (2010). *Learning to communicate in science and engineering: Case studies from MIT.* Cambridge, MA: MIT Press.
Powell, A. (1985). A chemist's view of writing, reading, and thinking across the curriculum. *College Composition and Communication*, *36*(4), 414–418.
Prelli, L. J. (1997). The rhetorical construction of scientific ethos. In R. A. Harris (Ed.), *Landmark essays on rhetoric of science: Case studies* (pp. 87–104). Mahwah, NJ: Lawrence Erlbaum.
Quammen, D. (2000) Megatransect: Across 1,200 miles of untamed Africa on foot. *National Geographic Magazine.* Retrieved June 2, 2016, from http://ngm.nationalgeographic.com/ngm/0010/feature1/fulltext.html.
Reynolds, J. A., Thaiss, C., Katkin, W., & Thompson, R. J. (2012). Writing-to-learn in undergraduate science education: A community-based, conceptually driven approach. *CBE-Life Sciences Education*, *11*(1), 17–25.
Rich, J., Miller, D., & DeTora, L. (2011, June 27). From concept to application: Student narratives of problem-solving as a basis for writing assignments in science classes. *Across the Disciplines, 8*(1). Retrieved June 4, 2016, from http://wac.colostate.edu/atd/articles/richetal2011.cfm.
Robinson, M., Stoller, F., Costanza-Robinson, M., & Jones, J. K. (2008). *Write like a chemist: A guide and resource.* New York: Oxford University Press.
Simpson, S. D., Meekan, M. G., McCauley, R. D., & Jeffs, A. (2004). Attraction of settlement-stage coral reef fishes to reef noise. *Marine Ecology Progress Series*, *276*(1), 263–268.
Simpson, S. D., Meekan, M., Montgomery, J., McCauley, R., & Jeffs, A. (2005). Homeward sound. *Science*, *308*(5719), 221–221.
Stanford, J. S., & Duwel, L. E. (2013). Engaging biology undergraduates in the scientific process through writing a theoretical research proposal. *Bioscene: Journal of College Biology Teaching*, *39*(2), 17–24.
Stanford University, Undergraduate Program in Writing and Rhetoric. (2017). "About the Notation in Science Communication." Retrieved June 24, 2017, from https://undergrad.stanford.edu/programs/pwr/notation-science-communication/about-notation-science-communication.
Sullivan, D. (2008). Extending the rhetoric of science into the first-year composition classroom. *The Review of Communication*, *8*(3), 292–295.
Turunen, T. (2013). *Introduction to Scientific Writing AssistaNt (SWAN) - Tool for evaluating the quality of scientific manuscripts.* M.Sc. thesis, School of Computing, University of Eastern Finland.
United States National Institutes of Health. Center for Scientific Review. "NIH Grant Review Process YouTube Videos." (March 21, 2014). Retrieved June 2, 2016, from https://public.csr.nih.gov/aboutcsr/contactcsr/pages/contactorvisitcsrpages/nih-grant-review-process-youtube-videos.aspx.
University of Maryland. College of Behavioral and Social Sciences. (2016). "Alan Alda Center for Communicating Science" Retrieved June 24, 2017, from https://bsos.umd.edu/event/alan-alda-center-communicating
University of Rhode Island, College of Environmental and Life Sciences. (2017). "SciWrite." Retrieved June 24, 2017, from http://web.uri.edu/cels/sciwrite/.
University of Wisconsin-Madison Graduate School. (2017). "Dissertation Writing Camp." Retrieved June 24, 2017, from https://grad.wisc.edu/currentstudents/dissertation/writing_camps/

Vázquez, A. V., McLoughlin, K., Sabbagh, M., Runkle, A. C., Simon, J., Coppola, B. P., & Pazicni, S. (2012). Writing-to-teach: A new pedagogical approach to elicit explanative writing from undergraduate chemistry students. *Journal of Chemical Education*, *89*(8), 1025–1031.

Washington State University Global Campus. (2017). "Online Professional Science and Technology Writing Certificate." Retrieved June 24, 2017, from https://online.wsu.edu/cert/scienceTechnologyWriting.aspx.

Watson, F. L., & Lom, B. (2008). More than a picture: Helping undergraduates learn to communicate through scientific images. *CBE-Life Sciences Education*, 7(1), 27–35.

Zerbe, M. J. (2007). *Composition and the rhetoric of science: Engaging the dominant discourse.* Carbondale, IL: Southern Illinois University Press.

9 Confronting the Objectivity Paradigm

A Rhetorical Approach to Teaching Science Communication

Maria E. Gigante

The Alan Alda Center for Science Communication at Stony Brook University is one of many new resources for improving the communication of science to non-expert public audiences. Alda's Center is concerned with serving society through the improvement of scientists' communication skills with acting lessons and other non-traditional methods (Alda, n.d.). In addition to innovative and unconventional resources like Alda's, more traditional resources—handbooks on scientific communication—have begun circulating online relatively recently. Two eBooks in particular have been sent to me by colleagues in the sciences who had heard that professors at other universities were using them in their graduate courses: *English Communication for Scientists* (2014), published by *Nature* and written by Jean-Luc Doumont, an engineer with a PhD in physics, and *Communicating Science* (2014), published by *Rogue* and written by Roy Jensen, an author of chemistry textbooks. Although they have differing approaches, both handbooks share the overarching purpose of helping scientists become more "effective" communicators. However, these books are both lacking a rich, rhetorical approach to scientific communication, which makes them insufficient for preparing students to communicate for various audiences, purposes, and situations.

What I am suggesting is not so much that these two handbooks are representative of the kind of communication instruction taking place in scientific communities or graduate programs, but that they have the potential, with their broad circulation as eBooks, to set the standards for scientific communication. This chapter proposes a set of learning outcomes for a course in science communication and advocates for taking a rhetorical approach to reach those objectives—an approach that aligns with scholarship that merges the rhetoric of science with composition theory and communication. Despite the fact that there are different schools of thought about what future scientists need to know in order to communicate effectively, a rhetorical approach is necessary for students to learn to be both effective and

DOI: 10.4324/9781315160191-9

civically-engaged communicators. In the *Journal of General Education*, Jeanne Fahnestock (2013) addresses common misunderstandings of rhetoric and clarifies the role that the rhetorical arts play, not only in higher education but also in civic matters. Securing cooperation, which is achieved through the art of persuasion, is, as Fahnestock puts it, "a necessary component of a successful polity" (p. 14). Thus, the inclusion of rhetorical training in science education can assist future scientists in understanding their role in society and in developing facility in communicating their work to the citizens (with varying levels of interest) who are inevitably affected by scientific research.

Before I delve into the central concepts of the course, I should note that I do not offer assessment criteria or results. Rather, the aim of this chapter is to propose a curriculum that is concerned with improving the communication of science to non-expert publics as well as assisting future scientists with their career objectives. The set of learning outcomes proposed in this chapter is geared toward not only improving the status of scientific research and development but also furthering the aims of a democratic civil society. In order to serve both scientific communities and society, the curriculum for an advanced science communication course, which ought to be transparent to students, incorporates three pairs of opposing or contrary concepts: historical/current discourse, internal/external science communication, and analysis/production of discourse. In what follows, I will briefly characterize the approaches to science communication taken in the two eBooks to distinguish them from the learning outcomes for my course. I will then explain the integration of the three contrary concepts into the curriculum to make a case for the centrality of rhetoric in the training of future scientists.

Course Foundations and Objectives

Universities are responsible for providing their students with the tools they need in order to be successful in their future careers; where the sciences are concerned, the ability to communicate—both within the scientific community and externally, with public audiences—is a necessary skill. In the principal article on internal and external scientific communication, "Accommodating Science" (1986), Fahnestock expresses the necessity of students learning how to address both expert and non-expert audiences *in the same course*: "only in such a course," she argues, "will [students] experience the problems, moral as well as technical, of accommodating information for different genres, audiences, and purposes" (p. 348). Now thirty years later, the type of course proposed by Fahnestock remains obscure (see Chapters 8 and 10, for such a course design).

Also integral to a rhetorically-based science communication course is Susanna Priest's (2013) contention regarding what nonscientist publics need to know about science in "Critical Science Literacy." Priest's argument that nonscientist publics have a right to know "how things work" in the sciences—e.g., how new knowledge is agreed upon, how experiments are done, how scientists communicate with each other—aligns with the notion that transparency in science communication can contribute to a more democratic society (Priest, 2013, pp. 138–139). The concept of "critical science literacy" is in contradistinction to current, but outdated, beliefs in the scientific community that non-expert publics would support scientific research if they only had more factual knowledge of science.

Bombarding nonscientist publics with facts about science is not helpful to the scientific community, nor is it helpful to society. Critical science literacy, according to Priest, would enable citizens to discern between trustworthy and untrustworthy information so as to make informed decisions about scientific issues. In a response to Priest's article, I have argued that science students in addition to nonscientist publics ought to be trained to be "critically literate" in science—meaning that they, too, need to understand how science works (Gigante, 2014).

A rhetorically-grounded science pedagogy and Priest's notion of critical science literacy are mutually supportive: it is rhetorical training that can sensitize students to different audiences and expose them to the moral concerns that arise in various situations. Rhetoric deals with case-specific analysis, and, as Fahnestock (2013) notes, it "always stresses the accommodation of an argument to a particular audience at a particular time and under a particular set of circumstances" (p. 22). The science communication course presented in this chapter asks students to become rhetoricians of science as well as science communicators.

I administered a survey on scientific communication to science graduate students at my university—a large, Midwest public research institution. Specializing in biological sciences, geological sciences, or physics, 25 students responded to the survey. Of those 25 students, the majority (92%) indicated that they had never taken a class that is completely dedicated to writing in the sciences (See Figure 9.1).

In accordance with the lack of discipline-specific writing instruction, the results also point to a lack of rhetorical awareness about the scientific enterprise—that is, how arguments are constructed and how new knowledge is formed: some students (40%) were ambivalent about or outright disagreed with the statement that scientists make persuasive arguments to advance new knowledge (See Figure 9.2).

Because I teach an undergraduate science writing and communication course and am in the process of developing a graduate version of the course, my colleagues in the sciences drew my attention to the two eBooks mentioned in the introduction to this chapter. They thought

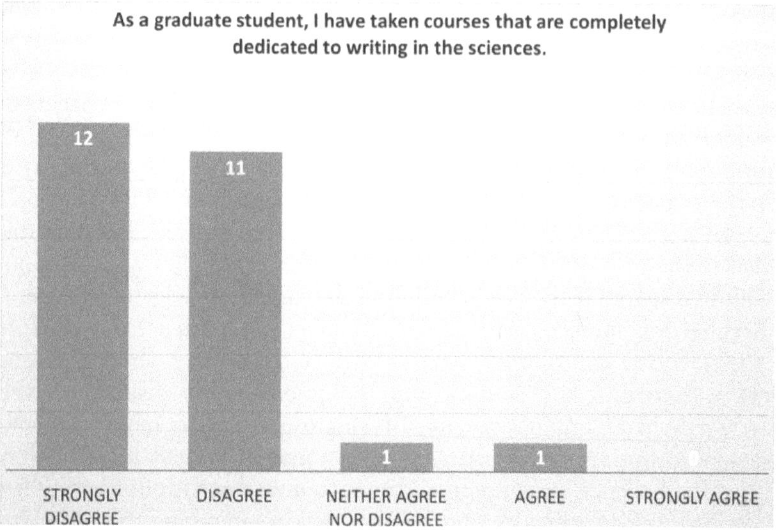

Figure 9.1 The majority of students polled have never taken a graduate science writing course.

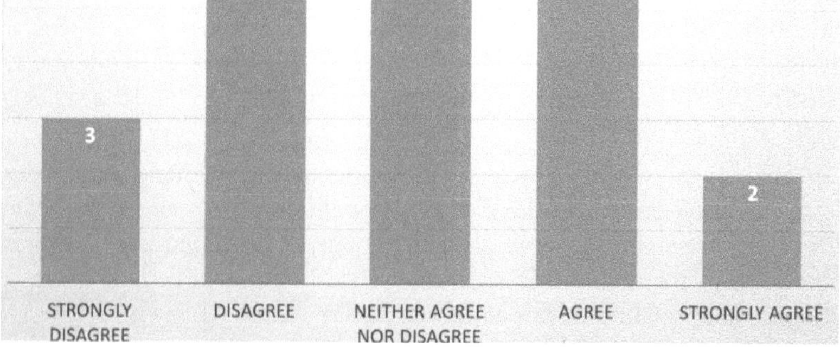

Figure 9.2 Less than half (40%) of students polled believe that scientists make persuasive arguments. Many students polled (32%) believe that scientists do not make persuasive arguments, and many others (28%) were ambivalent.

that I might be able to use them as required reading in my classes. The eBooks seem to be geared toward non-native speakers of English, but they also market themselves toward scientists and students who want to learn to communicate more effectively, especially graduate students. Although the books attempt to make discourse conventions accessible to students, they are less than suitable for teaching scientists how to be effective communicators for a few reasons.

First of all, like Alda's Center for Science Communication, these eBooks claim to have the objective of teaching scientists how to communicate more effectively to public audiences as well as to other scientists; in reality, however, neither eBook treats communication with non-specialist publics with as much depth as they do communication with expert audiences. Second, the handbooks offer a cursory overview of communication—for example, Doumont offers only two possible tools for communicating with non-specialist audiences: analogies and visualizations. Jensen, on the other hand, seems completely unaware of the fields of communication and rhetoric, as he conceives of the "Fundamentals of Communication" (the subject of the first chapter) as style, tone, and grammar (Jensen, 2014, p. 10, 13). Finally, although Doumont seems to be at least somewhat aware of rhetorical principles, like audience and purpose, neither author provides a resource with insights into both analyzing and constructing scientific arguments.

In other words, the training provided by these resources is missing a dimension that is explicitly rhetorical. Resources like Doumont's and Jensen's science communication eBooks are simple, concise, and perhaps easy to implement in pedagogical settings, but their objectives are insular—to serve the scientific community—and their methods are to present cursory overviews of communicative techniques rather than an in-depth awareness of how scientific communication works and whose interests are at stake.

Scientists cannot merely "translate" scientific information for "the public" (Jenson, 2014, p. 204); rather, future scientists should be held accountable for engaging with societal values to further the aims of a democratic civil society. As the title of this chapter indicates, rhetoric is essential to teaching current and future scientists how to communicate. Rhetorical training enables students to both analyze and produce discourse tailored to specific contexts, audiences, and purposes. Moreover, rhetorical training encourages students to consider their responsibilities to communicate science to non-specialist publics in a way that is respectful and mindful of the fact that science is socially situated, as opposed to existing above societal concerns (see, e.g., Perrault, 2013; Priest, 2013; Chapter 3; Chapter 11).

Pedagogically speaking, students most likely need to read and analyze several examples of communication, effective and ineffective, in a particular disciplinary genre before they practice producing discourse in that genre. Rhetoricians and scholars of scientific discourse have written extensively on the ways in which scientific communities negotiate knowledge formation (e.g., Bazerman, 1988; Gross, 1990; Prelli, 1989). Scholars have also explored and explained the intricacies of scientific argumentation, style, and structure (e.g., Fahnestock, 1999; Gross, 1990; Halliday & Martin, 1993), and have studied the issues that arise when science enters the public sphere (e.g., Harris, 1997; Myers, 2003; Paul, Charney, & Kendall, 2001). Another growing body of scholarship merges the rhetoric of science with writing pedagogy. Many of these pedagogically based articles emerge from or seek to complicate a movement in composition studies referred to as "writing to learn" (e.g., Bazerman, 1989; Carter, Ferzli, & Wiebe, 2007; Keys, 1999; Moskovitz & Kellogg, 2005; Zerbe, 2007), which holds that the writing process is a means of engendering critical thinking in a discipline. Taking that idea a step further, Carter et al. (2007) use theories of "situated learning," which posit that "by participating in the ways of doing that define a community, a newcomer learns its ways of knowing" (p. 284). Sharing the results of their interviews with students who were assigned a significant written task in their science courses, Carter et al. argue that writing assists science majors in joining the scientific community. In line with the scholarship mentioned here, the poll I conducted at my university showed that our graduate students want to learn to communicate more effectively (See Figure 9.3).

The idea that the act of writing in a field's professional genres is a means of socializing students into a discipline is significant to science

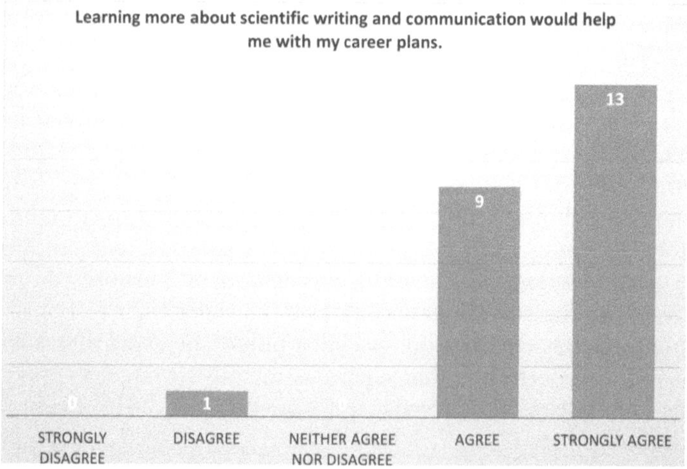

Figure 9.3 Most students agreed that a science writing and communication course would be beneficial to them.

communication efforts. Scholars of writing and rhetoric have noted that professors—not just science professors—have been initiated or socialized into their discourse communities without necessarily being mindful of the ways in which arguments are structured, meaning is negotiated, and knowledge is formed (see Bazerman & Paradis, 1991; Duff, 2010). The initiation process is likely to render discourse conventions all but invisible, then, to young scholars and majors advancing in the field. Indeed, this "taken-for-granted knowledge" (Priest, 2013, p. 144) needs to shift to the forefront of future scientists' communicative practices.

Beyond discipline-specific knowledge, students should learn about the cultural viewpoints ingrained in the way in which science is communicated. In the landmark essay, "Contesting the Objectivist Paradigm," Brasseur (1993) advocates for a curriculum that makes transparent the fact that:

> ...while traditional discourse models in technical and professional writing may contribute to successful communication within an organization, they may also promote enculturation to a kind of communication which diminishes peoples' voices, disinherits them from power and, thereby, limits the capacity to affect change.
>
> (p. 115)

Ultimately, Brasseur argues, students should be given opportunities "both to critique problematic communication models and contribute their own visions to a redefinition of the discourse" (p. 120). Surely, students would not be presented with this opportunity by reading eBooks such as Jensen's or Doumont's mentioned above.

Rather, a more nuanced rhetorical approach will make persuasive communicative choices transparent to novices and grant them access to the discourse community (Carter et al., 2007; Duff, 2010). The first step is to assist students in becoming more aware of their disciplinary conventions. Beyond that, students learn to apply discourse conventions, bearing in mind context, purpose, and audience. The learning objectives that I am proposing for a communication-intensive course are for students to:

1. Demonstrate an ability to rhetorically analyze various genres of scientific discourse.
2. Understand how communicating science impacts and is impacted by political, social, cultural, economic and ethical concerns.
3. Produce original projects that reflect knowledge of the fields of science communication and the rhetoric of science.
4. Effectively accommodate research interests for both expert and non-expert audiences through various modes of communication (e.g., written, visual, digital).

In order to accomplish the proposed objectives, I advance a pedagogy based in contrary concepts—assisting students with communicating science both *internally* and *externally*, developing facility in both *analysis* and *production*, and gaining knowledge of *historical* and *current* scientific discourse. Regarding the first pair of concepts, students learn how to communicate for both expert and non-expert audiences in the same course, and they learn how to communicate in different genres for different purposes, bearing in mind the ethical factors involved in each unique situation (Fahnestock, 1986, p. 348). Second, these objectives stress the necessity of rhetorical analysis to the successful production of discourse. In other words, students must learn how to analyze discourse in whatever genre they are attempting to reproduce prior to beginning the composition process (Carter et al., 2007; Fahnestock, 1986; Keys, 1999). Lastly, and toward that end, students can learn about the evolution of scientific communication to better understand the conventions authorized by the scientific enterprise today. Purposefully structuring the course around these contrary concepts allows students to understand the complex relationships between scientific communities and non-expert publics and can better prepare students to be engaged in civic matters than a course that only focuses on one or two of these components.

A Pedagogy Based in Contrary Concepts

I have argued elsewhere for the inclusion of rhetoric in undergraduate science curricula (Gigante, 2014) and the rationale for taking a rhetorical approach: to encourage students to engage deeply with scientific argument structures and to critically analyze the conventions that are in place. Like undergraduate science majors, graduate science students can benefit from a rhetorically-based communication course that not only makes disciplinary conventions transparent but also emphasizes the fact that science is a socially-situated, human endeavor that is not above persuasion and argumentation. Based on the results of the survey that I administered, some graduate students are under the impression that science is above societal concerns and is not subject to human biases (See Figure 9.4).

A pedagogy based in contrary concepts can help change these perspectives. This pedagogy emerged out of my experiences teaching an undergraduate science communication course, which I created after I talked to my colleagues in the sciences about their students' research and writing abilities. For example, they noted that students lack confidence when searching for and analyzing information, synthesizing their research findings in their own reports, and setting up their own hypotheses. Unlike undergraduate science majors, who are still learning

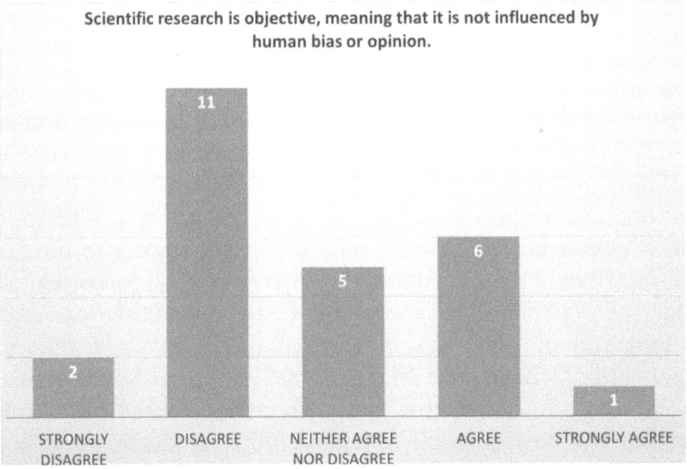

Figure 9.4 Approximately half of the students polled (52%) do not believe in the notion of objectivity, but many students (28%) believe that scientific research is objective, and many others (20%) were ambivalent.

how to navigate academic life, graduate students obviously require an elevated set of expectations. However, the science faculty at my university have indicated that graduate students are as in need of training in communication as undergraduates. A textbook that is appropriate for both advanced undergraduates and graduate students is Penrose and Katz's *Writing in the Sciences* (2010). One of the most helpful aspects of their book is the inclusion of several sample papers, reprinted from reputable scientific journals, which elucidate the major concepts presented in earlier chapters. Penrose and Katz's book is also flexible enough to accommodate my curriculum, as I could use their chapters out of their intended order without disorienting my students. Although I did not construct the course with a pedagogy based in contrary concepts in mind, I realized that that underlying framework was, in part, what made the course successful.

The notion of a pedagogy based in contraries dates back to Quintilian, who was known for his pedagogy of *controversia*, which required students to be able to speak on multiple aspects or "sides" of an issue (see Mendelson, 2001). Whereas Quintilian's pedagogy concerned contraries pertaining to the subject matter of argumentation, here I am suggesting a macro-version of the concept, pertaining to the structure of the course as a whole. Below, the contrary concepts are elaborated, beginning with historical/current binary, which merits consideration because it provides a foundation for the other pairs of concepts: internal/external communications and analysis/production of discourse.

Historical/Current Scientific Discourse

Beginning the course with a brief unit on the history of science is useful for a couple of reasons. First, most students have received only a cursory overview of the history of science and have not been required to read historical scientific papers. According to historians of science Bowler and Morus (2005), students' knowledge about the history of science is generally limited to important figures (e.g., Galileo and Newton) and their astonishing discoveries. Science textbooks tend not to do justice to the complexities of scientific history, especially with respect to the myth of the "ah-hah!" moment of epiphany or discovery: "All too often, it turns out that the conventional stories are vastly oversimplified—they are myths that 'tidy up' the messy process of controversy surrounding any new innovation" (Bowler & Morus, 2005, p. 2). Although it is nearly impossible to do justice to this messy process in a brief historical unit, it is, at the very least, possible to lend depth to students' understanding of the evolution of science in society over time. For example, both producers of and audiences for scientific discourse were much different in the seventeenth through the nineteenth centuries, prior to the birth of mass-market publications and the democratization of education.

The second reason that a historical unit is helpful is that it sheds light on how discursive conventions have changed drastically between the seventeenth century and the present time. One could certainly make the argument that the course should begin the history unit much earlier. In particular, I began in the seventeenth century because that is when the *Philosophical Transactions of the Royal Society of London* came into existence, the first major scientific journal, which gave my students a clear indication of what scientific writing was like at that time. Discussions of historical scientific papers and practices naturally facilitate a comparison/contrast scenario with modern discursive practices, making this unit an effective set-up for later discussions that address such concepts as audiences for scientific communication, production technologies, organizational conventions, and language and style. Students can work their way through seventeenth-century articles from the *Philosophical Transactions of the Royal Society of London* and identify all of the ways in which these articles differ from current scientific papers. An excellent model for this comparison/contrast process appears in Gross, Harmon, and Reidy's *Communicating Science* (2002), which students can read alongside the primary texts.

The explicit contrast between historical and current science communication gives students a vocabulary for talking about discourse conventions and for discussing the changes in audiences for scientific communication that have occurred since the seventeenth century. The contrast allows instructors to complicate the notion of the "gap" between science and society that is so often lamented today, as students engage with subjects such as scientific specialization, the democratization of education, and technological advancements, all of which contributed to the "gap."

Internal/External Science Communication

The second pair of concepts that gives structure to the course pertains to the types of communication science students must learn to do in order to be successful in their future careers—that is, they must learn to communicate with other scientists using internal rhetorics of science (Ceccarelli, 2004) as well as with broader, non-specialist publics using external rhetorics of science (Ceccarelli, 2004). Although this dichotomy could be seen as reductive (for example, it excludes mixed audiences, such as those for grant proposals, which generally require writing for both specialist scientists and non-specialist scientists), it is useful nonetheless for helping students to conceive of distinctive audiences and purposes for communicating science. Regarding "internal" communication, students can be introduced to rhetorical studies, broadly, and to the rhetoric of science more specifically. The notion that scientists make arguments—and, in fact, are particularly good at making arguments—struck my undergraduate students as surprising, as most of them had assumed that science is "objective." A significant number of the graduate science students that I polled believe the notion that science is free from human error and bias (See Figure 9.4). A helpful introductory reading is Randy Allen Harris's introduction to *Landmark Essays in the Rhetoric of Science* (1997), which tactfully debunks the notion of objectivity without maligning the scientific enterprise.

Regarding "external" communication, students will have already been primed to discuss concepts of audience and discourse conventions from the historical unit, so they can begin to differentiate between the types of communication in which they will engage. Students learn about the dramatic changes in purpose and mode of address when there is a change in audience for scientific information. Fahnestock's "Accommodating Science" (1986) can be assigned reading at this point in the course to emphasize the importance of being able to write for a scientific audience as well as for an audience of non-experts. To facilitate the process of writing for different audiences in the same course, students can use their own research topics in their specific fields of study to write a literature review, and then they can transform the information in the literature review into a brief accommodation. Their accommodations could take the form of an article suitable for publication in a newspaper or a popular magazine, a video, a blog, a website, or a podcast.

Projects that cross boundaries between experts and non-experts can help students navigate the ethical issues that arise in various situations as well as navigate different genres; even at the graduate level, students may not be fully aware of the extent to which communication is context-dependent. If students only learned about internal science communication, they would be in a position to view science as operating in a vacuum, as opposed to being socially-situated. Likewise, if students only learned about how to communicate with non-expert publics, they would be at a disadvantage because they would be missing a necessary component: namely, learning how to communicate their research in a way that makes them credible to their peers. Having exposure to different types of science communication in the same course prepares students to be "critically literate" scientists first, who can then learn to communicate with broader, non-specialist publics.

Analysis/Production of Discourse

The last binary that is essential to this course fulfilling its objectives is particularly dependent upon rhetorical training. As rhetoricians and compositionists have argued, students must first practice analyzing examples of texts before attempting to produce texts, if they are to become successful rhetors (Carter et al., 2007; Fahnestock, 1986; Keys, 1999; Chapter 13). Their arguments date back to the ancient rhetorical tradition—for example, the *progymnasmata* were exercises that students had to complete prior to speaking out in real-world situations. Put into the context of an advanced undergraduate or graduate-level course, students must first learn to seek out the motivations for science communication and the accepted methods of persuasion in their discourse community. To do this, students analyze scientific papers, learning their constituent parts. In addition to identifying the arguments of other scientists and the ways in which those arguments are deployed, students can also become more aware of stylistic qualities of scientific writing.

The major objective of analysis is to help students understand the argumentative and stylistic hallmarks of scientific discourse that have a tendency to make scientific writing esoteric to audiences outside of that area of expertise—including other scientists who specialize in different subfields. Simply having access to the knowledge of what makes scientific writing esoteric can be empowering to students. Linguists Halliday and Martin's work on scientific style can assist in exploring such concepts as the use of nominalization to make concepts concrete and give them fact-like status, hedging to strategically fit experiments into the dominant paradigm, and passive construction to remove the human element from the research project. John Swales' (1990) Create a Research Space (CARS) model illuminates the argumentative moves that scientists make in introductory sections of articles to reference previous work, identify gaps in that work, and establish relevance for the new research.

After students gain confidence rhetorically analyzing scientific discourse, they can apply their knowledge to the production of their own projects in scientific genres. Completing a literature review, for example, is well-suited to this curriculum because it can be done without setting up an experiment and collecting data. Rather, students become knowledgeable about a particular topic of interest, collect primary literature on that topic, and write a review of the relevant, current literature that would be useful both for their own future pursuits and for other scientists in their subfield. To demonstrate to students that writing is a process, the assignment can be broken down into its constituent parts such that students are required to submit an annotated bibliography and a rough draft prior to submitting a finished version of the literature review. Separating the project into different stages provides the instructor with an opportunity to meet with students one-on-one to discuss their individual concerns about writing.

With an understanding of the conventions of scientific argumentation, students can then move back into the analytical process to assess the changes that occur when scientific information moves out of the scientific community and into the public sphere. For the project requiring an accommodation of their findings for non-expert publics, students collect examples of accommodations in whatever field they are working in and can take note of the difficulties and pitfalls of the accommodation process so that they are inspired to take on the responsibility of communicating their own research in an engaging, but still accurate, way. Prior to producing accommodations, students should become familiar with the scholarship on science communication and the debates therein. For example, different approaches to science communication have been less successful, such as the "deficit" model, which presumes a one-way flow of information from experts to non-expert publics and fails to consider non-experts' values. There is a complicated and ongoing shift from a model called "Public Understanding of Science" to one called "Public Engagement with Science" (see Davies, 2013). The former presumes, like the deficit model, that the public is knowledge-deficient and would support scientific research with a better understanding of scientific facts, whereas the latter actively engages public opinion and values in conversations about science. Priest's "Critical Science Literacy" represents the most current model, and a similar model is represented in Perrault's *Communicating Popular Science* (2013). In this connection, continuous discussion about the ethics of communicating science to non-expert publics is essential during this part of the course (e.g., Goodwin & Dahlstrom, 2014; Goodwin, Dahlstrom, & Priest, 2013). Students can evaluate several examples of communication, particularly on controversial issues, to learn about how these cases have been handled and to debate how they might have been more effectively addressed (Goodwin, Dahlstrom, Kemis, Wolf, & Hutchison, 2014). After analyzing several examples of accommodations on their topics and becoming familiar with science communication models, students can then create their own accommodations, which, as I mentioned in the previous section, can be produced in genres as disparate as magazine articles, videos, or podcasts.

Conclusions

In this chapter, I have proposed objectives and outlined a curriculum for a science communication course that would be suitable for graduate students and advanced undergraduates. For students to be both effective and civically-engaged communicators, they should rhetorically analyze scientific discourse and accommodations; produce examples of scientific communication for different audiences and purposes; and determine how science communication and scientific communication impact and are impacted by political, social, cultural, economic, and ethical pressures. In order to accomplish these aims, the curriculum can implement the three pairs of contrary concepts elaborated above: historical/current scientific discourse, internal/external scientific communication, and analysis/production of discourse.

There are some limitations to proposing a curriculum of this nature. Rhetoricians of science are in relatively short supply, and graduate programs in rhetoric cannot be found at every university. Because the course requires students to produce several written assignments, it would not be possible to run it as a large lecture or workshop, as the instructor would not be able to give students one-on-one attention (or provide substantive feedback on their projects). Courses like the one I am proposing do exist, but they often go unpublicized; as Ceccarelli (2013) has noted, "barriers to publication of pedagogical reflections are high in the academy" (n.p.). The purpose of this chapter is not to be overly prescriptive about assignments and lectures, but rather to encourage more scholarly discussion about the best possible ways of teaching future scientists how to be effective *and* civically-engaged communicators.

Whether or not consensus is reached among rhetoric and communication scholars regarding how a science communication course should be taught, what is of utmost importance is to acknowledge that eBooks such as Jensen's and Doumont's exist to be used in lieu of a more nuanced and thorough approach. Despite their claims to teach students how to communicate more effectively, these handbooks are not designed to immerse students in writing in the sciences—they would not provide students with a foundation in the ways in which knowledge is produced and circulated. Likewise, such quick-fix handbooks certainly could not do justice to the type of socially- and culturally-aware curriculum for which Brasseur (1993) has advocated. Only a curriculum that is rooted in the rhetorical tradition can provide future scientists with the necessary foundation to become responsible and civically-engaged communicators both within and beyond their professional communities.

References

Alda, A. (n.d.). Our mission. *Center for communicating science at Stony Brook University*. Retrieved March 5, 2016, from www.centerforcommunicatingscience.org/our-mission-2/.

Bazerman, C. (1988). *Shaping written knowledge: The genre and activity of the experimental article in science*. Madison: University of Wisconsin Press.

Bazerman, C. (1989). *The informed writer: Using sources in the disciplines* (3rd ed.). Boston, MA: Houghton Mifflin.

Bazerman, C., & Paradis, J. (Eds.). (1991). *Textual dynamics of the professions: Historical and contemporary studies of writing in professional communities*. Madison: University of Wisconsin Press.

Bowler, P. J., & Morus, I. R. (2005). *Making modern science: A historical survey*. Chicago, IL: University of Chicago Press.

Brasseur, L. E. (1993). Contesting the objectivist paradigm: Gender issues in the technical and professional communication curriculum. *IEEE Transactions on Professional Communication, 36*, 114–123.

Carter, M., Ferzli, M., & Wiebe, E. N. (2007). Writing to learn by learning to write in the disciplines. *Journal of Business and Technical Communication, 21*, 278–302.

Ceccarelli, L. (2004). Rhetoric of science and technology. In C. Mitchem (Ed.), *Encyclopedia of science, technology, and ethics*, vol. 3: L-R (pp. 1625–1629). Detroit, MI: Macmillan Reference.

Ceccarelli, L. (2013). To whom do we speak? The audiences for scholarship on the rhetoric of science and technology. *Poroi, 9*(1). Retrieved May 20, 2015, from http://ir.uiowa.edu/cgi/viewcontent.cgi?article=1151&context=poroi.

Davies, S. R. (2013). Constituting public engagement: Meanings and genealogies of PEST in two U.K. studies. *Science Communication, 35*, 687–707.

Doumont, J. (2014). *English communication for scientists*. Scitable, by Nature Education. Retrieved September 5, 2015, from www.nature.com/scitable/ebooks/english-communication-for-scientists-14053993.

Duff, P. (2010). Language socialization into academic discourse communities. *Annual Review of Applied Linguistics, 30*, 169–192.

Fahnestock, J. (1986). Accommodating science: The rhetorical life of scientific facts. *Written Communication, 3*, 275–296.

Fahnestock, J. (1999). *Rhetorical figures in science*. New York: Oxford University Press.

Fahnestock, J. (2013). The rhetorical arts of cooperation. *Journal of General Education, 62*(1), 11–27.

Gigante, M. E. (2014). Critical science literacy for science majors: Introducing future scientists to the communicative arts. *Bulletin of Science, Technology, & Society, 34*, 77–86.

Goodwin, J., & Dahlstrom, M. (2014). Communication strategies for earning trust in climate change debates. *WIREs: Climate Change, 5*, 151–160.

Goodwin, J., Dahlstrom, M. F., & Priest, S. (Eds.). (2013). *Ethical issues in science communication: A theory-based approach* (Proceedings of a symposium at Iowa State University).Charleston, SC: CreateSpace.

Goodwin, J., Dahlstrom, M., Kemis, M., Wolf, C., & Hutchison, C. (2014). Rhetorical resources for teaching responsible communication of science. *Poroi, 10*. Retrieved May 20, 2015, from http://ir.uiowa.edu/poroi/vol10/iss1/7/.

Gross, A. G. (1990). *The rhetoric of science.* Cambridge, MA: Harvard University Press.

Gross, A. G., Harmon, J. E., & Reidy, M. (2002). *Communicating science: The scientific article from the 17th century to the present.* Oxford: Oxford University Press.

Halliday, M. A. K., & Martin, J. R. (1993). *Writing science: Literary and discursive power.* Pittsburgh, PA: University of Pittsburgh Press.

Harris, R. A. (Ed.). (1997). *Landmark essays on the rhetoric of science.* Mahwah, NJ: Lawrence Erlbaum.

Jensen, R. (2014). *Communicating science.* Rogue Publishing. Retrieved September 5, 2015, from www.roguepublishing.ca/content/Communicating_Science.

Keys, C. W. (1999). Revitalizing instruction in scientific genres: Connecting knowledge production with writing to learn in science. *Science Education, 83*, 115–130.

Kuhn, T. (1996). *The structure of scientific revolutions.* Chicago, IL: Chicago University Press.

Mendelson, M. (2001). Quintilian and the pedagogy of argument. *Argumentation, 15*, 277–293.

Moskovitz, C., & Kellogg, D. (2005). Primary science communication in the first-year writing course. *College Composition and Communication, 57*, 307–334.

Myers, G. (2003). Discourse studies of scientific popularization: Questioning the boundaries. *Discourse Studies, 5*, 265–272.

Paul, D., Charney, D., & Kendall, A. (2001). Moving beyond the moment: Reception studies in the rhetoric of science. *Journal of Business and Technical Communication, 15*, 372–399.

Penrose, A. M., & Katz, S. B. (2010). *Writing in the sciences: Exploring conventions of scientific discourse* (3rd ed.). New York: Longman.

Perrault, S. T. (2013). *Communicating popular science: From deficit to democracy.* New York: Palgrave-MacMillan.

Prelli, L. (1989). *A rhetoric of science: Inventing scientific discourse.* Columbia: University of South Carolina Press.

Priest, S. H. (2009). Reinterpreting the audiences for media messages about science. In R. Holliman, E. Whitelegg, E. Scanlon, S. Smidt, & J. Thomas (Eds.), *Investigating science communication in the information age: Implications for public engagement and popular media* (pp. 223–236). Oxford, England: Oxford University Press.

Priest, S. H. (2013). Critical science literacy: What citizens and journalists need to know to make sense of science. *Bulletin of Science, Technology & Society, 33*, 138–145.

Swales, J. M. (1990). *Genre analysis: English in academic and research settings.* Cambridge: Cambridge University Press.

Zerbe, M. J. (2007). *Composition and the rhetoric of science: Engaging the dominant discourse.* Carbondale: Southern Illinois University Press.

10 Dissolving the Divide between Expert and Public

Improving the Science Communication Service Course

Kate Maddalena and Colleen A. Reilly

Although contemporary scientists are becoming better public communicators, professional academic genres like the research article remain difficult for non-expert audiences to grasp and use for important work in everyday living and policy-making. Good science writing takes those non-expert audiences into account, even in the context of a journal article or an academic conference. Indeed, scholarly articles that writing experts and scientists hold up as exemplary writing often come from top-tier, generalist journals such as *Science* and *Nature*, whose audiences are broad and disciplinarily diverse (Schimel, 2012). That writing for wider accessibility is commonly seen as better writing is no surprise—such writing's arguments and stories are more explicitly tied to bigger-picture impacts. We propose the advanced science writing classroom as one ideal context where we can address the problem of the public/expert audience divide by asking students to see professional science writing in the same ecology as the public communication of science. We begin our chapter by briefly reviewing the current literature related to defining the boundaries of science communication and training scientists to write in professional contexts. Next, we describe our recent experiences in designing several upper-level undergraduate and graduate courses specifically for emerging professional scientists and discuss detailed examples of assignments. Our example assignments include drafting research questions, writing literature reviews, and designing conference posters—three genres we see as prime places for rhetorically-minded interventions into the conventions of professional science communication. We include student-produced examples of and excerpts from each of these assignments in order to demonstrate how our approach informed the writing process and shaped submitted artifacts. In the final section, we reflect on our assignments and course constructs as we imagine a robust pedagogy of advanced professional science communication for the future, a pedagogy structured to produce rhetorically aware scientists with broad, not binary, conceptions of their professional contexts.

DOI: 10.4324/9781315160191-10

The Expert/Public Divide as an Impediment in Science Communication

A recently published NSF-funded report calls for enhanced training for advanced science students so that they are prepared to communicate persuasively and in narrative form to a range of constituencies, including policy makers, business people, and the public, in order to gain support and resources (Neely, Goldman, Smith, Baron, & Sunu, 2014). Such reports and similar scholarship readily acknowledge the rhetorical aspects of science communication when focused toward non-peer audiences. For example, Downs (2014) provides a detailed analysis of how to use narrative strategies to help public audiences grasp relevant scientific information; her study focuses on the use of narrative to "reduce adolescents' risky sexual behavior" (p. 13632). Other scholars note the scarcity of positions in some areas of science and encourage graduate students to learn to communicate with the public to obtain positions outside of academia (Guannel, Bruno, Grand, Lee, & Day-Miller, 2014). Kuehne et al. (2014) detail the societal benefits of providing science graduate students with practice in writing for a range of external audiences (see also Chapter 11 that focuses on civic science).

Despite a proliferation of new media and potential new audiences, the structure of scholarship by scientists for other scientists in terms of genre conventions and methods of argumentation has not moved far beyond Swales (1990). As other chapters included in this volume have noted (see especially Chapter 8), scientists continue to struggle to respond to complex rhetorical situations. The IMRAD model that tends to dominate scientific publications reinforces the idea that the scientists' data serve to carry the persuasive load, downplaying the importance of analyzing audiences' needs and consciously crafting arguments for communication with peers. Intriguingly, Sollaci and Pereira (2004) highlight the IMRAD structure's facilitation of "modular reading" (p. 366), allowing readers to seek particular information in designated sections and avoid processing the argument that authors present. IMRAD and related writing conventions, if they are taught as one-size-fits-all prescriptive containers for science's products, implicitly uphold positivistic views of the scientific endeavor and elide the messy complications of science as a social construct.

Difficulties in reconsidering the approach to genres used in peer communications stem partly from the view that the term "science communication" connotes communicative work that Fahnestock (1986) famously coined "accommodation," i.e. the re-interpretation of material by experts for non-expert audiences or publics. The rhetorical concerns for external audiences are not foregrounded in peer communications. In fact, in programs designed to teach writing for science peers, students learn that their scientific ideas outrank their writing abilities. As Ding

(2008) explains to the graduate students when instructing them about applying for NIH grants, the evaluators do not score the quality of the writing in assessing the applications (current guidelines omit this criterion, see NIH, 2016). She argues, "writing seems to function more as a tool to convey ideas and to inform and persuade the audience than as a component to be evaluated and scored" (Ding, 2008, p. 35).

Internalizing peer-to-peer science communication as a specialized discourse separate from and inherently inaccessible to public audiences perpetuates difficulties in scientists' communications with external audiences. Recent considerations of science's relationship(s) with news media, especially, have uncovered problems like weight-of-evidence misrepresentation, or the tendency of media to represent two sides of a politicized issue (like climate change) as equal when in fact one side has much more scientific support (Kortencamp & Basten, 2015). Manufactured controversy is a related practice that creates an argument where there is none to produce rhetorical exigence (Ceccarelli, 2011). These problems persist because they are reproduced by the communicative habits and assumptions of expertise. The very meaning of technical terms often becomes a point of controversy; "stem cell" (Leydesdorff & Hellsten, 2005), "climate change" (Nisbet, 2009), and "tipping point" (Russill & Nyssa, 2009) are examples of terms that have been used for different rhetorical purposes and with alternate semantic outcomes in the ostensibly separate spheres of research science, applied science, popular press, and policy-making. Terms taken up by the popular imaginary become meaningful as analogies, applied to entirely different fields, and/or associated with partisan politics and are thus divested of clear technical meaning. Scientists often (understandably) describe these divisive linguistic-discursive phenomena as frustrating. They feel unable to educate the public and/or deem them as being too science-illiterate to understand (Besley & Tanner, 2011; Horst, 2013). Such attitudes contribute to the deficit model of science communication that scholars have problematized for the past decade (Besley & Tanner, 2011; Miller, 2001; Sturgis & Allum, 2004).

One potential solution to these communication problems is to avoid reifying the divide between science and publics in the first place, which necessitates an intentional inclusion of peer-to-peer communication under the disciplinary umbrella of science communication. Some of the most compelling arguments for such a turn come from the scientific disciplines. In a 2001 essay in *Conservation Biology*, Robertson and Hull (2001) exhort researchers to incorporate the principles of public ecology into their research and writing from the outset:

> [A]ll the people who produce, review, and apply conservation research should evaluate the success of their knowledge according to its ability to influence conservation decisions. In addition to possessing

> conventional "scientific" attributes such as validity, generalizability, and precision, conservation knowledge must also possess qualities that make it effective in the political arena of decision making.
> (p. 970)

A more recent study in the field of science communication entreats scientists to be aware of how their speech and social interaction affect the quality of the communication and the attitudes of their listeners:

> Research may reveal how dialogue should be conducted in order to get these questions on the table, and how the cultural context can be conducive for having such dialogue. *Experts-in-training who have not yet completely internalized the discursive patterns and professional identities that help reproduce the hegemony of technical-scientific expertise may be of crucial importance here.*
> (Mogendorff, te Molder, van Woerkum, & Gremmen 2016, p. 47, emphasis ours)

In other words, scientists' approaches to peer communications affect how they engage publics, and instigating changes in communication practices may be best done among burgeoning scientists. One obvious place to intervene in the cycle of internal/external viewpoint (re)production is in the professional and/or academic science writing classroom. On our own campus and at other universities across the country, teachers of writing are working with teachers of science to imagine interdisciplinary pedagogies (see especially Saitta, Zemliansky, & Turner, 2015) in the form of writing across the curriculum (WAC) initiatives and service courses.

Scholars in writing studies, including rhetoric, composition, technical communication, and professional writing, have been studying and providing instruction in science writing for almost a century. Most notably, Penrose and Katz's (1998) textbook, *Writing in the Sciences*, now in its third edition, is designed for science writing in academic contexts, focuses on research genres, and applies rhetorical perspectives to important but challenging genres like the grant proposal. Penrose and Katz's work was groundbreaking because explicit instruction in this kind of writing in upper-level science curricula was rare in the late 1990s. Since then, service courses like those we describe here have become more common and have served as models for our efforts.

Institutional Contexts: New Courses, New Audiences

Our department has historically offered one science writing course at the undergraduate level. The course was designed for and taken by English majors in the Professional Writing track and intentionally covered a broad range of science-related literacy practices. In response to current trends in the field and demand on our campus (specifically, requests from professors of environmental sciences and chemistry), we have expanded our science writing offerings to include upper-level undergraduate and graduate courses that focus on writing in professional science fields: an undergraduate writing course required of all environmental science majors, an online writing course for graduate students and professional scientists, and a graduate-level writing course in the environmental sciences that serves as a research writing workshop.

Although designing the courses was exciting, we knew that we could not approach courses like these—service courses by definition—without very careful, critical attention to learning objectives. By designing courses for specific majors, ostensibly to prepare them for certain fields, we could easily fall into the trap of prescribing disciplinary writing and thus help to re-create the very silos that we problematize at the outset of this chapter. As the long and controversial history of the field of composition shows, service courses present a double danger, potentially undermining our own disciplinary expertise as scholars of language and robbing our students of agency and voice within the academy (see, for example, Crowley, 1995).

We responded to these important pedagogical and ethical concerns in multiple ways. First, at the logistical level, *we refused to limit seats in our service courses to only the science majors and professional scientists they were designed to serve.* This choice addressed several departmental and pedagogical needs; we wanted to continue to serve our own majors, draw as many student instructional hours as possible, and encourage a diverse range of perspectives in every class. We knew that limiting the class to the service of one scientific field would automatically delimit the territory of expertise to one discipline, which is exactly what we want to argue that professional science writing pedagogy should strive to avoid. There were immediate and interesting results from keeping the courses relatively open: biology, business, coastal policy, communications, and physics majors enrolled in the new courses without prompting.

Second, at the level of course materials, *we designed our assignments as learning experiences to* ***promote rhetorical awareness*** *rather than experiences in discipline-specific training* (see also Chapter 9). Although we encouraged students to use the courses and assignments to enhance their professional development, the true content of each course (and our expertise) are rhetoric and writing. As such, each class's disciplinary profile is different, emergent, and exploratory. Students' interests guide and produce class content. For example, the course materials for our upper-level undergraduate course include a science writing textbook that serves as a rhetoric and a bibliography of examples compiled by students. The bibliography's contents are compared, categorized (as secondary and primary research, for instance), and returned to as context-specific exemplars throughout the semester.

Third, *we used discipline-specific examples as samples to* ***critique and re-imagine****, rather than to uncritically emulate.* Student-found exemplars, including the articles in the class bibliography above, are not taken unquestioned as standard, but seen as single instantiations of genres, produced by the social contexts to which they respond (see Miller, 1984). Although critique and revision of published science may at first sound slightly antagonistic to the goals of our colleagues in the scientific disciplines who asked us to contribute to their curricula, in fact they are in keeping with their expressed desire for science majors to become better writers. When we met to imagine the courses, the science professors expressed general dissatisfaction with students' written work, but also with the professional written products of peers in their field—especially articles and conference posters. Our science-minded colleagues' characterization of communication problems in their own discourse communities made our desire for rhetorical and critical approaches a refreshingly easy sell.

Despite being asked by our colleagues to aid their students to improve their science communication abilities, we are aware that we approach this endeavor as non-scientists whose ethos can be called into question, particularly when we are advocating critical departures from the genres central to the students' and their instructors' disciplines. As Harwood and Hadley (2004) note, advising students to question and deviate from the norms of their respective disciplines puts them at risk; they can be subject to critique and earn lower grades from their disciplinary faculty. As a result, we decided that departures should be presented in that context; students should be aware that all rhetorical choices come with certain risks. For example, using the passive voice, common in some science disciplines, may help the writer conform to norms but may result in their ideas being expressed less clearly and succinctly. Students should be empowered to understand that there are rhetorical choices to be made and, likewise, consequences for all choices.

To further situate our rhetorical expertise, particularly in the online graduate course, we located instructional materials by scientists that advocated rethinking scientific communication in ways that aligned with our values. Instructional texts by Meredith (2010) and Schimel (2012), both scientists, bring such insider perspectives. For example, in a chapter about communicating research results to peers, Meredith (2010) encourages the use of "thrifty" versus "expensive" words to convey complex ideas more clearly, directly, and efficiently (p. 72). He also stresses the importance of lively prose and active voice for all genres of writing. Similarly, Schimel's (2012) entire approach to peer-directed writing centers on crafting compelling narratives as vehicles for memorable communication. Drawing upon these arguments from science experts helps to situate the rhetorically-oriented arguments that we make, especially when they conflict with guidelines that students have previously encountered.

Interventions: Pilot Assignments

The three example assignments that we describe here focus on traditionally "internal" scientific genres, but we re-vision them to highlight nuanced audience awareness and rhetorical engagement. Two main metaphors for written discourse in context—the narrative and the conversation—guide our treatment of these genres and encourage students to focus on rhetorical goals. These metaphors are well tested in the context of academic writing instruction and writing in the disciplines. Narrative is a common prescription for public-facing science communication and is exhorted as a tool for making technical expertise more palatable for non-experts (Dahlstrom & Ho, 2012; Downs, 2014). Encouraging a focus on narrative in peer-to-peer writing could be a step towards more public-friendly professional science communication. The narrative metaphor—the idea of "letting the data tell a story" or "finding the story in the data"—is already salient in the professional science communities our writing courses are designed to serve, especially the environmental and biological sciences. In fact, when students in our undergraduate pilot course interviewed researchers in their respective fields about their approaches to the writing process, more than one researcher described writing as storytelling. Our students' findings fit well with the chosen text for the 300-level undergraduate course, Schimel's (2012) *Writing science: How to write papers that get cited and proposals that get funded,* which highlights storytelling as an academic author's primary mode. Schimel's attention to connecting research to audiences and stakeholders beyond one lab, question, and/or discipline (a feature he calls a story's "stickiness") is also in keeping with our goal of increased rhetorical awareness.

In addition to the Schimel text, we added additional reading material from the fields of education and rhetoric and composition. Specifically, we introduced students to Montuori's (2005) consideration of the literature review as a creative means of engaging in dialogue and Swales's (1990) "Creating a Research Space (CARS)" model from his influential genre analysis of research articles. Both of these texts forward a conversational model for disciplinary discourse that finds its most famous roots in Burke's parlor. More recently, Graff, Birkenstein, and Durst (2006) have articulated the conversation metaphor very effectively for the composition classroom in their rhetoric, *They Say/I Say.* It is fitting that we add a conversational model (from rhetoric) to the narrative one (from science) since we are asking students to expand the parlor, to imagine more and various interlocutors. Such a task requires us to go "back to basics" with the elements of convincing argument.

The three example assignments that we discuss below come from a set of semester-long projects that require students to choose a topic, develop a research question, and produce an article-length paper that includes a brief literature review as well as a conference poster that highlights key findings. We received authorial consent to include the examples in the discussion below in order to represent students' voices and illustrate the efficacy of our approach. A student in Maddalena's ENG 315: Writing in the Academic Sciences course from Fall 2015 agreed that comments from her informal course evaluation be included without her name. Students in Reilly's ENG 551: Professional Science Writing course from Spring 2016 provided written permission to quote from their literature reviews. A group of students in Maddalena's ENG 315: Writing in the Academic Sciences course from Fall 2015 provided written permission to include their poster. At their request, we have removed students' names to preserve anonymity. Finally, Kelsey Potlock, an honors student who graduated in Spring 2016, provided written permission to include a draft of her honors project poster that was critiqued by Maddalena's ENG 315: Writing in the Academic Sciences course from Spring 2016. She indicated that she wished to keep her name on the draft.

The Research Question: Method Design as Writing Task

Research questions can ultimately determine the precision, pertinence, and knowledge-making power of an empirical project. In science texts, research question development is rarely framed explicitly as a writing task (this differs greatly from approaches in rhetoric and composition; see for example Miller-Cochran & Rodrigo, 2013). Moreover, in science curricula, research questions are folded into considerations of method. Given that students in the sciences learn methodology and corresponding methods in labs, courses, and in the field through what Ding (2008) terms as a process of social apprenticeship (p. 8), direct instruction in how to draft, revise, and incorporate research questions into their writing process is not guaranteed in a scientist's education. From the professional science writing teacher's perspective, however, research question writing is embedded in the larger task of writing and revising method; method design is exceptionally well-supported in a rhetoric and composition-style pedagogy, as it focuses on process and allows for iterative revision. Though Schimel's (2012) science writing text does not address research questions explicitly, it proposes a heuristic for research narratives to be S.imple, U.nexpected, C.oncrete, C.redible, E.motional, S.tories (SUCCES) (p. 17). For the purpose of teaching the research question as writing, we take the first three items—simplicity, unexpectedness, and concreteness—as criteria. Specifically, we maintain that "simple, unexpected, concrete" research questions make it possible for "credible, emotional [scientific] stories" to be written.

The first two of these criteria are often surprising to students and require unpacking. Science is technical—how can it be simple? Science is circumscribed by clear, strict, knowledge-oriented, conventional practices—how can it be unexpected? The intentionally rhetorical perspective of a writing class provides an excellent opportunity to consider these questions, as they are issues of context and audience. Simplicity functions in two ways: (1) to make the epistemological goals of the project clear to *both expert and non-expert audiences*, and (2) to make the implications of the project's findings navigable and applicable, *often in various contexts.* Unexpectedness is similarly complex, as it operates in both epistemic and rhetorical paradigms: (1) epistemically, it sets up the study to make "new" knowledge, giving the research disciplinary traction, and (2) rhetorically, it maximizes the effect of kairos and makes for an engaging, exciting read. The third criterion, concreteness, is the research question's link to the more material and pragmatic aspects of method, and we highlight that link by focusing on another writing task: operationalization by way of term definition. Research questions need to be clearly addressed by the method; specifically, each conceptual term must have a measureable, perceptible corollary. This instrumental connection is literally *made by writing.* Defining terms is of supreme importance to a scientist's knowledge-making work.

To show how the research question writing process works in our professional science writing classroom, we will include one set of research question draft plus research question revisions here and describe the epistemic and rhetorical reasons for the revisions in terms of our course's approach. The research questions initiated for this project focused on subjects about which the students, regardless of background, could conduct primary research; as a result they are not grounded in experimental science. Learning the process of research question development was paramount here.

Draft 1 of student research questions:

1 Do students use cell phones in socially acceptable ways?
2 Is cell phone use good or bad for students?

The group revisited this first iteration of their research questions after (1) receiving peer response and (2) amassing a preliminary list of sources for their literature review. Both of these key steps gave them new insights and helped to produce a more focused and interesting set of questions when they revised. From peer review, they learned that the term "socially acceptable" was much too complex (i.e. not *S.imple*) and would require too much work to operationalize via a method, especially since the question's context was much too vague (i.e. not *C.oncrete*). Likewise, the second question lacked any concrete context, and the terms "good" and "bad" would likely need too much work to make simple and concrete, as well.

But the group's first look at current literature about college student cell phone use, which they gathered in preparation for the literature review assignment, gave them a layer of knowledge that enabled them to hone their questions in even more sophisticated ways. First of all, they realized that their original questions were not *U.nexpected* at all. In fact, the broad topics of the social perceptions of cell use and the technology's benefits versus harms were well-trodden territory. Their questions, then, did not set up any potential for (epistemically) new or (rhetorically) interesting knowledge making. If the group wanted to keep their topic, they would need to adjust the focus of their questions to produce something that would be new to the social science and exciting and compelling for all audiences. As they discussed their sources for the literature review, the group noticed that, although student behaviors and attitudes were often the subject of surveys' inquiry, few if any seemed to get at what students deemed "acceptable" in their own micro-level cultures. The fact that the study was restricted to their own campus, they then realized, could lend itself to an *U.nexpected* consideration of a given campus's student culture being only one instantiation of what is socially acceptable. A concession to limitation, then, could invite further research and conversation—other studies on other campuses. The revised research questions (included below) manifest this new orientation by adding the concrete context of our campus as well as the sophisticated twist of asking about students' attitudes rather than their behaviors in comparison to external norms. The group also added a third question to address a trend they saw in their sources. The bolded terms are terms that the group knew they would need to operationalize as they drafted their method.

Students' revised research questions:

1. When is it **acceptable** to students on UNCW's campus to **use** a cell phone?
2. Do UNCW students find cell phone use **disruptive** or **beneficial**?
3. Do UNCW students **prefer** to text or call people, and why?

The research questions' iterative, constitutive relationship with the development of literature reviews seems relatively obvious. But the depth of students' simple, preliminary consideration of current literature and recognition of how their studies would contribute to a conversation was frankly revelatory to us in terms of claiming a "value added" from our course. During an informal evaluation of the first student group, one environmental science major mentioned that our attention to drafting research questions had directly benefitted her in the next semester's methods course:

> Being able to draft research questions and see exactly how they were used to inform the methods section of a paper allowed me to have better insight into every future scholarly article I read for research. Learning how to draft the actual questions in [this] course ... gave

> me an advantage in the methods course because that knowledge [of research question's relationship to method] was there already.
> (Personal communication, 22 Jun. 2016)

Responses like these are rewarding and promising.

The Literature Review: Practicing Creative Inquiry and Dialogue

Typically, literature reviews are perceived and taught as surveys of published scholarship that represent the significant ideas from the research under consideration accurately, systematically, and thoroughly. As exemplified by the approach discussed in Cooper (1982), the methods for selecting the scholarship to include and the representation of the ideas from that scholarship should hold up under scrutiny as objective and systematic. The writer searches for texts that will be perceived as most relevant in the eyes of the reader, generally the instructor, and represents or, as Montuori (2005) notes, reproduces the ideas from the surveyed scholarship faithfully and succinctly. Strikingly, even recent scholarship centers on helping students to systematically survey and coherently represent the extant scholarship (Luederitz et al., 2016); little is said about empowering students to insert their voices into their analyses. For example, although Luederitz et al. (2016) present their approach to literature reviews as empowering to student researchers, they continue to privilege systematic process over innovation and critique: "ensuring that the literature is being reviewed in a consistent, coherent, and reproducible manner" (p. 234). While some instructors ask students to make comparisons between the scholars' perspectives or include a summary paragraph that supports or critiques some of the ideas that they encountered, there is little room in this reproductive model for students to have agency, own how their perspectives shape their choices of scholarship, dialogue with the ideas they encounter, and engage in self-discovery.

Frustrated with the formulaic texts that we received based on the reproductive model of the literature review, we sought an alternate approach to the genre that empowered students to personally connect with the scholarship in their fields, engage in self-discovery through interrogating the work of others, and construct texts that are lively, interesting, and pleasurable to read by audiences of peers. Montuori's (2005) approach, emphasizing creative inquiry and dialogue, provides this generative alternative; as he explains, revising the genre to require innovative response instead of mimesis aids students to see the scholars whose work they survey as part of a "living community with a history, motivations, passions, conflicts, alliances, errors, dead ends, and creative outbursts" (p. 375). Students can thus identify with the scholars and see themselves as potentially contributing to and even shaping the discussion: "in the context of creative inquiry, we are actively participating in the community; we are in the discourse and engaging in inquiry in that context" (Montuori, 2005, p. 377).

In the guidelines for the assignment, we emphasized creativity, dialogue, and synthesis. We read and discussed Montuori's article, and students developed research questions related to disciplinary conversations. We also read sample literature reviews by students from previous classes and examined them for evidence of creativity and synthesis through posting responses to the online course discussion board dedicated to the topic. Finally, we encouraged students to insert themselves into their reviews through their language choices by using personal pronouns and active voice, if possible. As noted previously, prompting students to depart from the perceived dominant discourse of their fields by employing personal pronouns in their academic writing is a powerful intervention; even if in the end the students return to traditional norms, they do so knowing that they are making a conscious choice and that alternatives exist (Harwood & Hadley, 2004, p. 372).

Student responses to this revised assignment have been extremely positive; formulaic reviews are now a minority of the submissions that we receive. Since initiating this approach, the majority of the students' literature reviews demonstrate that they are truly grappling with the ideas in their sources, developing their own taxonomies to categorize and analyze their sources' ideas and arguments, and participating in the key discussions in their disciplines. For example, in a review that discusses the uses for knockout mutants, Lauren Scheetz explains, "While conducting my survey, I attempted to search for organisms occupying low and high order positions," which clearly asserts her agency in deciding that her analysis needs to survey a particular range of organisms used in these processes. Similarly, Hayley Grabner, who focused on how to cultivate marine citizenship in populations, locates what is for her the center of any debate surrounding this issue:

> The debate, also discussed later, occurs when scholars deliberate the prevailing mechanisms and barriers to inciting the desired individual action, as well as the types of future policies they suggest. I suspect that these question[s] persist because cultivating a sense of marine citizenship is—now, more than ever—of particular interest to environmentalists, community leaders, and graduate students like myself.

In both examples, the students direct the survey of scholarship to suit their needs; they are not being dragged along by the weight of the approaches that have preceded them. Molly Gabler not only surveys a range of sources relevant to her research, she also categorizes them by the methodologies that they employ: "Currently, there are many methods used to address the area of N_2-related DCS [decompression sickness] in the mandibular fats of odontocetes. These methods include biochemical, anatomical, behavioral and modeling techniques." Creating taxonomies and organizing the discussions of sources around them leads to increased synthesis in these literature reviews as students then discuss related sources in tandem and avoid summarizing each text consecutively.

Additionally, students make novel connections between sources, employ metaphor and other figurative language, and make astute critical comments about their sources. Grabner, for instance, remarks that "unique and important concepts percolated to the surface." Scheetz makes this imagistic comparison in a discussion of knockout mutants: "'intrinsic resistome,' a map of natural resistance to treatments." Gabler critiques an assumption that she repeatedly encounters in the literature that underpins the models used to support the conclusions of other scholars:

> All of these models consider blood, brain and muscle to be fast tissues and fat to be slow. I fully support using models to predict certain animals that may be at risk for DCS; however, fat is considered a 'slow' tissue based on the low solubility values derived from experiments using plant lipid (e.g., olive oil).

Juli Hood also provides an insightful critique of the data provided by her sources: "Some of the studies reviewed gave vague information regarding sample collection (4), exact ages of patients (4), and unclear results and conclusions (5)...". Surprisingly, some students analyzed both the content and form of their sources; for instance, Grabner explains that

> [a]lthough not all articles were reporting on an original study, those that did typically used a familiar IMRaD format: starting with an introduction and following with methods, results and a discussion. In doing so, each study drawn upon in this review proposed a hypothesis and then used survey research and subsequent quantitative analysis to evaluate their findings.

Furthermore, students were not afraid to include some sources asserting unique or even radical positions because our assignment guidelines encouraged them to look for and evaluate alternative approaches. For instance, Gabler emphasizes the importance of recent developments in her field:

> However, there is recent data suggesting that nitrogen solubility in the mandibular fats of these animals is even higher than the blubber (Lonati et al., 2015). Integrating these values into the models should be the next step in assessing DCS risk in these animals.

She confidently aligns herself with this sort of innovative approach.

Finally, because the assignment instructions highlight the importance of the students' voices in their reviews, more students used personal pronouns and active voice, as shown in the quotations above, making their reviews more lively, and, as a result, more enjoyable to read. For example, Hood asserts:

> Several studies reviewed used semi-automated instrumentation (1, 3, 7) while others used visual interpretation (2, 6, 7). Based on my work with diagnostic test strips, slight variations in visually read results can occur between operators and the interpretation of color is subjective.

The students' engagement with the ideas in their sources, as exemplified in the excerpts included here, shaped their research papers, the next assignment in our graduate courses. Because students were encouraged to consider what the scholarship of others has revealed to them about their positions, they transitioned more easily to writing their research papers and used portions of their literature reviews to propel their writing forward.

The Conference Poster: Solving an Audience Problem

Posters were included as a top priority on a list of genres that advanced undergraduate and graduate students should be prepared to produce in a science writing course. According to our teaching collaborators in the environmental sciences, the scientific community recognizes that the poster as a genre has a communicative problem: too often, they just don't *work*. Presenters patch together an abstract, a few figures, and implications, and any sense of argument or story is lost in a perceived need to cover all of the science. Words like "awful," "ugly," "embarrassing," and "unreadable" were used to describe professionally-produced conference posters in our brainstorming meetings. We suspected that the problem might, again, be directly linked to issues of context, particularly audience. Though research projects themselves are often narrow and specialized, poster sessions are huge, festivalesque contexts; scientists stand in a large hall full of peers and strangers. The audience is suddenly broader and less well known. The professional conference is a context that necessarily requires communicators to reject the false binaries of internal/external and expert/public. Making a poster, for many, means reaching beyond the scientific skill set into the intimidating fields of document design, graphic design, and public speaking.

Our pilot assignment, therefore, required that students first collect examples of conference posters and critique them before making their own, explicitly asking them to improve the genre through their own contributions to it. In addition to the rhetorical perspectives we had been developing throughout the semester, we added a brief introduction to the most basic of document design principles to give them a starting point for critiquing poster layout. The collected examples and ad-hoc genre analysis produced two types of information for students to consider as they began their own designs: (1) a set of professional *conventions*, or expectations, that students knew they had to take seriously before deciding to ignore or reject and (2) a set of *critiques* of both individual poster design and those same community conventions to serve as a rationale for how they would tackle the poster problem and propose to improve the genre from the inside. Their attitudes towards community conventions was informed, of course, by their experience with the literature review and our own explicit problematization of the conventions of professional science communication.

The students' solutions to the poster problem varied. Some designs diverged more radically from convention than others. There was a general agreement among students and across classes, however, that neither narrative nor conversation models for good science writing were served by the typical conference poster and that any nuanced sense of audience was altogether lacking. The typical poster was much too busy, incorporated too much text, and basically tried to do the same work as an article, but in a different (and inappropriate) context. Some basic conventions that students discovered in their samples included an echo of the trifold science fair poster in the form of three or four columns, maintenance of the IMRAD structure on the poster in the form of headings, prominent graphic representations of funding sources (i.e. NSF) and institutional affiliations, and data presented in the form of figures almost exclusively. Most students agreed that poster designers needed to choose a less complicated communicative goal for their posters than for their articles. They chose to focus on research context (in terms of public stakeholders), findings, and implications for that same public context. This observation was interesting to us, of course, since we had been trying to get students to include non-expert readers as part of their audience from the outset.

The two student examples we include below chose to preserve and/or challenge conference poster conventions in different ways. The group who designed Figure 10.1 took a conservative approach, upholding most genre conventions (but we should note that their document design is superior to many examples). The conventions they chose to resist were well considered. First, they rejected the three-column layout in order to highlight their context (Introduction) and results. They also decided to include a figure of their full survey instrument in their Method section to let the audience see more quickly and clearly what they did. The most interesting aspect of this poster, however, is something we would call a failure in design that resulted from our own exhortations to resist convention. Class discussion about how to represent context without depending on too much text led to a suggestion that context be represented visually, in the form of a picture. This group liked that idea, and they tried it by including a beautiful picture of the pier that was featured in their study. But in trying to preserve other genre conventions and maintain the neat, clean balance of their layout, they mislabeled the image as data by including it in the "Data and Results" section. Such issues are testament to the difficulty of designing a document that observes convention while also trying to innovate.

The next example, produced for one of our classes and later incorporated into an honors project, is a more radical divergence from a conventional conference poster. Kelsey Potlock took her fellow students' proposal to let images take the place of text to heart and let an image (literally the location and local context for the study) serve as the backdrop for the poster. Photos also "tell" the viewer about the various conservation methods being tested by the study. The most radical decision that Potlock made, however, was to forego presenting the actual results or data on the poster. Her explanation for this decision was that an

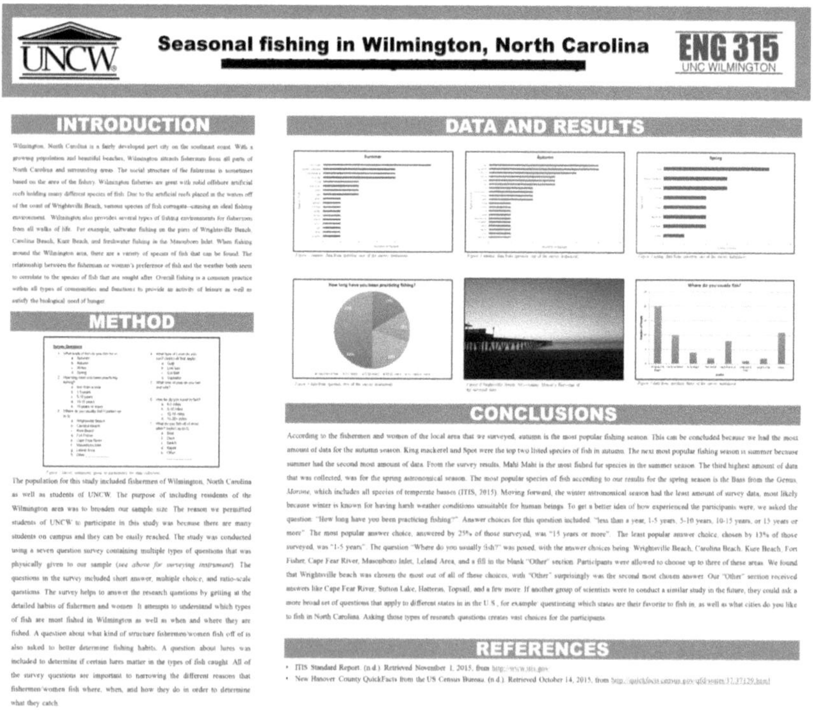

Figure 10.1 Student-produced poster for Writing in the Academic Sciences: A more conservative approach. Color figure in plate section.

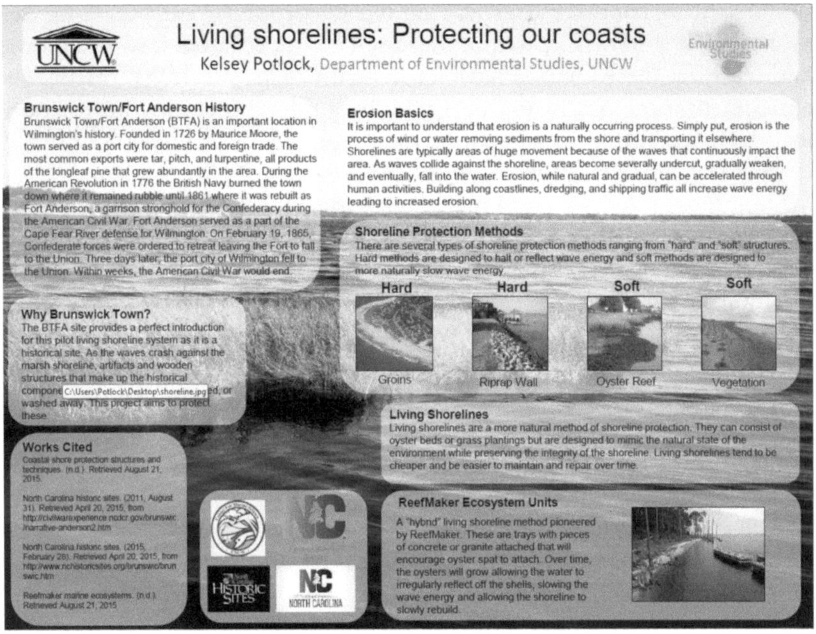

Figure 10.2 Student-produced poster for Writing in the Academic Sciences: A more radical divergence from convention. Color figure in plate section.

expert audience would be more interested in the location and historical context of the interventions being tested and would ask about results if necessary. The most important takeaway for non-expert audiences would be the mechanics of the technologies. Though this draft of the poster has a few design issues, we think she balanced the consideration of such diverse audience needs very well (Figure 10.2).

The conference poster project is a natural extension of the critical perspectives that students developed through writing the literature review. The literature review as a genre is not as readily accessible as the conference poster—new scholars need to be told, via Montuori (2005), about problems with conventional approaches. In contrast, conference poster samples are evidently problematic to students, and because of our gradual building of a critical perspective in this course, students come to the project doubly ready to employ their new skills. Conference posters are also excellent sites to foreground complex audiences for science.

Conclusions: Future Directions

The professional science writing courses and assignments described here are still quite new, and we do not yet have full impressions or data to speak to their effects on students' responses to academic and professional writing contexts after leaving our classrooms. Our year-long experiences of partnering with colleagues in the sciences and designing and teaching these courses have left us with clear directions for further experimentation locally and beyond. Our findings and our future goals may have interesting implications for the larger disciplinary landscape of science communication.

On a local level, we want our classes to serve more students and departments in the sciences while simultaneously growing and deepening our own department's identity on campus, demonstrating the central function of rhetoric in effective science communication. Our working partnerships with environmental sciences and chemistry will continue, but we are actively seeking interest and partnerships in other disciplines. In fact, in early 2017, our department received approval for a new post-baccalaureate certificate in science and medical writing. Through this endeavor, we plan to partner with the health sciences, a growing area on our campus. Another way to showcase the deliverables of our courses and increase their impact is to find platforms through which to promote and publish student work. From the first cohort of students, two from the undergraduate course have gone on to explicitly incorporate professional science communication elements in honors capstone projects.

In a larger context, we want to emphasize three important ideas that we argue should be paramount in discussions of science communication for students and professionals alike: (1) crafting narratives about scientific discovery for clarity, readability, and enjoyment is equally important for peer audiences; (2) genre conventions need to flex to serve rhetorical ends; and (3) models should be seen as starting places, important for critique as well as understanding genre expectations.

Rethinking the expert/public audience binary and broadening notions of science communication are in line with other important developments in science and technology taking place worldwide. Most notably, the written genres and contexts of science are escaping the borders of traditional institutions, implying that the scientist of the future will require a more nuanced perspective about expertise. What Kelly (2014) calls "parascientific genres"—blogs, crowdsourced data production, public crowdsourced funding proposals (i.e. Kickstarter), and citizen science projects—are becoming required communicative tools in the postmodern scientist's belt. Scientific content is also fast becoming a tradable currency on social media; the further content is removed from the original researcher, the more likely that layers of accommodation will shift the focus and even warp the message, as in a game of telephone. Our integrative approach to science communication responds to such evolving pathways for the dissemination of information by preparing students and professional scientists for a reality in which the distinctions between internal and external audiences, scientists and non-scientists, public and private, and a myriad of other established boundaries are blurred and even erased.

References

Besley, J. C., & Tanner, A. H. (2011). What science communication scholars think about training scientists to communicate. *Science Communication, 33*(2), 239–263.

Ceccarelli, L. (2011). Manufactured scientific controversy: Science, rhetoric, and public debate. *Rhetoric and Public Affairs, 14*(2), 195–228.

Cooper, H. M. (1982). Scientific guidelines for conducting integrative research reviews. *Review of Educational Research, 52*(2), 291–302.

Crowley, S. (1995). Composition's ethic of service, the universal requirement, and the discourse of student need. *JAC, 15*(2), 227–239.

Dahlstrom, M. F., & Ho, S. S. (2012). Ethical considerations of using narrative to communicate science. *Science Communication, 35*(5), 592–617.

Ding, H. (2008). The use of cognitive and social apprenticeship to teach a disciplinary genre: Initiation of graduate students into NIH grant writing. *Written Communication, 25*(1), 3–52.

Downs, J. S. (2014). Prescriptive scientific narratives for communicating usable science. *PNAS, 111*(4), 13627–13633.

Fahnestock, J. (1986). Accommodating science: The rhetorical life of scientific facts. *Written Communication, 3*(3), 275–296.

Graff, G., Birkenstein, C., & Durst, R. (2006). *They say, I say: The moves that matter in academic writing.* New York: W.W. Norton.

Guannel, M. L., Bruno, B. C., Grand, M. M., Lee, N., & Day-Miller, E. A. (2014). In Hawaii, a pilot course in professional development fulfills an unmet need in graduate education. *Bulletin: Limnology and Oceanography, 23*(3), 56–59.

Harwood, N., & Hadley, G. (2004). Demystifying institutional practices: Critical pragmatism and the teaching of academic writing. *English for Specific Purposes, 23*, 355–377.

Horst, M. (2013). A field of expertise, the organization, or science itself? Scientists' perception of representing research in public communication. *Science Communication, 35*(6), 758–779.

Kelly, A. R. (2014). *Hacking science: Emerging parascientific genres and public participation in scientific research.* (Unpublished doctoral dissertation). North Carolina State University, Raleigh, NC.

Kortencamp, K.V., & Basten, B. (2015). Environmental science in the media: Effects of opposing viewpoints on risk and uncertainty perceptions. *Science Communication, 37*(3), 287–313.

Kuehne, L. M., Twardochleb, L. A., Frischie, K. J., Mims, M. C., Lawrence, D. J., Gobson, P. P., … Olden, J. D. (2014). Practical science communication strategies for graduate students. *Conservation Biology, 28*(5), 1225–1235.

Leydesdorff, L. & I. Hellsten (2005). Metaphors and diaphors in science communication: Mapping the case of stem cell research. *Science Communication* 27(1): 1–36.

Luederitz, C., Meyer, M., Abson, D. J., Gralla, F., Lang, D.J., Rau, A., & von Wehrden, H. (2016). Systematic student-driven literature reviews in sustainability science—an effective way to merge research and teaching. *Journal of Cleaner Production*, *119*, 229–235.

Meredith, D. (2010). *Explaining research: How to reach key audiences to advance your work*. Oxford: Oxford University Press.

Miller, C. R. (1984). Genre as social action. *Quarterly Journal of Speech*, *7*, 151–167.

Miller, S. (2001). Public understanding of science at the crossroads. *Public understanding of science*, *10*(1), 115–120.

Miller-Cochran, S. K., & Rodrigo, R. R. (2013). *The Wadsworth guide to research*. Boston, MA: Wadsworth, Cengage Learning.

Mogendorff, K., te Molder, H., van Woerkum, C., & Gremmen, B. (2016). Turning experts into self-reflexive speakers: The problematization of technical-scientific expertise relative to alternative forms of expertise. *Science Communication*, *38*(1), 26–50.

Montuori, A. (2005). Literature review as creative inquiry: Reframing scholarship as a creative process. *Journal of Transformative Education*, *3*(4), 374–393.

Neely, L., Goldman, E., Smith, B., Baron, N., & Sunu, S. (2014). *GradSciComm report and recommendations: Mapping the pathways to integrate science communication training into STEM graduate education*. Retrieved March 6, 2017, from www.informalscience.org/gradscicomm-report-and-recommendations-mapping-pathways-integrate-science-communication-training.

NIH: National Institutes of Health: Office of Extramural Research. (2016). *Definitions of criteria and considerations for research project grant (RPG/X01/R01/R03/R21/R33/R34) critiques*. Retrieved March 10, 2017, from https://grants.nih.gov/grants/peer/critiques/rpg_D.htm.

Nisbet, M. C. (2009). Communicating climate change: Why frames matter for public engagement. *Environment: Science and Policy for Sustainable Development*, *51*(2), 12–23.

Penrose, A. M., & Katz, S. B. (1998). *Writing in the sciences*. New York: St. Martin's Press.

Robertson, D. P., & Hull, R. B. (2001). Beyond biology: Toward a more public ecology for conservation. *Conservation Biology*, *15*(4), 970–979.

Russill, C., & Nyssa, Z. (2009). The tipping point trend in climate change communication. *Global Environmental Change*, *19*(3), 336–344.

Saitta, E. K., Zemliansky, P., & Turner, A. (2015). A model for program-wide assessment of the effectiveness of writing instruction in science laboratory courses. *Journal on Excellence in College Teaching*, *26*(4), 149–172.

Schimel, J. (2012). *Writing science: How to write papers that get cited and proposals that get funded*. New York: Oxford University Press.

Sollaci, L. B., & Pereira, M. G. (2004). The introduction, methods, results, and discussion (IMRAD) structure: A fifty-year survey. *Journal of the Medical Library Association*, *92*(3), 364–367.

Sturgis, P., & Allum, N. (2004). Science in society: Re-evaluating the deficit model of public attitudes. *Public Understanding of Science*, *13*(1), 55–74.

Swales, J. (1990). *Genre analysis: English in academic and research settings*. Cambridge: Cambridge University Press.

11 A Rhetorical Approach to Scientific Communication Pedagogy in Face-to-Face and Digital Contexts

Carleigh Davis and Erin A. Frost

As this collection shows, scientific communication has been somewhat neglected within the field of technical and scientific communication. Further, there is evidence of a disconnect between theories and practices of working scientific communicators and what happens in our classrooms. There is growing emphasis among practitioners of scientific communication on the application of civic science (see Chapters 1, 6, and 7) but little coordinated effort to reflect this emphasis in our classroom practices. This chapter seeks to fill that gap by offering rhetorical approaches to scientific communication pedagogy. Our approach is explicitly culturally inflected; we often use civic science as a way of conceptualizing the cultural implications of scientific communication for students. This introduces complexities both because culture is an intellectually challenging subject to teach and also because one of us (Carleigh) was faced with undertaking this already challenging task in a virtual classroom. Thus, we seek to answer a two-part question in this chapter: What are some best approaches to teaching scientific communication, and what are the affordances and challenges of teaching scientific communication in online versus face-to-face contexts?

In trying to answer the two-part question above, we recognize that "best" will mean productive and useful but also flexible and permeable—we seek to offer ideas, not to standardize or stabilize approaches to teaching science communication. We recognize that many universities do not offer this specific course and that student populations differ; however, our hope is that we can offer approaches here that others might adapt or draw inspiration from to suit their own needs. In addition, with the proliferation of online course offerings, this chapter provides a rare example of how to implement a historically face-to-face curriculum in an online space.

We base our reflections and suggestions on our work teaching English 3820: Scientific Writing at East Carolina University (ECU). We have taught this course a combined total of six times, two sections of which were co-taught and one of which was online. This was the first time English 3820 had been offered online at our university. Working under IRB approval, we share some of the insights from students in the course,

DOI: 10.4324/9781315160191-11

using pseudonyms and omitting identifying information. To accomplish this work, this chapter begins by explaining the context of the course; next, we offer an overview of the course design, including specific details of the assignments we used. We make the following pedagogical recommendations: (1) use the concept of civic science, (2) incorporate group interaction, (3) include specific terminology and examples, and (4) make it personal. Finally, we offer more specific recommendations for incorporating these recommendations in distance education (DE) classrooms before offering suggestions for future work.

As numerous scholar-teachers have established (Arola, Shepherd, & Ball, 2014; Eldred & Toner, 2003; Kalmbach, 1997; Selfe, 2003; Wysocki, Johnson-Eilola, Selfe, & Sirc, 2004), digital delivery affects the content of a course; thus, the online format meant that Carleigh had to transfer our rhetorical pedagogy (with its attendant foci on social responsibility, belief systems, and reflection) to a digital learning environment where collaborative, reflective spaces look different than they do during face-to-face interaction.

In teaching all of these classes, we purposely introduced "failures of science orthodoxy" (for example, the reclassification of Pluto), breakdowns of scientific ethics (for example, the Tuskegee experiment), and schisms between public and scientific communities (for example, the thoroughly disproved autism-vaccine link) in order to get students to interrogate the specifics of established practices in scientific writing as well as the social and individual contexts that surround them. It was our hope that focusing on this particular kind of critical thinking skill would better prepare students to participate conscientiously in scientific practice in the future. As an example, pre-medical students sometimes arrive in our classrooms believing their future patients bear the responsibility for looking up risks involved in medical practice; such students are therefore resistant to exercises in communicating those risks and generally unfamiliar with creating genres that would allow them to do so effectively. These students are used to looking up information themselves and often have not yet interrogated how such information comes to be so readily available to them or what barriers might prevent others from accessing and understanding it. These expectations illustrate one of the ways that rhetorical education can demonstrate its importance for science students. Likewise, the rhetorical pedagogy detailed in this chapter offers some ways in which we can encourage students to think of their roles as scientists and as communicators as mutually inclusive.

Contexts of ENGL 3820: Scientific Writing

Scientific communication courses have followed some of the same patterns as technical communication courses. Namely, scientific communication courses are often service courses, and they tend to be pragmatically focused in that they are designed to increase job prospects and performance for students. We recognize the importance of service courses (and their connections to interdisciplinary teaching and scholarship), and we believe culture and rhetorical studies to be a foundation for scientific communication both in pursuit of pragmatic goals and in broader theoretical contexts.

Several scholars have offered writings useful in understanding the larger context of scientific writing and thus of our course (Garfield, 2002; Gibson, 1982; Longo, 2000; Penrose & Katz, 2010; Rossiter, 1986). However, we wish to begin our discussion with Francis Bacon's approach, which "reformulated scientific knowledge as contingent, rather than being absolute" (Longo, 2000, p. 49). Bacon did not have the last word on the subject, though, as science communicators after him (notably Thomas Henry Huxley) continued to claim that "pure" science should be "unconcerned with [and] ... not responsible for social outcomes" (Longo, 2000, p. 57). Despite the claims of Huxley and others who share his views, more recent scholarship (Feyerabend, 1987) and the modern rise of civic science demonstrate increasing traction for the idea of science as socially mediated. Further, the largest science funding agency in the nation—the U.S. government—has long been persuaded that science is connected to social outcomes (Abbott et al., 2014). These interactions shed light on a longstanding (and continual) "tension between private and public science" (Longo, 2000, p. 29). This ongoing debate regarding attention to social and cultural concerns of the public in the history of scientific communication is a major element informing our pedagogical approach to the subject.

At least some learning goals for scientific communication courses should be based in the intersection between this uncertainty regarding the place of public concern in scientific practice and the rich history of social advocacy in technical communication and technical communication pedagogy. As one productive example, scholars such as Gurak and Bayer (1994) emphasize the influence that feminist approaches to technical communication can have on science and technology fields. Segal's (1995) work, likewise, is suggestive that fostering classroom space in which students are able to critically evaluate genre conventions and norms of writing in their fields better prepares them to work in a variety of scientific fields. Also essential to these courses is the idea, prominent in technical communication scholarship, that critical awareness of the rhetorical foundations and social implications attendant to both technical action (in this case, practicing science) and technical writing can improve students' abilities to engage successfully in both processes (Durack, 1997; Haslanger, 2002; Jung, 2007).

Despite this attention, scientific communication is often subsumed under the umbrella of technical communication. According to the program list published by the Council for Programs in Technical and Scientific

Communication (2011), only five out of 82 institutions offer programs that explicitly name science or scientific communication as a focus (see Table 11.1).

Our institution is not one of the above five, but we do take up the science-as-culture question extensively in English 3820: Scientific Writing. ECU, a large public Higher Research Activity doctoral university in Greenville, NC (The Carnegie Classification of Institutions of Higher Education, 2015), has offered some version of English 3820 for nearly 50 years. This is an important and unusual history, made all the more significant because of our university's position as the nation's largest producer of medical professionals who practice in underserved areas. In our local context, English 3820 was distinct from other technical, business, and professional writing classes from the beginning because of its directed service to medical pre-professionals, and it has remained distinct primarily because of the disciplinary orientation of students who continue to enroll. For more on the historical situation of this course, see Combs, Frost, and Eble (2015).

Table 11.1 Programs with Focus on Science/Scientific Writing/Communication

Institution	*Title of Program(s)*
Ferris State University	BS in technical and professional communication with a concentration in science and medical writing
James Madison University (Institute of Technical and Scientific Communication)	• BA in technical and scientific communication with concentrations in online publication, publications management, or technical and scientific communication in the public sector • BS in technical and scientific communication with concentrations in online publication, publications management, or technical and scientific communication in the public sector • MS in technical and scientific communication • MA in technical and scientific communication
Michigan Technological University	BA in scientific & technical communication
North Carolina State University	English minor in technical and scientific communication
University of Minnesota, Twin Cities	• BS in scientific and technical communication • Graduate minor in rhetoric and scientific and technical communication • MS in scientific and technical communication • MA in rhetoric and scientific and technical communication • PhD in rhetoric and scientific and technical communication

The current undergraduate catalog charges English 3820 with providing students with "practice in assimilation and written presentation of scientific information." The course offers humanities training to STEM majors (primarily biology and chemistry, but also geography, sociology, and a smattering of others) in order to prepare them for public communication scenarios. English 3820 is a writing intensive (WI) course that contributes to a 12-hour WI requirement for all students; most students who enroll are biology students with intentions of going into medicine or dentistry because the course is also one element of the research skills requirement for their majors. The inclusion of this rhetoric-based writing course as a mandatory part of the biology major comes as a result of the pre-professional medical programs here recognizing that in order to communicate with a diversity of patients, medical professionals produced by ECU must have practice in writing and communication.

Demographically, about half of the students who enroll in English 3820 are female and one-third identify as minority, as compared to the university's makeup of about 60% female students and 20% minority students (Combs et al., 2015). In this sense, English 3820 is actually more reflective of regional demographics—we draw from a rural region with an ethnically diverse population—than is the university population at large.

Despite intense demand for English 3820 and the presence of vibrant online curricula in our department, English 3820 had never been offered as an online course prior to 2015. We offered this course online for the first time in summer 2015. Student responses to the F2F courses that we have been involved in have been overwhelmingly positive, whereas students in these new DE sections reported significantly less overall satisfaction with the course as a whole as well as their own learning throughout. Part of our current work (and part of the exigence for this chapter) is to reflect on the DE experience and theorize ways to improve student learning in that context.

Course Design

The students who enroll in English 3820 are bright and engaged, but they have had little experience or training in translating their (newly earned) STEM expertise in ways that will allow them to communicate it productively with potential audiences outside their field. Thus, the learning goals for this course were developed primarily to meet students at their current level of competency, which allows them to communicate well with other scientists, and to help prepare them for careers in which they will need to communicate about scientific principles with non-expert audiences. One way that we respond to this public responsibility is by focusing English 3820 on civic science and the development of transformative leadership abilities. Civic science is "scientific inquiry that offers opportunities for participants to develop their capacity to work across differences, create common resources, and build a democratic way of life" (Abbott et al.). In other words, it is science done in a participatory way, with accountability to the public. We also asked students to think about the role of leadership (Astin & Astin, 2006) in performing civic science. Given our situation at a public institution, this kind of engagement is essential.

With these values in mind, we complemented the five process- and genre-based learning outcomes that are standard to our university writing-intensive courses with three additional outcomes that were developed for this course (see list below and Table 11.2). Through a combination of student-centered, writing-enhanced, civic science-oriented, and leadership-informed principles, we developed an effective and useful pedagogy for teaching scientific communication. This pedagogy is not without flaws, and for that reason we offer an extended reflection below on the affordances and challenges of this approach so that other teachers of scientific communication might benefit from our experiences.

Assignments and Implementation

In designing and implementing this course, we placed a tremendous amount of value on class discussion, relying on only short periods of lecture or presentation in order to encourage student interaction. We also made group work a priority along with peer review, emphasizing the need for instructors and students to learn from each other as we progress through complex scientific and rhetorical topics. As such, students were always practicing process-based writing strategies, engaging in collaborative learning, and developing more nuanced understandings of the rhetorical situations they were being called upon to address. We found that this kind of engagement better prepared students to think through the demands of any given writing situation and respond flexibly, avoiding the somewhat prescriptive genre-based approach that we were concerned might narrow their understandings of what it means to be a scientific writer.

The major writing assignments in our course focused on three specifically designed learning outcomes:

- Learning Outcome 1: Consider the situated nature of particular contexts of scientific writing and produce actual examples of scientific writing for various purposes.
- Learning Outcome 2: Examine theories, methodologies, and ideologies that undergird scientific writing with an eye to perfecting both critique and imitation of scientific styles.
- Learning Outcome 3: Theorize a variety of reasons, using rhetorical language, for why a responsibility to the public is important for working scientists and medical professionals in order for their writing practice to be useful and effective.

Table 11.2 Course Assignments

Learning outcome 1	Personal statement A personal statement regarding students' career goals (i.e., for medical or dental school, job, graduate school, or internship applications).
Learning outcome 1	Teaching and learning science A collaboratively produced, curriculum-driven website. Students were divided into groups based on the stated needs of a local client.
Learning outcome 1	Writing for specified publics A project that communicated a scientific message of students' choice to an audience for whom their message would be significant by using an appropriate medium.
Learning outcome 2	Journal project Feedback in a cohesive, persuasive letter to authors of manuscripts for an issue of the journal *Communication Design Quarterly* that focused on health and medical rhetorics.
Learning outcome 2	Artifact analysis A rhetorical analysis of a scientific "artifact" discussing how the artifact conveys information to its intended audience.
Learning outcome 3	Debate A scripted debate among 4 to 5 students on the various perspectives surrounding an issue of contention among scientists and/or other interested publics.
Learning outcome 3	Exams <u>Midterm:</u> An in-class examination designed to test a student's ability to understand and apply rhetorical principles used in scientific communication. <u>Final Exam</u>: A multi-option take-home exam. Options included a "scavenger hunt" related to course concepts in practice, designing an examination for future students, creating a website to communicate a scientific concept with a specific target audience, or completing a mock interview. Exam options were designed collaboratively with students.
Learning outcome 3	Extra credit A reflective cover letter about the course as well as future applications of material; OR a Public Service Announcement regarding course content and expectations.
Learning outcome 3	Writing portfolio A revised project of the student's choice from the semester along with a reflection on that project and its future applications.

We utilized several assignments for each learning outcome (see Table 11.2) and made sure to distribute the assignments evenly among the specified course outcomes so as to place equal weight on each goal.

In our experience, this structure has been very successful. Students typically report a great deal of satisfaction with the course and we have been satisfied with each class's ability to achieve the course goals by the end of the semester. While students often find the assignments themselves to be challenging and unusual, we have found that this kind of adjustment to their expectations of what the class will be like gives them a more complete understanding of what it means to be a scientific writer by the end of the semester.

Pedagogical Recommendations

In the following sections, we offer suggestions for course concepts and structural elements that, based on our reflections and observations, were beneficial to our teaching of scientific writing. Namely, we suggest that scientific writing instructors use the concept of civic science to encourage students to think about scientific writing from various perspectives, incorporate group interaction to help students learn from each other, rely on specific terminology and examples when dealing with complex rhetorical concepts, and encourage students to engage with their writing as individuals rather than exclusively as scientists. Our observations and suggestions are local, couched in the circumstances surrounding our particular courses and students, including both traditional and online courses. However, if used well, they can be instructive in a variety of contexts. As such, the following discussion also attempts to bridge the pedagogical gap between traditional and online classrooms to offer productive suggestions for both. We first discuss what each of these recommendations might look like in a traditional classroom, and we then offer suggestions for implementing them in digital learning environments.

Recommendation 1: Use the Concept of Civic Science

Emphasizing civic science not only encourages students to actively think about ethics and the social repercussions of scientific writing, it also provides them with an inroad for comparing and contrasting the application of rhetorical principles in writing done for various publics, both scientific and nonscientific. Such an awareness encourages them to think through the whys and hows of common scientific writing practices by emphasizing the connection between form and function in writing. Through this process they begin to bridge the gap between civic, industrial, and academic careers in science fields.

Many students enter into the class with the expectation that much of the course time will be spent learning the nuances of advanced scientific publication. As such, it can be difficult to help them find value in projects that ask them to communicate with public audiences. To prepare students for this kind of work, we introduced the concept of civic science early as a part of class readings and discussion. In our experience, students engaged well with the idea in theory, but they were resistant to feeling an obligation to public audiences, particularly in relation to their own research projects. Given our position at a public university and her experience in previous semesters, Erin wanted to provide students some frameworks for thinking about responsibility to the public as well as some examples of how this might happen. An understanding of science as symbiotic and community oriented, i.e. "I am a scientist because I want to help people," allows students to adjust to the dissonance they experience in a writing class when compared to their more clinical scientific courses, which often focus on a more isolated view of scientific practice. We addressed this by incorporating discussion questions inquiring about the purpose of "science" and the motivations that frame students' interest in particular projects. In response, students in our classes have been nearly unanimous in the understanding that science exists "to help people," especially through discovery and understanding. From this point, discussions with us and with their classmates allowed students to bridge the gap between "helping people" and "public action" with regard to communication. The teaching and learning of science and the writing for specified publics assignments allowed students to practice putting these principles into action while also making apparent the value that their work has outside laboratory or academic contexts. The debate assignment required a similar rhetorical move by asking students to take on the perspective of a particular group with a vested interest in the topic and craft an argument based on the particular concerns of that group.

As an example of this practice, during one F2F class most students took a very firm stance on the subject of climate change and were initially appalled that policies could be in place that tolerate practices that have a negative impact on the planet's climate. Accepting, then, that these negative effects were a valid concern, the class constructed a scenario revolving around a factory with practices that violate environmental ethics codes. When asked to represent the view of factory employees whose job is the only means of supporting their families, several students crafted eloquent and persuasive arguments claiming that financial stability for their families took precedence over the likelihood of climate change in the future. The conversation then centered around how the class, acting as climatologists at this point, might work to persuade community members that they should take the steps supported by climatology research while taking into account their financial and personal concerns. Through this kind of discussion and others like it, students began to draw clear connections between their actions as science writers and the broader ramifications of those actions in various communities.

Recommendation 2: Incorporate Group Interaction

We recommend that both traditional and online scientific writing courses place significant emphasis on group work and collaborative writing, as well as collaborative interactions such as peer review and communication with stakeholders who are not members of the class (e.g. clients in service-learning projects, local reporters, members of parallel course sections or members of topically-related classes, and scientists and specialists who work with the public via social media; also see Chapter 13). We have found this to be useful not only because it helps students to engage with multiple perspectives throughout the course but also because it more closely represents the writing scenarios that these students will participate in later in their careers, which they often have limited opportunity to practice ahead of time (See Chapter 2's treatment of authorship practices in scientific communication). For example, in the teaching and learning science assignment, students adapted curricula from various fourth and fifth grade science textbooks to create a website that would suit the needs of a local elementary school. Specifically, the site was intended to be a supplementary source that would engage the elementary school students with scientific topics to keep them interested and help them master the required concepts. When working directly with the needs of this particular elementary school, which has a number of English as a Second Language students as well as community ties to our university, students had to find ways to not only communicate the information in a way that would be age-appropriate, but also to appeal to this particular group of users. As such, they had to collaborate both within the class and outside to negotiate the presentation of their content in a way that would meet the needs of the school.

Through discussions in which students often embodied their roles as individuals, community members, and scientists simultaneously, students in the traditional classes were able to remind each other of the real-world implications of scientific actions and, therefore, the interplay between scientific practice and nonexpert audiences on an ongoing basis. This kind of discussion allowed them to give productive feedback to each other as they worked on assignments. We believe that students in the traditional classes reported a great deal of benefit from these projects in part because the format of the course facilitated student cross-talk.

A key element in this kind of diversified communication strategy is an understanding that all writing, even scientific writing, is value-laden. As such, it was essential for students to interrogate their understanding of objectivity and understand that a variety of perspectives/approaches can be equally valid. In this context, we found it especially useful during our co-teaching semester for students to see two different instructors using different teaching styles, both to positive effect. By witnessing our interactions, students were able to complicate their understanding of teaching as an objective imparting of fact-based knowledge and see our value judgements come into play. The interdisciplinary worldviews of both student and instructor participants in the class helped to demonstrate the importance of multiple perspectives, further reinforcing the importance of audience awareness and rhetorical training in scientific communication. We believe similar (and perhaps even more productive) results could be achieved through various kinds of co-teaching—pairing a rhetoric instructor with a biology instructor, for example, or a science-writing specialist with a research-productive geneticist—so long as the co-teachers have sufficiently diverse approaches.

While this dynamic was a guiding force throughout our co-taught semester, it was important to us to find ways both to allow students to engage in this kind of interplay themselves and to incorporate assignments in future classes that would allow students to witness this type of interaction when the added benefit of co-instruction was not available. The debate assignment accomplished this goal by encouraging students to analyze the ways in which their positionality might influence their communicative practices with various audiences both inside and outside scientific communities. In order to successfully fulfill the terms of the debate assignment, students had to show an advanced understanding of their scientific topic as well as the ability to articulate relevant points of that topic to various interest groups and effectively communicate concerns of and to those groups. As such, students were able to demonstrate rhetorical dexterity by understanding and re-aligning their perspectives to suit a variety of audiences in a given situation.

Recommendation 3: Include Specific Terminology and Examples

Students in our courses had a great deal of experience learning through cross-disciplinary naming and identification. They have spent years committing scientific terminology to memory and are often used to course structures that rely on vocabulary as a foundation for understanding connections between concepts. In our experience, they tend to be comfortable approaching rhetoric through a similar process—by identifying key terms representing metalinguistic and rhetorical concepts (process, appeals, audience, etc.) and then applying those terms together to act as a formula for communication (we wouldn't call this a fundamental characteristic of the minds of science students, but we do think this is the kind of practice that they have with "doing school."). As such, continually returning to the names of rhetorical terminology across projects helps students to engage reflectively with these concepts in order to better apply them in their own frameworks. At the same time, engaging specific terminology from students' home disciplines helps them to see how this process works similarly across academic contexts. Formulas only work when they contain discrete representations; so, too, with the communicative formulas these students create. Encouraging students to develop precise understandings of elements of rhetorical situations such as audience (e.g. fourth-graders and their parents rather than "the public") and to alter their definitions of those elements appropriately helps them to develop a more practical understanding of the application of rhetorical ideas.

To begin this process, we introduced the canons and modes of rhetoric to help students develop a common language for talking about the issues that they were encountering in scientific communication. Having been used to a very fluid "humanities"-like approach to teaching rhetoric, it was strange to Carleigh to observe students taking a formulaic and methodical approach to implementing rhetorical principles once they were introduced. While jarring to her, it seems like their more linear, question-based approach (for every assignment: "Who is the audience? Why am I writing this? What is my message?") made students much more rhetorically aware and helped them to consistently apply the principles more so than other classes she had taught. These questions and the understanding that students gained from them likewise provided a reference point for students working on projects throughout the course as well as a means of productive critique in peer review of later projects.

In keeping with this approach of terminology and process elements functioning as course content, we chose to incorporate formal examinations in our course structure. This was a new experience for Erin, who had never given tests in this course when teaching it before. We did this for three main reasons: (1) to help ensure students were doing the reading and learning to use the appropriate terms for the concepts we were talking about; (2) the population in these classes are more comfortable with tests than papers, and we wanted to be responsive to their experiences and also to value more than just a single learning style; and (3) despite anti-test sentiment (justifiably resulting from the governmental handling of K-12 benchmarking) in academia/education right now, we believe that a well-written and appropriate test can still be a valuable measure of and incentive for learning. We did our best to design the tests, which included multiple choice, true/false, short answer, and short essay questions, in ways that would both feel comfortable and familiar to students who are used to being evaluated in this format and also allow them to demonstrate the process-oriented skills that are central to the course. This meant that in addition to fundamental vocabulary and recall questions, we also included questions that asked students to identify concepts at play in specific scenarios and then alter those concepts to suit a different purpose. For example, one short essay question offered an explanation of a chemical reaction that was originally written for college students. To respond to this question, students were asked to identify the audience and purpose based on textual cues and then offer the original author advice on how to change the passage to make it useful for an audience of fourth graders.

Recommendation 4: Make It Personal

Drawing on the concept of civic science, we found a great deal of value in encouraging students to consider their perspectives as individuals rather than simply as scientists. This allowed them to engage more productively with the concept of bias in scientific research as well as increase productivity when working with a variety of collaborators and audiences. Such practices were evident in the personal statement assignment, which when completed at the beginning of the course and processed through peer review allowed students (and the instructor) to get to know each other better and reference personal and research interests of others throughout the semester.

Part of our work in this course is to encourage students to question their understanding of objectivity (and therefore inherent "logic" or "correctness") in the hard sciences to better prepare them to communicate effectively with different kinds of audiences and to interrogate the factors that might hinder that communication. One way we tried to get at the idea that scientific "truth" depends upon perspective and interpretation was by introducing the personal—in this case, the personal statement—as an important element of scientific writing. In class discussion we introduced this concept by first interrogating the concept of "objectivity." Many students entered our classes with the understanding that objectivity is a state of mind that frees a researcher from bias; however, during our discussions, we characterized objectivity as a style of writing reflected in scientific papers that contributes to the ethos of the author, implying that she did not intentionally or carelessly skew the results or representation of her research. We simultaneously stressed the impossibility of removing oneself entirely from the research process, emphasizing the influence that personal and field-wide paradigms have on any scientist's research question, methods, and conclusions.

While we expected students to be at least somewhat resistant to these ideas, given the emphasis placed on objectivity in their home disciplines, they generally were not—at least not overtly. They seemed eager to engage with the social construction of expectations both in scientific writing and in research methods. However, students in each class did struggle with the leap between understanding the concept of objectivity defined as such on an intellectual level and applying it to the "logical" link between research methods and analysis. Students had trouble with the idea that analysis itself even takes place, using the term in closer alignment with "observation" than anything else—that is, scientific conclusions are true ("logical") because they can see them, not because they are applying an interpretive lens. This disconnect was an area we worked on constantly throughout all classes.

Because re-evaluating objectivity is one of the toughest concepts we took on, we felt that it was useful to align it with a very pragmatic personal statement assignment; most students in this course are headed for some kind of graduate school and thus are very pleased with the opportunity to work on a document that will help them during the application process. To be successful in the personal statement, students must use their existing knowledge of their discipline to help them recognize the belief systems that shape the kind of writing that discipline values. Likewise, in order to discuss their experiences and goals in the way that a personal statement requires, they must identify the ideas and beliefs that overlap, conflict, and guide them as they operate in different (but porous) communities as scientists, individuals, and in their other identities. For example, we have helped more than one pre-dental student refine a personal statement about the personal and professional ramifications of having a nice smile—these students recognized the multiple but overlapping communities affected by a person's smile/appearance and discussed ways to navigate these communities. In articulating these underlying ideologies, students practiced recognizing the ways in which their various identities shape their scientific practice. As science students writing for science teachers, these students had already received years of training in following methodical and logic-based patterns to communicate with audiences who have also been disciplined into those same logical patterns. Expanding upon this training in the personal statement helped to prepare them for projects in which they would examine the ways belief systems differ from person to person to improve communication across various social and professional positionalities.

Adapting Recommendations for DE Classrooms

Recommendation 1: Incorporating Civic Science in DE Classrooms

In the online course, it was much more difficult for the connections between scientific practice and civic responsibility to become apparent. We suspect that, because the course was organized into weekly units, the topics from each unit were less likely to carry over from week to week in the discussion boards than they were in traditional classroom discussion. In the construction of this course, Carleigh was very resistant to the inclusion of much lecture material, having seen how F2F students in traditional classes developed nuanced understandings of significance by interacting with each other. However, in the future, we would recommend introducing specific scenarios in an online lecture mode and modeling the value of civic science for scientists as well as the other interactants, making sure to emphasize the connections between scientific writing practices that have been discussed previously and how those practices connect with our understanding of civic responsibility. For example, DE students might first move through a module that utilizes video of the instructor talking about the Tuskegee experiment, followed by references to historical and modern documentation about similar transgressions and subsequently developed ethical guidelines (e.g. the Belmont Report, specific institutional IRB guidelines and the formation of their committees). From that point, students could discuss the implications of two to three other scenarios in a more concrete way that allows them to put the concept into practice rather than developing hypothetical scenarios on their own and trying to postulate through discussion boards.

Recommendation 2: Incorporating Group Work in DE Classrooms

Group interaction is particularly important in online iterations of the class. While both students and instructors may be more resistant to the idea of group projects due to time and geographic constraints, it is the kind of interpersonal interaction facilitated by group projects, as well as peer reviews and discussions, that is most limited by DE course structures that rely solely on repository tools (we used Blackboard, but anecdotally the same is true of Moodle and Wordpress courses as well). These interactions are necessary to help students to complicate their identity as scientists to include roles as educators and community members, but they are difficult, if not impossible, to achieve through traditional discussion boards or blog posts. We encourage instructors to implement synchronous meeting/collaborative tools such as facilitating Google Hangouts, Skype sessions, or Saba meetings in order to get class members talking more interactively. As we discuss below, we found the lack of synchronous learning space to be a challenge. However, we also recognize that asynchronous coursework has many affordances for students. Part of the reason our institution produces so many physicians working in underserved areas is that those physicians start out as students from rural and economically depressed communities; these students often work full time and thus the affordances of an asynchronous online course are important in our educational context. We therefore found it necessary to explore options other than synchronous discussion.

Careful incorporation of social media platforms and other online community-based resources (e.g. Twitter, Facebook, Today's Meet, Instagram, or participation in massive open online courses (MOOCs)) could be one option. Social media tools can be used to sponsor conversations that blur the line between synchronous and asynchronous communication while also allowing for smooth incorporation of resources and access to larger/public conversations.

Recommendation 3: Incorporating Specific Terminology in DE Classrooms

As part of incorporating specific terminology in teaching scientific communication, we demonstrated to students that the discipline of rhetoric has specialized terms just like biology and that writing about science involved particular mechanics. This means students were learning both a new set of terminologies as well as learning about the rhetorical function of technical language. This multi-layered learning process was, unsurprisingly, more difficult to manage in online contexts.

For example, we began a co-taught traditional class by having students do actual work in a science writing discipline—students got hands-on experience editing articles that were forthcoming in a special issue of *Communication Design Quarterly* focused on health and medical rhetorics (Meloncon & Frost, 2015). This assignment gave students practice in dealing with a lengthy manuscript, familiarized students with the format, and introduced students to science writing as a discipline beyond their individual specialties. These articles were not necessarily in their specialties, so they had to play the role of the nonexpert reader. Thus, we immediately got students focused on the mechanics of writing as a process before moving back to larger-picture disciplinary conversations.

To parallel this kind of learning in a DE class, we recommend having students complete reviews of pre-selected published scientific articles. To replicate the group dynamic described above, students might complete (draft) reviews of these articles individually but then also read and comment on each other's reviews. This approach can help students to consider the situated nature of particular contexts of scientific writing while also giving them practice in identifying and analyzing theories, methodologies, and ideologies that undergird scientific communication more generally.

Recommendation 4: Making DE Classrooms Personal

As previously discussed, we recognize the value of the personal statement assignment and recommend its inclusion early in the semester regardless of the course format. However, in DE classes it is difficult for the characterization of each student that is present in these documents to remain intact throughout the course, particularly because students will only read each other's statements closely if they are involved directly in peer review. Even in this case, it is unlikely that the knowledge gained about their classmates would carry through subsequent weeks. As such, in online iterations of the course, we advise the creation of student profiles displaying their personal and research interests as well as images and other interactive content. This allows the digital space to embody the personal as well as the course work so that students might see connections between the two.

To help carry these concepts through the course, we also suggest assigning students (or allowing students to assign themselves) particular roles to fulfill in class discussion from week to week, thereby encouraging them to embody their roles as scientists on some occasions but other roles at other times. During the DE course, Carleigh felt that the discussion board method of interaction limited the ability of the class to engage with social topics. The removal of the interactions from a physical space made it more comfortable for students to see themselves exclusively in their role as a scientist (rather than acknowledging social positioning in terms of gender and race, for example, which would be more visible in a physical classroom). Carleigh concluded that the online experience lacked the opportunity for students to have live discussions and work through advanced concepts, like the community-based discussion of climate change referenced above, and to consider multiple perspectives in writing projects. Encouraging students to speak from different perspectives through various modes of communication would allow for the representation of different interest groups and, hopefully, help to avoid the mono-lined discussion boards that hindered this particular DE class.

Conclusions: Directions for the Future

Based on the experiences described above, we argue strongly for a rhetorical approach to scientific communication pedagogy that takes positionality, social science, and culture into account and focuses on the needs of students. To offer some evidence of the efficacy of such an approach, we give below specific examples of students' intellectual inquiry based around one particular assignment. At the end of our semester of co-teaching, we assigned students a debate project in which they worked in groups to present multiple sides of a controversial public science issue. During this project we saw concrete evidence not only of students using rhetorical principles to think about science, but of them doing so in interdisciplinary and contextually relevant ways. Some examples:

- In response to a debate about vaccinations, Justin asked, "Is it fair to tell people what scientific studies they should believe?" While students in the class generally agreed that a long-discredited study linking autism and vaccinations (Deer, 2004) was unethical and false, they nevertheless had a lively discussion about the rights of the public in defining for themselves which scientific studies are useful.
- Leslie stretched across disciplines to ask, "How would a tort case [about vaccinations] find patient zero?" She wanted to know the processes by which a "patient zero" might be identified, and she also wanted to know what rhetorical moves a civil lawsuit might make in establishing this.

- In a debate about radio-frequency identification (RFID) chipping of adults, students asked for more specification about the legal, ethical, and social boundaries of this proposed practice. Amanda asked, "Is this debate about mandatory chipping or voluntary? And, what about the tension between consent issues and lowering crime rates?" and Elizabeth asked, "Who's in charge of this? The government, or some organization? What [organization] regulates?" Students were as concerned with the cultural situation and implications of such a practice as they were with the science of the chip itself.

To conclude, we employed a pedagogical approach that encouraged students to develop a critical awareness of their own belief systems and those of others, as well as an understanding of how those belief systems influence communication practices. A quotation from one of Carleigh's online students who came to understand science as culture demonstrates the importance of such pedagogical approaches to scientific communication:

> I had to rethink a few of the opinions I had formed before hearing other views on topics such as the Scientific Method and bias in science. For example, I have never thought of objectivity as being a writing style. ... I believe that this will allow me to be more honest in my work and allow my research to be taken more seriously.

This student's understanding of objectivity as enculturated will have, as she points out, both theoretical and practical effects on her future work. We believe the recommendations we offer above may help move more students toward this understanding and the benefits it offers. We hope that in the future, instructors of scientific communication might examine, implement, revise, and critique the recommendations we offer. We know that not all experiences will parallel our own, and only by assembling a body of knowledge around the teaching of scientific communication can we, as a field, begin to address the unique needs of students in this area.

References

Abbott, S., Boyte, H., Jordan, N., Ottinger, G., Peters, S., Spencer, J. P., ... Zelazo, P. (2014, October 2–3). *A call to action: Civic science and the grand challenges of the 21st century* (A White Paper for the National Science Foundation Workshop). Washington, DC: National Science Foundation Civic Science Workshop.

Arola, K., Shepherd, J., & Ball, C. (2014). *Writer/Designer: A guide to making multimodal projects*. New York: Bedford/St. Martin's.

Astin, A. W., & Astin, H. S. (2006). *Leadership reconsidered: Engaging higher education in social change*. Greenville, NC: East Carolina University.

Combs, D. S., Frost, E. A., & Eble, M. F. (2015). Collaborative course design in scientific writing: Experimentation and productive failure. *Composition Studies, 43*(2), 132–149.

Council for Programs in Technical and Scientific Communication. (2011). Program list. In *Council for Programs in Technical and Scientific Communication website*. Retrieved February 21, 2017, from www.cptsc.org/programlist.html.

Deer, B. (2004). Revealed: MMR research scandal. *The Sunday Times: Health*. Retrieved February 21, 2017, from www.thetimes.co.uk/tto/health/article 1879347.ece.

Durack, K. (1997). Gender, technology, and the history of technical communication. *Technical Communication Quarterly, 6*(3), 249–260.

Eldred, J. C., & Toner, L. (2003). Technology as teacher: Augmenting (Transforming) writing instruction. In P. Takayoshi & B. Huot (Eds.), *Teaching writing with computers: An introduction* (pp. 17–32). Boston, MA: Houghton Mifflin Company.

Feyerabend, P. (1987). *Farewell to reason*. London: Verso.

Garfield, E. (2002, April 1). Highly cited authors. *The Scientist, 16*(7), 10.

Gibson, S. S. (1982). Scientific societies and exchange: A facet of the history of scientific communication. *The Journal of Library History (1974–1987), 17*(2), 144–163.

Gurak, L., & Bayer, N. (1994). Making gender visible: Extending feminist critiques of technology to technical communication. *Technical Communication Quarterly, 3*(3), 257–270.

Haslanger, S. (2002). On being objective and being objectified. In L. Antony & C. Witt (Eds.), *A mind of one's own: Feminist essays on reason and objectivity, 2nd ed.* (pp. 209–253). Cambridge, MA: Westview Press.

Jung, J. M. (2007). Textual mainstreaming and rhetorics of accommodation. *Rhetoric Review, 26*, 160–178.

Kalmbach, J. R. (1997). *The computer and the page: The theory, history and pedagogy of publishing, technology and the classroom* (New directions in computers and composition studies). Norwood, NJ: Ablex.

Longo, B. (2000). *Spurious coin: A history of science, management, and technical writing*. New York: SUNY Press.

Meloncon, L., & Frost, E. A. (Eds.) (2015). Charting an emerging field: The rhetorics of health and medicine and its importance in communication design. [Special issue]. *Communication Design Quarterly, 3*(4).

Penrose, A. M., & Katz, S. B. (2010). *Writing in the sciences: Exploring conventions of scientific discourse* (3rd ed.). New York: Pearson Longman.

Rossiter, M. W. (1986). Women and the history of scientific communication. *The Journal of Library History (1974–1987), 21*(1), 39–59.

Segal, J. (1995). Is there a feminist rhetoric of medicine?: Toward a critical pedagogy for scientific writing. *Textual Studies in Canada, 7*, 109–116.

Selfe, D. (2003). Techno-pedagogical explorations: Toward sustainable technology-rich instruction. In P. Takayoshi & B. Huot (Eds.), *Teaching writing with computers: An introduction* (pp. 17–32). Boston, MA: Houghton Mifflin Company.

The Carnegie Classification of Institutions of Higher Education. (2015). *About Carnegie classification*. Retrieved April 3, 2015, from http://carnegie classifications.iu.edu/.

Wysocki, A., Johnson-Eilola, J., Selfe, C. L., & Sirc, G. (2004). *Writing new media: Theory and applications for expanding the teaching of composition*. Salt Lake City: University of Utah Press.

12 MetaFeedback

A Model for Teaching Instructor Response to Student Writing in the Sciences

Lindsey Harding and Liz Studer

Despite the importance and prevalence of the writing scientists do, often science courses lack writing instruction and instructors lack training in writing pedagogy. When science-and-writing courses are offered, they take on many forms depending on their departmental, institutional, and instructional contexts (see Chapter 8). Undergraduate students may take writing courses as part of their general education requirements, but these courses often fail to cover primary scientific communication (Moskovitz & Kellogg, 2005) and prepare students for the writing assigned in their other major courses (Kutney, 2008). Technical and scientific communication courses may be offered by writing programs or English departments (see Chapter 11 for such a course). While these courses enable students to practice scientific writing, they often exist apart from the required program of study for science majors. Writing across the curriculum (WAC) and writing in the disciplines (WID) programs attempt to insert necessary writing instruction into undergraduate science courses by integrating technical and scientific communication curricula into content-driven courses. Research on the effectiveness of these initiatives attests to their success at introducing students to scientific writing (Brownell, Price, & Steinman, 2013; Holstein, Steinmetz, & Miles, 2015; Paszkowski & Haag, 2008), engaging students more fully in course material (Moskovitz & Kellogg, 2011; Chapter 13), and encouraging writing as a process (Lee, Woods, & Tonissen, 2011). Our study focuses on specific aspects of writing-intensive science curricula and their pedagogical foundations and reports our strategy for professionalizing science lab instructors so that they are prepared to provide effective instruction in scientific communication.

Feedback is a hallmark of writing instruction for the positive impact it has on student learning (Mory, 1992). As a written discourse, feedback can take the form of instructor commentary, peer review, and self-assessment. Our goal with this research was to develop a methodology for improving feedback training for instructors; as a result, both peer review and student self-assessment are beyond the scope of the current study.

DOI: 10.4324/9781315160191-12

In this chapter, we focus specifically on the comments and marks composed by graduate instructors in response to student writing. In the literature, feedback concerned with addressing sentence-level writing issues and correcting errors is characterized as local, micro, and lower-order; feedback concerned with targeting larger, more abstract matters, such as the organization of the text or the writer's sense of purpose and audience, is characterized as global, macro, and higher-order. Historically, writing studies research has found instructors primarily use feedback for justifying grades on finished pieces of writing (Connors & Lunsford, 1993). In courses across the disciplines and upper-division science courses, instructors tend to focus their feedback on lower-order concerns, rather than attend to higher-order issues relevant to scientific communication (Stern & Solomon, 2006; Szymanski, 2014). Such research calls for training instructors to be aware of what they are doing when they assign and respond to student writing (Connors & Lunsford, 1993; Szymanski, 2014). Other feedback research considers the perceived helpfulness of written commentary to identify best practices for writing instructors and peer reviewers and describe characteristics of effective feedback (Blair, Curtis, Goodwin, & Shields, 2013; Cho, Schunn, & Charney, 2006; Patchan, Charney, & Schunn, 2009). Scholars also investigate feedback as an ongoing dialogue between student and teacher that extends across drafts and assignments and supports writing as an ongoing process (Barker & Pinard, 2014; Scott, 2014; Sommers, 2006). This diverse and growing body of research presents a strong, if underutilized, foundation for training writing instructors in the sciences.

In contrast, relatively little has been written about feedback practices in introductory science courses (Morgan, Fraga, & Macauley, 2011). Moreover, we found no study to date that has examined the kind, extent, and goals of the pedagogical training discipline-specific instructors receive to prepare them to provide feedback to student writers. Thus we know little about how feedback works in introductory science courses and how to prepare instructors to engage in effective feedback practices to improve students' understanding of and engagement in scientific communication. Yet the need for such training is acknowledged in the literature. Elton (2010) argues that feedback may be unclear and vague because instructors are not taught how to respond to student writing and because a discipline's writing conventions tend to be viewed as implicit knowledge and therefore unable to be taught (see also Chapter 9). Sadler (2010) found feedback to be largely ineffective even when instructors provide substantial commentary to students unless students are taught to understand feedback and to use it to improve their writing. This research calls for pedagogical training that focuses on feedback as a method for teaching writing in the disciplines. More research on feedback practices and feedback training may encourage programs to develop writing-intensive opportunities for students in the sciences and better support instructors who teach scientific communication.

Specifically, this chapter reports a training program designed for graduate students who teach writing-intensive introductory biology lab courses. A variety of timely and regular interventions in the feedback process yielded a series of snapshots of how feedback works in pedagogy and practice to teach scientific communication. This study not only investigates the relationship between feedback strategies and their perceived effectiveness, but also evaluates a pedagogical training model designed to prepare instructors to provide feedback that will help students learn the conventions and principles of scientific discourse, understand course content, and improve writing skills.

Institutional Context

The training program and study discussed in this chapter took place at a public land-grant institution. At this university, students who enroll in introductory biology lecture courses must also enroll in writing-intensive biology labs, which are taught by graduate lab assistants (GLAs) in the sciences and supported by the university's Writing Intensive Program (WIP), a writing-in-the-disciplines initiative that trains graduate students to serve as writing coaches. The graduate students teaching these courses are often unaccustomed to writing pedagogy and struggle to articulate effective writing feedback to guide the development of their students. To address this challenge, all GLAs who teach writing-intensive labs are required to take a one-credit writing-in-the-disciplines pedagogy seminar (Writing Intensive Program Pedagogy [WIPP] 7001), which is offered by WIP every fall and spring. The course meets for 50 minutes each week for 15 weeks and addresses a broad range of topics in addition to its focus on feedback strategies, including writing-to-learn, classroom assessment techniques, and professionalism. Because the GLAs regularly provide feedback to students on a number of writing assignments throughout the semester, responding to student writing is the cornerstone of the pedagogical training they receive.

We initiated this project when we redesigned WIPP 7001 in the summer of 2015 in preparation for assuming instructional responsibilities for the course in the fall. This co-instructional design and co-teaching environment enabled us to combine one instructor's expertise and experience in writing pedagogy with the other instructor's expertise and experience in scientific communication and teaching introductory science. The feedback strategy workshops and the associated research project were the product of this collaboration, developed with the goal of strategically improving our approach to teaching response to student writing. By considering the perceived helpfulness of feedback composed by GLAs from multiple perspectives—self, students, and peers—our goal at the outset was to gain insight into the effectiveness and utility of the strategies we taught. In fall 2015, we introduced and conducted four feedback strategy workshops in two sections of WIPP 7001, which served a total of 27 GLAs.

Feedback Strategy Workshops

Feedback strategy workshops involved the presentation of a new feedback strategy, followed by peer review of written comments on recently assigned student work. As well, GLAs were expected to bring a self-assessment of their feedback to the workshop. During peer review sessions, GLAs discussed the best ways to use the new strategy presented in the workshop and reflected on how previous strategies had impacted their teaching experience. GLAs seemed to enjoy these sessions and learn from their peers' teaching experiences and suggestions.

We designed the workshop schedule based on a number of key considerations (see Table 12.1). We selected strategies that supported WIP's mission to teach writing as a process; a mode of learning; and a discipline-specific way of creating, sharing, and vetting knowledge. We also selected these particular strategies because they were practical, and we predicted that the GLAs would be able to readily learn, adopt, and use them, especially within the structure of their lab courses. For example, we introduced feedback as dialogue around the time GLAs began conducting conferences with students. Further, we kept in mind the progression of writing assignments from low-stakes tasks to high-stakes projects to ensure that the strategies we discussed could be used on a variety of assignments. Finally, we sequenced the workshops so the strategies would increase in complexity and build upon each other throughout the semester. While the strategies themselves are not new or innovative, particularly for writing teachers, their application in a training regimen for GLAs who are brand new to writing pedagogy is game-changing. That is, these practices help GLAs approach the task of responding to student writing with purpose and focus and understand feedback as a teaching tool.

Table 12.1 Description of Each of the Four "Feedback Strategy Workshops" Presented to the GLAs during the WIPP 7001 Course

Feedback Strategy Workshop	*Key Concepts*
Workshop 1: The compliment sandwich	Praise – critique – praise
Workshop 2: Feedback as dialogue	Conversational moves, questions, conferences, cover letters, and revision plans
Workshop 3: Formative vs. summative feedback	Emphasis on guidance for improvement vs. grade justification
Workshop 4: Macro vs. micro comments	Emphasis on conventions and communicative strategies vs. sentence-level errors

Workshop 1: The Compliment Sandwich

We began the semester with the "compliment sandwich" method to give GLAs a structured strategy that emphasizes the value of praise over criticism. Research shows that students perceive praise to be helpful, yet graduate students tend to include more criticism in their written comments (Cho et al., 2006). The "sandwich" comprises three total statements: a compliment or encouraging message, a comment that directs the student toward improvement, and another compliment to offer further encouragement. This strategy—much like the peer and public feedback associated with the Wikipedia writing assignment discussed by Carmichael & Klock in the next chapter—motivates students to pay attention to feedback and positively disposes them toward the process of providing and receiving feedback. As well, the compliment sandwich offers a simple heuristic that GLAs with no prior teaching experience could begin using immediately.

To further support this strategy, we emphasized that less is more. Overloading student writing with comments and marks makes it hard for students to know what to focus on in revision or for their next writing assignment. Since too much feedback can overwhelm students (Walvoord & Anderson, 1998; White, 2007), the compliment sandwich reminds GLAs to limit the number of comments they write.

Workshop 2: Feedback as Dialogue

For the second workshop, we presented the concept of "feedback as dialogue." We wanted to introduce GLAs to a process-oriented approach to feedback, as too often in higher education, feedback transmits information in one direction, from instructor to student (Nicol & Macfarlane-Dick, 2006). To this end, we emphasized composing feedback as questions that asked students to think through issues on their own. We also encouraged in-class dialogues and conferences about written feedback, as well as opportunities for students to respond to or reflect on the feedback they received. As one example, we presented a cover letter assignment as a formal practice GLAs could employ to require students to participate in a conversation about their writing. For this assignment, students share responses to feedback with their instructors the same way that scientists respond to reviewers and editors after they submit articles to be considered for publication. Our emphasis on dialogical methods encouraged GLAs to see feedback as an ongoing discourse that extends beyond written remarks composed on a single, isolated draft.

Workshop 3: Formative vs. Summative Feedback

During Workshop 3, we discussed the value of formative feedback to help students understand and engage in writing as a process. Formative comments provide guidance in the midst of the writing process to facilitate revision, learning, idea development, and improvement moving forward (McGarrell & Verbeem, 2007). Summative comments, on the other hand, address a finished piece of writing and provide an evaluation of the final product's quality. We stressed the importance of formative remarks to encourage students to rethink their ideas, re-see their writing, and discover opportunities to build their writing skills over time, as opposed to isolated commentary applicable only to a single assignment. Simple suggestions like "Titles of scientific articles should be formal and informative but just like in other writing, catch readers' attention" and "Results sections should always be written in past tense" give students information pertinent to all science writing experiences. Beyond directive feedback on scientific writing conventions, many GLAs also used formative feedback to instruct students on the scientific process (e.g., "Your question was clear but more details are needed on your experimental design. Remember, the readers should be able to replicate your experiment based on details you provide here.").

Workshop 4: Macro vs. Micro Comments

We saved the most challenging strategy for the final workshop. Understanding macro and micro comments became especially important for GLAs when providing feedback on drafts of high-stakes writing projects, as many GLAs were starting to do 10 weeks into the semester. We described macro comments to GLAs as remarks that address global concerns with a text (e.g., comments that suggest ways to improve the organization of a scientific article or ask the writer to further introduce and more clearly explain research goals). We defined micro comments as references to local, or sentence-level, concerns (e.g., comments or marks that highlight punctuation errors, citation issues, awkward phrases, word choice, and grammar/spelling issues).

In our experiences with a writing-in-the-disciplines initiative, instructors in all disciplines regularly express frustration that students ignore their comments and fail to revise. While this frustration may indicate a number of complex situations, students may ignore feedback from their instructors because they are inundated with a barrage of written comments. In addition, "red pen shock" flattens any hierarchy among comments and marks, so students ultimately see problems with sentence structure, grammar, content, and scientific writing on the same level. If students attend to feedback, they typically edit linearly and locally, fixing all the "micro" issues, while larger problems go unacknowledged, unquestioned, and unrevised. For example, if an instructor leaves a comment asking for more evidence to support a hypothesis along with marginal remarks throughout the text such as "citation here" and "fix citation format," the student may feel that the evidence comment has been addressed when they add and edit citations. In this scenario, the student may consider citing background research appropriately and supporting a hypothesis as one and the same.

To reduce the risk of misleading or confusing students, we recommended that GLAs read student work without a pen in hand. We hoped this suggestion would help them avoid the temptation to compose micro comments throughout the text. In addition, we outlined a simple three-step process to further nudge GLAs toward leaving macro comments: (1) read the entire text, (2) identify the three to five most important areas of concern or suggestions for improvement, and (3) compose a final or attached note that points out what the student does well and what the student needs to work on. This process, we stressed, can require less time, as it guides GLAs to compose a limited number of high priority comments that encourage growth and development, rather than mark-up the paper line by line and comment on everything. We reminded GLAs that their job was to coach students on their writing, not copy-edit their prose.

Assessment of GLA Training

Our goal with this study was to assess whether the time spent rehearsing the above feedback strategies translated into helping GLAs provide effective feedback and, in turn, helping undergraduates improve their scientific communication skills and build scientific knowledge over time. As a result, we developed a loop among instructional training, best practices in response to student writing, the perceived effectiveness of feedback in practice, and feedback research.

The training took place in the WIPP 7001 course and feedback workshops outlined above (see Table 12.1). The research component of this loop involved surveys completed by GLAs and their undergraduate students on the perceived effectiveness of the feedback generated by GLAs, as well as compilation and analysis of the feedback itself. We chose to study perceived helpfulness in order to compare perceptions from students evaluating the feedback they received, GLAs reflecting on the feedback they composed, and GLAs assessing the feedback of their peers. Over the course of the semester, we cycled through four iterations of GLA self-assessment, feedback workshop, feedback peer review, and student survey. The recursive nature of this model enabled the GLAs to incorporate new strategies into their approach to feedback over time. After the semester, we looked at the comments GLAs composed in response to student writing to assess strategy use. Together, the surveys and the feedback itself provided data for us to evaluate the effectiveness of our feedback training sessions and assess GLAs' abilities to learn and put into practice the strategies presented during the feedback workshops.

Survey Methodology

We developed three IRB-approved surveys to evaluate the perceived effectiveness of feedback on student writing. GLAs completed the self-assessment survey prior to participating in each of the four feedback workshops. During the workshops, GLAs completed peer review surveys. Undergraduate students were asked to complete the student survey four times during the semester after they received feedback on their writing assignments. Data from these surveys were then compiled and compared across groups.

The surveys were structured to provide quantitative and qualitative data. Quantitative data were received for each of five mirrored statements in the form of one to five scores (1-Strongly Disagree to 5-Strongly Agree) by all groups (students, GLAs, and peers) (see Table 12.2). Students and GLAs were also asked to provide written commentary in response to the statement, "Please provide any thoughts or other comments about the feedback given/received." GLAs during the peer review workshops were asked additional questions (Table 12.3) to provide further perspective to their fellow graduate students. These qualitative data were compiled and used in conjunction with final reflections submitted by GLAs at the end of the semester.

This study was designed in accordance with human subject ethical research practices and was reviewed and approved by the University of Georgia Institutional Review Board and Independent Ethics Committee. All groups involved were invited to participate in the study, and participation was voluntary. Those involved had the purpose, study design, projected outcomes, and benefits explained verbally and/or in writing by the researchers prior to the start of the study. All persons who elected to participate signed a consent form in which the details of the study were explained. Finally, student texts reviewed in the peer sessions were made anonymous by removing all names associated with the work, and any names referred to in this chapter were changed to maintain confidentiality.

Results and Discussion

Graduate Lab Assistants' Perceptions of Feedback

GLAs consistently scored themselves highly on all survey statements throughout the course, though responses were higher on some statements than others (see Table 12.4). One-way ANOVA results showed that GLAs scored significantly higher on the statements "I gave useful feedback on the current assignment" and "My students can apply this feedback to future assignments" than the three following statements: "The feedback I gave on this assignment will help my students improve

Table 12.2 Quantitative Data from Mirrored Survey Statements Modified for GLAs, Peer GLAs, and Undergraduate Students. Scored on 1-Strongly Disagree, 2-Disagree, 3-Neither Agree Nor Disagree, 4-Agree, 5-Strongly Agree

	GLA *Self-Assessment* Survey Statements	GLA *Peer-Assessment* Survey Statements	Undergraduate *Student* Survey Statements
Statement 1	I gave useful feedback on the current assignment.	The instructor gave useful feedback on the current assignment.	I received useful feedback on the current assignment.
Statement 2	My students can apply this feedback to future assignments.	Their students can apply this feedback to future assignments.	I believe I can apply this feedback to future assignments.
Statement 3	The feedback I gave on this assignment will help my students improve their understanding of scientific concepts and ideas.	The feedback they gave on this assignment will help their students improve their understanding of scientific concepts and ideas.	The feedback I received on this assignment will help me improve my understanding of scientific concepts and ideas.
Statement 4	The feedback I gave on this assignment will help my students improve their science writing skills.	The feedback they gave on this assignment will help their students improve their science writing skills.	The feedback I received on this assignment will help me improve my science writing skills.
Statement 5	My feedback skills are improving due to the feedback strategies I've learned and incorporated into my teaching.	My (reviewer) feedback skills are improving due to the strategies I've learned from my peer teachers through discussion and peer-review.	

their understanding of scientific concepts and ideas," "The feedback I gave on this assignment will help my students improve their science writing skills," and "My feedback skills are improving due to the feedback strategies I've learned and incorporated into my teaching" ($p < 0.05$, $F = 6.83$, JMP version 13.0.0).

While the averages of the latter three were all above 3.8, a significantly higher response for the first two survey statements indicates a difference in perception of feedback between more immediate benefits

Table 12.3 Additional Qualitative Questions for GLAs Participating in Peer Review of Their Fellow GLAs' Feedback to (Anonymous) Students

	Additional GLA *Peer-Assessment* Written Survey Questions
Question 1	Can you see evidence of WIP principles – where/why not?
Question 2	Read as a student – are the comments clear? Do you know what you need to do to improve?
Question 3	Read as an instructor – can students apply this feedback to future assignments? Is this helping their understanding of scientific concepts? Etc.
Question 4	Please provide any other comments.

Table 12.4 Overall Survey Results from all Groups with Standard Deviation and Sample Size (N) Per Response. Statements Specific to Each Group Surveyed can be Seen in Table 12.2

		Mean Survey Score	St. Dev.	N
Statement 1	GLA	4.19	±0.54	74
Instructor gives	Student	4.25	±1.09	221
useful feedback.	Peer	4.34	±0.80	65
Statement 2	GLA	4.34	±0.69	74
Students can apply	Student	4.20	±1.12	221
feedback.	Peer	4.15	±0.87	65
Statement 3	GLA	3.91	±0.64	74
Feedback helps	Student	4.00	±1.20	221
students learn science.	Peer	4.00	±0.85	65
Statement 4	GLA	3.85	±0.84	74
Feedback helps	Student	4.11	±1.16	221
students learn scientific communication.	Peer	4.02	±0.94	65
Statement 5	GLA	3.85	±0.81	74
Feedback skills are improving.	Peer	4.02	±0.90	64

and impact on broader writing concepts and scientific knowledge. This difference suggests GLAs were more confident in the short-term benefits of their feedback as opposed to long-term outcomes. This is likely an indirect reflection of GLA confidence. GLAs may feel more comfortable leaving feedback about the content covered and writing assigned in their own class, while they may feel less confident using feedback as a tool to teach science and scientific communication more generally. A number of factors could account for this distinction, including GLAs' academic status as apprentice scholars and a lack of familiarity with the content and writing assignments of other courses.

When considering scores over the course of the four workshops, we had anticipated either an increase in scores to reflect GLAs' increasing confidence in their feedback as they developed their skills and knowledge or a decrease in scores to reveal increasing uncertainty as new knowledge broadened their horizons. Contrary to our expectations, GLAs did not significantly change their overall scores on any individual question as the semester progressed. Consistent high scores could indicate the value of regular and timely feedback workshops to prepare GLAs for the student writing they were responding to, such that GLAs maintained a high level of confidence in their ability to provide feedback. Indeed, in casual conversations and in-class discussions, GLAs demonstrated greater awareness of effective responding practices as the semester progressed. Some GLAs informally reflected that material presented in our training course opened their eyes to the complexities of pedagogy generally and responding to student writing specifically. The fact that these opinions were shared with us informally but not reflected in our survey data highlights the simplicity of the survey tool and the inadequacy of perceived helpfulness to show the development of teaching philosophies and practices over time. Regardless, other patterns did emerge from the use of survey data.

Graduate Lab Assistant vs. Peer Perceptions of Feedback

One of the goals of our study was to give GLAs tools through the feedback workshops that they could practice and further explore within peer groups. Average scores for all quantitative statements on the peer review survey were above four, making "strongly agree" and "agree" the most common responses; thus, peers generally found their fellow GLAs to be utilizing feedback strategies we discussed and composing clear comments focused on helping students improve their writing and learn course material. GLAs also responded highly (Average 4.02) to the statement, "My (reviewer) feedback skills are improving due to the feedback strategies I've learned from my peer teachers through discussion and peer-review," indicating that they valued the peer-review model implemented in our class.

When commenting on each other's feedback in the qualitative portion of the peer review survey, GLAs complimented their fellow instructors and reflected on what they learned from each other. For example,

> [She] is a genius! Reading her feedback is always helpful to give me new ideas and perspectives for my own responses to students.
>
> I really liked ... that you told them why you gave them the comments you gave. This gives me ideas on how I can improve as a science writer specifically.

By participating in feedback peer review, GLAs were exposed to feedback in practice, which helped reinforce the strategies we presented in workshops and motivate GLAs to continue developing their commenting practices.

As well, peer survey results indicated that GLAs found evidence of peers employing feedback strategies covered in workshops when responding to student writing. A GLA remarked about a peer's formative, future-oriented comments, "Solid feedback. The students should be able to incorporate this in their lab reports/future writing." Another GLA pointed out how a peer used questions to dialogue with a student and encourage active learning: "Comments always in the form of a question that led the student to think more about the assignment. They will probably need to go back and look things up in a textbook—helping to further their knowledge." In the following comment, a GLA discussed a peer's use of praise: "Anna utilized the [compliment] sandwich technique. She was not overly critical and provided informative feedback."

GLAs also regularly reflected on what their peers' feedback means for students and their writing and learning processes. For instance,

> Many students will appreciate the feedback. Some students may not read the comments because they are in depth. However, more feedback is better than little. Her students will benefit from her advice.
>
> I feel that the TA devotes a large amount of time to grading each assignment, which is something that I would like if I was a student in his class.

These remarks—which highlight the potential for students to appreciate, like, and learn from feedback—suggest that GLAs believed feedback could positively contribute to student learning.

At the same time, GLAs pointed out opportunities for their peers to improve their feedback—both in terms of practices utilized and content. In providing constructive criticism to their peers, GLAs hedged with words like "maybe," "guess," and "perhaps." This pattern could be evidence of hesitancy on behalf of GLAs or an attempt to provide facilitative, rather than directive, feedback to their peers. Instead of identifying a precise solution to a problem they saw in the comments, GLAs preferred to pose suggestions to improve feedback practices and enhance response content:

> I guess providing a rubric would have provided guidance to the students. The goals of the task may have been too non-specific.
>
> Maybe could improve feedback that improves [students'] understanding of scientific concepts, like the proper way to record units?
>
> I think you could ask more big picture questions to get [students] to think more about topics ("formative").
>
> Maybe explain why [students] need to further describe data, etc.
>
> "A little revision" is a vague directive. Also, saying things "sciency" may not mean something to a non-major.

The above comments reinforced for us the value of the peer review sessions as opportunities for GLAs to practice identifying feedback practices in context and see comments as both a form of discourse and an instructional space.

Graduate Teaching Assistant vs. Student Perceptions of Feedback

Finally, the primary objective of this study was to examine the relationship between GLA perceptions and their students' perceptions of feedback on writing. Interestingly, students scored their instructors significantly higher than GLAs scored themselves in all categories (ANOVA, $p < 0.05$, $F = 2.39$) except statement 2, "I believe I can apply this feedback to future assignments" in which they scored 0.14 lower but not significantly (see Table 12.4). These results suggest that despite the struggles graduate students face when giving helpful feedback and the tendency graduate students have to be more critical in their commentary (Cho et al., 2006), their students responded positively to it. Positive student response was also evident in the comments students shared about the feedback they received:

> My GLA very specifically pointed out what I did well and what I needed to improve upon and how.
>
> My feedback was very thorough and easy to understand. I completely agreed with all my TA's comments and since this was my first assignment, I know what to improve upon next time.
>
> Extremely helpful information. I plan to incorporate this feedback into my future assignments.

For other students, however, feedback was not as helpful or instructive. As the comments below illustrate, students recognize the potential to learn from the feedback they receive, and, just as importantly, they see a lack of constructive feedback as counterproductive to their learning:

> He didn't have any critiques of what I wrote, just said it was correct. So there isn't much to learn from his feedback.
>
> Only received 4/5 points on assignment with no other feedback other than "looks good." I wish I knew what I could improve on.

Because students participating in the study completed surveys on a voluntary basis, we acknowledge that biases likely exist. Student participants were either positive about or unsatisfied with their GLAs and the feedback they received. Few students with moderate opinions participated in the study. Thus, while our results do not offer a full spectrum of student perceptions, we are reporting the range of responses we received. We noted that regardless of the negative reviews, students were more positive in their responses than the GLAs were.

Qualitative Results and the Use of WIP Principles

Throughout the feedback given to students by GLAs and in their students' responses to it, we found evidence of GLAs incorporating the principles presented in the feedback workshops into their teaching.

Workshop 1: The Compliment Sandwich

The compliment sandwich strategy introduced in our initial feedback workshop was adapted and used frequently in GLA feedback on a variety of assignments and throughout the semester. The following feedback examples from GLAs to biology students demonstrate effective implementation:

> Very good! Love the way you presented your points. You might want to elaborate on the social classes idea to provide a better context for your evidence. I also like the idea of including subjects from different age groups as it will provide more conclusive results.
>
> I like the question, but I think it needs a bit more before it would be biologically testable, right? Try to drill down to really get a testable question that still relates to the sessile nature of Scenedesmus. I also like the way you made the link between zooplankton predation and phytoplankton life history strategies! Great work.

Many GLAs commented in their surveys that they found the compliment sandwich technique enabled them to establish expectations at the beginning of the semester and build rapport with students through feedback as the semester progressed. For example, a GLA reflected, "I believe that, by giving this kind of feedback on a minimal assignment, students will be more conscious that I will thoroughly inspect every writing activity." Other GLAs found it hard to find a balance between giving thorough feedback and managing their time. As one GLA commented, "Potentially my feedback is too extensive and takes me too long to be practical." But the time invested in this strategy seemed to be well spent, according to its positive reception among students:

> Really good about noting the strengths and weaknesses and providing helpful ways to improve.
>
> Very thorough feedback with reasoning behind the positives and negatives of my work. Much appreciated!
>
> I did not expect any feedback [on a short assignment] but was surprised to receive it. My GLA gave positive feedback regarding my assignment too!

Our results from GLA and student comments on this strategy support prior research suggesting that students appreciate positive reinforcement

(Patchan et al., 2009) and that instructors believe such commentary to be helpful for student writers (Connors & Lunsford, 1993; Hyland & Hyland, 2001).

Workshop 2: Feedback as Dialogue

While we presented a variety of implicit and explicit methods for achieving a dialogical approach to feedback, questioning was the only method discussed in the surveys and evident in GLA comments (See Workshop 2: Feedback as Dialogue above). No mention of other dialogic strategies may be indicative of the survey statements and prompts used to elicit comments from GLAs. Specifically, the surveys did not ask about verbal interactions, nor did they inquire into the use of feedback response assignments, such as the cover letter. Still, a few students commented on the use of in-class discussions to bolster their understanding of the feedback they received. On student reflected, "He gave helpful feedback, and in class made sure we understood what we were asking and that it had an abiotic and biotic element that were measurable." We see remarks like this one as evidence that these verbal conversations were taking place to clarify or further emphasize points expressed and material covered in written feedback. Similarly, a GLA used feedback as a conversation starter: "Really think about how/if you want to put your shortcomings in so blatantly. What do they add to your argument? What do they make your reader think about your conclusions? Let's talk about it next class!"

As well, several GLAs highlighted the effective use of questions by their peers:

> Made strong comments with questions following to help students think through their writing.
>
> Asked questions, didn't just leave comments like "incorrect."
>
> [The GLA] wrote directed questions that can lead the students to analyze what they did wrong and come up w[ith] what they need to do to fix it.

Finally, we saw many instances of GLAs incorporating questions into their feedback statements to encourage students to continue thinking through scientific concepts as well as the effective written expression of scientific material. For instance,

> In other words, why [is] knowing about vascular efficiency of Coleus relevant?
>
> Are you saying because termites have high B-12 concentration they might react similarly to 5hr. Energy? Not quite clear about what you are trying to say …
>
> So what does this tell us about how well the starch is working?

These results suggest that GLAs readily adopted and recognized the value of inquiry-based feedback to prompt students to think more carefully about concepts and how they explain those concepts in writing. At the same time, however, we did not see evidence of GLAs using questions to encourage students to think more about scientific communication conventions or writing more generally. Thus GLAs seemed far more likely to use questions to create a dialogue with students about *what* they were writing, rather than about *how* they were engaging in scientific communication.

Workshop 3: Formative vs. Summative Feedback

Providing formative rather than summative feedback was easily accepted by GLAs in theory but proved more difficult to implement correctly. GLAs recognized the importance of orienting feedback toward process-related considerations to enable "students to think about how they are structuring and presenting their data and then connect it to the larger concepts they are learning and testing," as one GLA commented. Out of all the feedback strategies, students responded most often and most favorably to comments that offered formative assessment:

> Extremely helpful information. I plan to incorporate this feedback into my future assignments.
>
> I love getting feedback on my assignments because it helps me to improve my future assignments before handing them in.
>
> Some of the comments I use for my other classes as well and they helped a lot! Thanks!

Such comments highlight the value of formative feedback with a global focus to encourage students to see scientific writing as a skill they need to continually develop over time, rather than a series of assignment-specific requirements, which is often the overarching message of summative feedback.

Workshop 4: Macro vs. Micro Comments

Our emphasis on using macro comments rather than micro comments was the most memorable and thoroughly discussed of the strategies by the GLAs, both positively and negatively. Predictably, this concept was met with some contempt. Many GLAs argued that micro comments were necessary, as they believed (and most likely had been taught themselves) that submitted work should be error-free and mistakes were highly unprofessional. While we endorsed the desire for clean prose, we stressed that correcting students' sentences does not help students learn to correct their own errors, and micro comments can encourage students to elide the difference between correct writing and effective scientific communication. Still, many GLAs commented that they "stand by micro-comments, it is essential in science," despite our presentation of the literature on the subject. Others tried their best to implement macro comments. One GLA remarked, "it was hard, but I made an effort not to comment on any grammar or language issues so that they could focus on the important stuff." Another GLA referred students to the lab manual so they could improve their citations. Still other GLAs made a point to only comment on micro issues if the larger concepts had been addressed. For instance, one GLA wrote, "I tried to provide comments about content and concepts first before providing any feedback on their essay structure or grammar." Students commented on the use of this strategy by pointing out feedback focused on global issues and feedback connected to higher-order thinking. One student commented, "I was able to better organize my article because of the feedback," and another reflected, "He made some comments that I think will aid my critical thinking for the next assignment." Such remarks show students engaging meaningfully with feedback that asks them to think deeply when writing and revising.

Conclusions

After compiling and analyzing survey data and feedback, we noted the limitations and affordances of our methodology. The four iterations of the surveys presented four distinct snapshots of feedback training and practice but did not demonstrate any trends or trajectories as the semester progressed to chart pedagogical development. That is, the surveys provided us with a series of evaluations of specific feedback on specific assignments at specific moments in the semester. At the same time, we found the survey methodology to be a compelling teaching tool, as it offered us a way to hold GLAs accountable for their training. The GLAs knew they would be evaluated by themselves, their peers, and their students, and this knowledge motivated GLAs to implement the strategies we covered in class and work hard to compose effective, meaningful feedback. In addition, we saw the survey instrument as a way to increase investment in feedback training and practice. Specifically, the self-assessment survey guided GLAs through a reflective analysis of their feedback strategies, which in turn helped direct their attention to the purpose of feedback: to help students build capacity in scientific knowledge and writing. Further, the survey instrument also cultivated GLAs' confidence in their teaching ability. By engaging in peer review and receiving feedback on their practices, GLAs were able to experience pedagogy much like their students were experiencing writing: as skills and knowledge applied in the ongoing pursuit of certain goals. The support and praise they received from their peers especially helped them see response to student writing as something they could do well.

We have been pleased with the success of this project, which trained GLAs to provide better feedback and documented the high level of perceived helpfulness of that feedback by their students. A future study could analyze the feedback GLAs are actually composing and attempt to measure gains in student writing and learning as a result of that feedback. Another future investigation could extend the existing literature on peer feedback (Cho et al., 2006; Patchan et al., 2009; Topping, 1998) and consider methods for training students to respond effectively to their peers' written work. Such research is necessary to continue exploring feedback in the sciences as a diverse metadiscourse and identify disconnects among pedagogy research, training, and practice.

Recommendations for Instructors and Program Administrators

Scientists must write to publish their research and, especially in the digital age, connect to the public. Therefore, training for science students needs to include opportunities for them to learn, practice, and engage in scientific communication. For those opportunities to be successful, training for those who teach undergraduate science students needs to highlight the importance of effective feedback. The positive response from GLAs involved in our research validates extensive feedback training; we saw increased confidence in our GLAs, who paid more attention to their feedback and likely increased their motivation to respond intentionally and thoughtfully to their students' writing. From these outcomes, we recommend the following practices for integrating writing and pedagogical training into other science programs and departments:

1 Train graduate and faculty instructors through coursework, professional development initiatives, and/or mandatory workshops focused on discipline-specific writing pedagogy.
2 Combine pedagogical training with professional development to increase incentive and motivation to participate in writing-related training.
3 Assess effectiveness of training and participants' accountability for training through tools like surveys, peer review, evaluations, and/or reflection.
4 Assess the perceived effectiveness of the written feedback students receive through surveys.

References

Barker, M., & Pinard, M. (2014). Closing the feedback loop? Iterative feedback between tutor and student in coursework assessments. *Assessment and Evaluation in Higher Education*, 39(8), 899–915. doi:10.1080/02602938.2013.875985.

Blair, A., Curtis, S., Goodwin, M., & Shields, S. (2013). What feedback do students want? *Politics*, *33*(1), 66–79. doi:10.1111/j.1467-9256.2012.01446.x.

Brownell, S. E., Price, J. V., & Steinman, L. (2013). A writing-intensive course improves biology undergraduates' perception and confidence of their abilities to read scientific literature and communicate science. *Advances in Physiology Education*, *37*(1), 70–79. doi:10.1152/advan.00138.2012.

Cho, K., Schunn, C. D., & Charney, D. (2006). Commenting on writing: Typology and perceived helpfulness of comments from novice peer reviewers and subject matter experts. *Written Communication*, *23*(3), 260–294. doi:10.1152/advan.00138.2012.

Connors, R. J., & Lunsford, A. A. (1993). Teachers' rhetorical comments on student papers. *College Composition and Communication*, *44*(2), 200–223. doi:10.1177/0741088306289261.

Elton, L. (2010). Academic writing and tacit knowledge. *Teaching in Higher Education*, *15*(2), 151–160. doi:10.1080/13562511003619979.

Holstein, S. E., Steinmetz, K. M., & Miles, J. D. (2015). Teaching science writing in an introductory lab course. *The Journal of Undergraduate Neuroscience Education*, *13*(2), A101-A109. doi:10.2307/358839.

Hyland, F., & Hyland, K. (2001). Sugaring the pill: Praise and criticism in written feedback. *Journal of Second Language Writing*, *10*(3), 185–212. doi:10.1016/S1060-3743(01)00038-8.

Kutney, J. P. (2008). Guaranteeing the failure of first-year composition: Four assumptions about writing expertise that support an unattainable standard for transfer. *International Journal of Learning*, *15*(8), 223–227. Retrieved March 13, 2016, from www.Learning-Journal.com.

Lee, S. E., Woods, K. J., & Tonissen, K. F. (2011). Writing activities embedded in bioscience laboratory courses to change students' attitudes and enhance their scientific writing. *EURASIA Journal of Mathematics, Science, and Technology Education*, *7*(3), 193–202. Retrieved March 12, 2016, from www.ejmste.com/v7n3/EURASIA_v7n3_Lee.pdf.

McGarrell, H., & Verbeem, J. (2007). Motivating revision of drafts through formative feedback. *ELT Journal*, *61*(3), 228–236. doi:10.1093/elt/ccm030.

Morgan, W., Fraga, D., & Macauley, W. J. (2011). An integrated approach to improve the scientific writing of introductory biology students. *American Biology Teacher*, *73*(3), 149–153. doi:10.1525/abt.2011.73.3.6.

Mory, E. H. (1992). The use of informational feedback in instruction: Implications for future research. *Educational Technology Research and Development*, *40*(3), 5–20.

Moskovitz, C., & Kellogg, D. (2005). Primary science communication in the first-year writing course. *College Composition and Communication*, *57*(2), 307–334.

Moskovitz, C., & Kellogg, D. (2011). Inquiry-based writing in the laboratory course: Writing lab reports in science classes can be more productive and engaging if the experience is structured well. *Science*, *332*(6032), 919–920. doi:10.1126/science.l200353.

Nicol, D. J., & Macfarlane-Dick, D. (2006). Formative assessment and self-regulated learning: A model and seven principles of good feedback practice. *Studies in Higher Education*, *31*(2), 199–218. doi:10.1080/03075070600572090.

Paszkowski, C., & Haag, M. (2008). Writing-to-learn in first-year biological sciences. *Collected Essays on Learning and Teaching, 1*, 132–137. Retrieved March 13, 2016, from https://eric.ed.gov/?id=EJ1055107.

Patchan, M. M., Charney, D., & Schunn C. D. (2009). A validation study of students' end comments: Comparing comments by students, a writing instructor, and a content instructor. *Journal of Writing Research, 1*(2), 124–152. Retrieved April 2, 2016, from https://doaj.org/article/cbd6071bca534215a10e3c41cd2516cc.

Sadler, D. R. (2010). Beyond feedback: Developing student capability in complex appraisal. *Assessment and Evaluation in Higher Education, 35*(5), 535–550. doi:10.1080/02602930903541015.

Scott, S. V. (2014). Practising what we preach: Towards a student-centred definition of feedback. *Teaching in Higher Education, 19*(1), 49–57. doi:10.1080/13562517.2013.827639.

Sommers, N. (2006). Across the drafts. *College Composition and Communication, 58*(2), 248–257.

Stern, L. A., & Solomon, A. (2006). Effective faculty feedback: The road less traveled. *Assessing Writing, 11*(1), 22–41. doi:10.1016/j.asw.2005.12.001.

Szymanski, E. A. (2014). Instructor feedback in upper-division biology courses: Moving from spelling and syntax to scientific discourse. *Across the Disciplines, 11*(2). Retrieved March 12, 2016, from http://wac.colostate.edu/atd/articles/szymanski2014.cfm.

Topping, K. (1998). Peer assessment between students in colleges and universities. *Review of Educational Research, 68*(3), 249–276. Retrieved March 2, 2017, from www.jstor.org/stable/1170598.

Walvoord, B. E., & Anderson, V. J. (1998). *Effective grading: A tool for learning and assessment.* San Francisco, CA: Jossey-Bass.

White, E. M. (2007). *Assigning, responding, evaluating: A writing teacher's guide.* Boston, MA: Bedford/ St. Martin's Press.

13 Incorporating Wikipedia in the Classroom to Improve Science Learning and Communication

Becky J. Carmichael and Metha M. Klock

Many people today consult Wikipedia to get answers to questions like these:

- How is the flu spread?
- Why is the sky blue?
- What is climate change?

Wikipedia, the digital encyclopedia, has approximately 15 billion page views a month (Anderson, Hitlin, & Atkinson, 2016; "Report Card," n.d.) and is a platform where editors worldwide collaborate to improve content on topics, including the questions above. For students, Wikipedia presents opportunities to collaborate with global editors, engage in discussion about topic presentation, and develop effective science communication skills. In this chapter, we provide an overview of Wikipedia to introduce the platform, outline ways students can contribute to the creation of articles, illustrate scaffolding of Wikipedia-based assignments, share faculty and student examples to highlight benefits and challenges of working with Wikipedia, and offer tips for students and teachers.

Wikipedia is a repository of increasingly reliable information, primarily due to implementation of strict guidelines for contributors. The Centers for Disease Control and Prevention monitor Wikipedia access logs to gauge interest in communicable diseases and forecast potential outbreaks (Generous, Fairchild, Deshpande, Del Valle, & Priedhorsky, 2014), and medical professionals consult Wikipedia articles for reference about particular diagnoses (Haigh, 2011; Heilman, 2011; Purdy, Thoma, Bednarczyk, Migneault, & Sherbino, 2015). This online, open-access encyclopedia bridges the knowledge gap between scientists and the public by providing science information in a comprehensible, neutral format ("Citing sources on Wikipedia," n.d.). Millions of editors contribute to Wikipedia, making scientific information broadly available to anyone with Internet access (Salvaggio, 2016c). Through its straightforward, user-friendly platform, Wikipedia increases public familiarity with science content and the scientific process (Horrigan, 2006; Moy, Locke,

DOI: 10.4324/9781315160191-13

Coppola, & McNeil, 2010). Wikipedia is a unique source of information not only for the public, but also for students learning to communicate scientific information.

Wikipedia supports science communication in several ways. It helps readers comprehend information and contributors clarify the meaning and implications of scientific knowledge. It provides an easily accessible source to research scientific topics and participate in the investigative process. By doing so, Wikipedia increases the general public's awareness, interest, and involvement in science (Burns, O'Connor, & Stocklmayer, 2003). Wikipedia depends on writers and editors who employ the standards of effective scientific communication. University students are well-suited to create and improve the quality of Wikipedia, expanding access to scientific content while developing their own communication skills. Since 2014, 645 students in 32 courses at Louisiana State University (LSU) have edited 912 articles and created 90 new articles on Wikipedia. Collectively, these articles have received over 32.7 million views ("Campaign: Louisiana State University," n.d.). Since 2010, 22,000 students in classrooms throughout the United States have contributed to ~35,000 Wikipedia articles ("How do you measure the difference that open knowledge makes?" 2015); these numbers continue to grow (Dewey, 2016). Students contributing to Wikipedia disseminate course content and share knowledge beyond their academic settings. By delving into scientific topics and publishing information through Wikipedia, students both learn and teach.

In this chapter, we apply the methodology of practitioner inquiry (Liggett, Jordon, & Price, 2011). Practitioner inquiry values the experiential knowledge of practitioners who use reflexive research and dialectical means to investigate and validate new knowledge. A reflexive practitioner critiques through encounters with others, including related literature and observation (Qualley, 1997). We apply practitioner inquiry to class observations and student samples from several semesters at LSU, showing how students can develop skills in science communication by contributing to Wikipedia. We share feedback from faculty and students who have participated in Wikipedia-based assignments (first names or pseudonyms were used when referencing course work, with permission of students and faculty). We share examples of assignments demonstrating how students develop an appreciation for and understanding of the sciences, develop self-confidence by participating in scientific conversations, and engage global audiences through Wikipedia content creation and collaboration.

The Wikipedia Platform

One daunting aspect for students (and perhaps teachers) who use Wikipedia is in the initial stage, familiarizing themselves with the Wikipedia interface. Wiki Ed is a non-profit organization that provides tutorials for professors and students to help increase their confidence in contributing to

Wikipedia. Wiki Ed aims to improve student learning in higher education by partnering with instructors, supporting student-driven Wikipedia contributions that enrich course learning objectives and content access (Wiki Education Foundation homepage, n.d.). Eileen, a student in the course Natural Disturbances and Society at LSU said,

> As I was beginning the assignment, I was far more than apprehensive. The editing tools on Wikipedia look a far cry from user friendly... [T]he workshops and online training were useful and necessary. These are certainly two or three hours that are necessary for becoming familiar with and mastering the editing process.

Providing an introduction to the Wikipedia platform, including the guidelines for Wikipedia use and publication, is a key step in helping student contributors. The guidelines are centered on Wikipedia's three core content policies: contributions must have a neutral point of view, be supported by verifiable sources, and include no original research ("Core content policies," n.d.). Design and layout consistency permits edits to any page, further facilitating Wikipedia's concept of open, crowd-sourced knowledge generation. Article pages, found in the Wikipedia mainspace, contain neutral, topic-specific information. Well-written article pages, devoted to notable topics, are focused, organized, and verifiable, and include appropriate graphics. Article pages are organized by a set of tabs, including Talk, outlined in the Anatomy of Wikipedia section.

The Talk page is an integral component of Wikipedia where students can collaborate with each other, as well as other editors, to discuss topics, offer advice, and resolve disagreements. Students new to Wikipedia can examine Talk pages on articles of interest to see how these online conversations help to shape or modify existing articles (Jenkins, Purushotma, Weigel, Clinton, & Robison, 2009). For example, editors may discuss page content or reference suggestions. Comments, questions, or ideas added to Talk pages require editors to "sign" their posts with four tildes (~), leaving a Username and timestamp. This exchange is recorded on the Talk page, providing students with artifacts of interactions and documenting differences and consensus in knowledge construction. Evaluating Talk pages allows students to develop a sense for specific guidelines on Wikipedia and gain experience in "netiquette" to be effective contributors (Brailes, Koskinas, Dafermos, & Alexia, 2015).

Summary and additional information about the Wikipedia platform, including descriptions of key features such as the Sandbox or Stubs, can be found in the list below.

Anatomy of Wikipedia

The anatomy of Wikipedia is symmetric, allowing for ease in contribution and discussion. Key terms used on the platform are defined below.

Article pages: Article pages are found in the Wikipedia mainspace. Well-written article pages, devoted to notable topics, are focused, organized, and verifiable; written in a neutral style; and include appropriate graphics.

Edit: The Edit tab enables an editor to add and modify content in Wikipedia. It allows the editor to input information similarly to word processing software. Editors concisely note contributions in an Edit Summary, where each addition is recorded lending to transparency. The *Edit Source* tab is another option for editing, allowing for edits to be made in wikicode. Editors will find access to formatting options such as bold, italics, and a citation wizard in both the *Edit* and *Edit Source* tabs.

Read: The Read tab provides a view of the article in its current state. Consumers of Wikipedia articles typically see this view.

Sandbox: Every Wikipedia User has a Sandbox in which to draft and organize contributions and test code. The Sandbox has fewer restrictions compared to the live article pages, though civility is still required because the contents can be viewed by anyone on Wikipedia. The Sandbox also has an associated Talk page, a useful space for providing peer and instructor feedback and critique before content goes live in the mainspace of Wikipedia.

Stub: A Stub is a short, undeveloped article on a notable topic that does not provide adequate coverage. Stubs are pages that students may choose to modify or enhance for a course assignment.

Talk pages: Talk pages are where Wikipedia editors discuss topics. Talk pages are associated with each Wikipedia Article, User, and User Sandbox pages where conversations between editors occur. Comments, questions, or ideas added to Talk pages require editors to "sign" their posts, leaving a Username and timestamp.

User pages: User pages provide space to organize new content and facilitate interaction with other editors. User pages have an associated Talk page where editors can converse about edits, ask questions, provide resources, resolve conflicts, and praise each other's work.

View history: The View history tab allows a user or editor to review the development of any Wikipedia page. This tab is particularly useful to examine how an article has evolved with updated information, research, etc. From this tab, page statistics can also be accessed, providing additional information about interest in the topic.

Wikipedia-Based Assignments

Wikipedia-based assignments range from making small edits, such as copyediting a series of science-related topics, adding citations, or inserting internal links to existing Wikipedia pages, to more substantial contributions, such as adding paragraphs of information to existing pages, updating content to convey research developments, creating new article pages, or adding visuals or audio. In this chapter, we offer three assignments used over multiple semesters at LSU that showcase the important role Wikipedia can play in science communication and illustrate the benefits to students. These assignments include exploring referencing and plagiarism in Wikipedia articles, contributing content, and critiquing content. These assignments can stand alone or, if assigned over the course of a semester, provide scaffolding for a major project. Such assignments familiarize students with how to contribute to Wikipedia as they build scientific knowledge. The assignments were designed to aid students in developing the following skills:

- Assessing accuracy of content
- Identifying needs of a target audience
- Using online technology and netiquette
- Applying and developing information literacy in the sciences
- Understanding science concepts
- Integrating information from various courses and sources
- Evaluating neutrality in resources and writing styles
- Generating, revising, and editing written communication
- Collaborating effectively with peers and editors

For each assignment, we briefly describe its objectives, include student examples, and indicate the benefits to students. The main course used to illustrate Wikipedia assignments is Natural Disturbances and Society, a science course for non-science majors taught at LSU by Dr. Becky Carmichael. The course is designed to introduce the principles of disturbance ecology, explore how natural disturbances shape ecosystems, examine ways humans affect and are affected by disturbance events, and introduce scientific methodology and principles. During the course, students selected several Wikipedia articles about natural disturbances or natural disasters, evaluated the articles' current state, edited the articles to improve clarity, and revised their contributions based on feedback from their peers, instructor, and global Wikipedia editors. Additional LSU courses that employed Wikipedia assignments are discussed to emphasize skills developed or show other assignment options.

Assignment 1: Exploring Existing Wikipedia Articles for Referencing and Plagiarism

Students new to science are often unfamiliar with how to find and cite peer-reviewed resources. Contributing to Wikipedia can help them develop these skills. Ideally, every sentence in Wikipedia should be verifiable and referenced ("Citing sources on Wikipedia," n.d.). Because such documentation is missing from many Wikipedia pages, students have several opportunities to identify statements needing verification. Through the processes of statement verification, students gain skills using tools such as Google Scholar or Web of Science. They also learn how to discern the differences among "gray" literature, peer-reviewed scientific articles, tertiary references, and online sources. Along with differentiating the value of sources, students can gauge the neutrality of information, sorting verifiable data from unsupported opinions.

In the course Natural Disturbances and Society, students are tasked to locate resources related to a chosen disturbance event. However, students have difficulty determining whether sources are appropriate and struggle to retain meaning of content without directly copying the original text. For example, one student located information on a recent hurricane event from an online source, but inserted the content almost verbatim without attributing text to the original author. Other students had difficulty ascertaining reliable content, selecting blog posts or advertisers over peer-reviewed scientific journals or reputable news agencies as references.

Challenges faced by students necessitate "just-in-time" instruction on reference reliability, content incorporation, and rules regarding plagiarism (including Wikipedia standards). In the Natural Disturbances course, students are provided with criteria for evaluating reliable sources. Students then assess the reliability of several sources, comparing popular news media outlets (BBC News, NPR), governmental agencies (National Aeronautics and Space Administration (NASA), National Oceanic and Atmospheric Administration (NOAA)), organizations (Greenpeace, Red Cross), and scientific journals (*Nature*, *Science*). Discussions typically center around accuracy of content, biases and neutrality, motive for publication, and intended audience. In one lesson, students ranked example sources from most to least reliable to gain an appreciation for source bias and reliability. Students also compared content among these sources, exploring how information was presented to different audiences. Next, students learn how to locate reliable scientific references. Many times, searches for scientific literature begin on Google, but during this in-class exercise, the search was expanded to library databases, such as Web of Science. With these resources at hand, students reviewed Wikipedia's criteria for paraphrasing, identifying what is considered ideal incorporation of new information, when to use direct quotation, and how to use appropriate citation metrics ("Citing sources on Wikipedia," n.d.). Such exercises initiate discussion on the ways publication guidelines differ across journals and disciplines.

After developing new skills for assessing source reliability, students assess a Wikipedia article for existing statements requiring citation. In a recent semester, students copied statements into a Google search and attempted to locate an original reference. Many students discovered that content on Wikipedia was repeatedly plagiarized. As a class, students scrutinized the existing pages, discussing how to paraphrase statements under Wikipedia guidelines. Every Wikipedia User has a Sandbox in which to draft and organize contributions and test code. When plagiarism or close paraphrasing was located, students drafted revisions in their Sandboxes and noted changes they made on the article's Talk page (see *Anatomy of Wikipedia*). The following revision example from the article "Pine processionary" (2016) is the work of Connor, who corrected plagiarized statements from the source, www.impactproject.eu. In the revised statement, Connor identified alternative ways to communicate information from this source.

> Before: "The typical cylindrical egg masses range in length from 4 to 5 cm."
>
> After: "The eggs of the Moth are laid in cylindrical bodies ranging from 4 cm to 5 cm in length."

Connor's revision conforms to Wikipedia guidelines for paraphrasing and use of quotations. Contributors to Wikipedia are encouraged to summarize an original author's work, limiting direct quotations to short statements. In the revision, Connor synthesized the necessary components and summarized the ideas in his own words, demonstrating his understanding of the original content and methods for removing plagiarism.

Understanding where to locate sources and how to evaluate information are integral components of building literacy in a field. Relevant, reliable sources are required to support statements and build arguments. Wikipedia assignments challenge students to locate appropriate scientific articles they can use to cite new content and translate ideas for the broader Wikipedia audience. Dr. Cameron Thrash at LSU, who uses Wikipedia for his course Prokaryotic Diversity, found,

> The primary challenge [for students contributing to Wikipedia] is identifying all the relevant information. This is the process I most want them to experience...because that's what we do as scientists both in writing papers to report our results, but also in creating background for our grant proposals.

Colleen, a student in Dr. Thrash's course, said that she doubted her ability to read scientific papers and apply their content to Wikipedia:

> I not only had to read the papers but read them quickly and understand what the researchers were trying to communicate. However, the more papers I read the better I was able to understand them and recognize key information. ... A large part of the process was absorbing the information from the scientific papers, then figuring out how to report [it] ... with proper citations to avoid plagiarism.

Wikipedia assignments require students to develop literacy in different styles and genres of scientific communication and help them to increase confidence in reading and translating scientific information. As another student explains, "I can use this in the future ... now I know how to find scientific sources through Wikipedia and check for validity and also be involved in the scientific community."

Assignment 2: Contributing to Wikipedia

Contributing to Wikipedia provides an opportunity for students to improve their writing skills. Writing engages students in the construction of coherent content through critical analysis of information and is one of the best ways to learn new material (Barkley, Cross, & Major, 2005). Purposeful writing assignments require that students conduct research to expand their knowledge and information literacy and develop an understanding of how experts in the discipline construct content and share it with an audience (Bean, 2011). When contributing to Wikipedia, students must consider course content, connect new information to familiar understanding, and evaluate novel ideas in the context of foundational disciplinary concepts. Traditionally, students have worked toward these goals through term papers and lab reports. By reframing the class term paper using Wikipedia contributions, students expand learning, evaluate what information to share, and engage in a global exchange of knowledge, informing a massive audience on specialized topics (Salvaggio, 2016b). The following three examples show how an assignment can be designed to involve different levels of content creation by students.

Small Contribution

Students in Natural Disturbances and Society are tasked to contribute content to a series of disturbance articles on Wikipedia based on research in primary literature. These small contributions consist of a few sentences that connect the science or mechanisms of how a disturbance occurred to a specific incidence and build available information on the disturbance type. Connor, the student quoted above, added the following excerpt to the article "Pine processionary" (2016). The contribution provides Wikipedia readers with details on the disturbance caused by pine processionary caterpillars, filling content gaps and supporting the information with citations from peer-reviewed scientific journal articles.

> The pine processionary caterpillar is responsible for most of the defoliation of southern Europe (Li, Daudin, Piou, Robinet, & Jactel, 2015). Although pines are most susceptible to the caterpillar,

> other trees such as larches are also vulnerable. The caterpillars can completely defoliate trees if large quantities are present.
>
> (Forestry Commission, 2017)

Another Natural Disturbances and Society student examined the article "2013 Colorado floods" (2013) and identified ways to expand knowledge of the event. The student noticed the article was missing information related to the United States federal government shutdown and its implications on relief efforts, a topic that had been discussed in class. The student added the following excerpt connecting content from the course with this event:

> … The [United States federal government] shutdown compromise signed on October 17, 2013 includes funding for Colorado relief efforts, specifically referencing Rep. Gardener's bill H.R. 3174; 113th Congress. The cap typically set at $100 million has been raised to $450 million in light of Colorado's current conditions. It is not uncommon for this cap to be raised for disaster struck areas such as those states hit by Hurricane Sandy or Hurricane Katrina.

In this excerpt, the student identifies omitted details and provides context for a reader to better grasp what occurred during the event. Further, the student recognizes the need to include a hazardous impact section describing the potential disruption to clean water due to flooding.

> Structures located in high risk flood zones were soon inundated. Sewage treatment plants affected by the flood waters released 20 million gallons of raw sewage as well as 150–270 million gallons of partially treated sewage, as estimated by the State health department. What resulted was higher levels of *E. coli*, some as high as 472–911 colonies per millimeter of water (126 colonies per millimeter of water is considered unsafe) (Denver Post, 2013). The Colorado Oil and Gas Conservation Commission (COGCC) reports that oil lines and containment facilities failed and leaked a total of 1,027 barrels of 43,134 gallons of oil. The COGCC is monitoring 13 substantial leaks as of October 8, 2013.
>
> (Colorado Oil and Gas Commission, 2013)

By adding content to Wikipedia, students become familiar with the editing process, observing how their written contributions are interpreted by a larger community of informed editors outside academia.

Substantial Contribution

Substantial contribution to Wikipedia can be as simple as locating and expanding a Stub, short undeveloped articles on a notable topic.

Students can select one of the many designated Stub articles from a variety of topics on Wikipedia. This encourages students to take ownership of information learned in class. Students must identify gaps in the content currently available, recognizing missing information and clarifying ideas. Although the exercise can be challenging, it has rewards for students and the online community alike. Creation of new content requires students to cover a topic comprehensively, identifying subsections, choosing citations, selecting or generating relevant images, while following Wikipedia guidelines for a neutral style.

Several students in the Natural Disturbances and Society course elected to expand existing Wikipedia Stub articles. This assignment required students to research their topics, seek updated references, consult the Article Talk page to access what additions were needed, and incorporate new content.

Brad elected to expand the Stub page on the Morris J. Berman oil spill. The article, created in January 2010, consisted of only 2,281 characters and two references (see Figure 13.1). Beginning in his Sandbox, Brad added more than 16,000 characters to existing content and expanded the article to address effects of the oil spill on the environment, tourism, and wildlife (see Figure 13.2). Brad's addition was not only substantial, but also earned a place in the "Did you know...?" section on the main page of Wikipedia, receiving 875 views in one day. This was rewarding for the student and demonstrated global readers' interest in the topic.

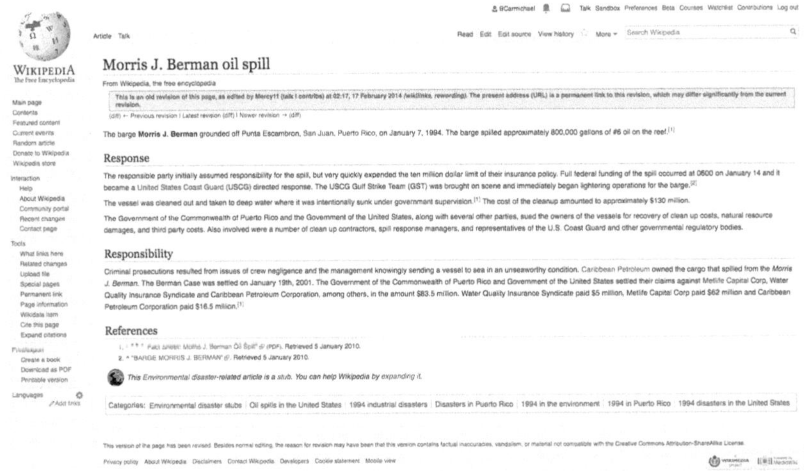

Figure 13.1 The "Morris J. Berman oil spill" Wikipedia article as a stub before Brad's contributions ("Morris J. Berman oil spill," 2014a). CC BY-SA 3.0. Wikimedia Foundation, Inc. (WMF) Marks are trademarks of WMF, the authors of this chapter are independent of WMF, and the WMF Marks are used under license.

Figure 13.2 The "Morris J. Berman oil spill" Wikipedia article after Brad's contributions. ("Morris J. Berman oil spill," 2014b). CC BY-SA 3.0. Internal images: "Condado Beach," (2012), CC BY-SA 2.0; "Brown booby," (2005), CC BY-SA 3.0. Wikimedia Foundation, Inc. (WMF) Marks are trademarks of WMF, the authors of this chapter are independent of WMF, and the WMF Marks are used under license.

New Article Creation

Dr. Alex Webb, University of Hong Kong, used Wikipedia assignments as an alternative to term papers in two geology courses (Plate Tectonics and Evolution of the Terrestrial Planets) while a professor at Louisiana State University. Creating a new Wikipedia article was a semester-long project that consisted of four sections: selecting a topic, drafting a contribution in the student's Sandbox, moving a revised contribution into Wikipedia (making it live), and interacting with the Wikipedia community. Each section was sequenced to allow for development of the article, instructor and peer review, and feedback from global editors. Students had to select topics relevant to the course at the beginning of the semester and have a polished draft within the first month to allow time for interaction with other students in the class and the global community.

Success of the assignment hinged on students selecting appropriate topics, providing feedback on each other's projects, and understanding audiences. Students identified topics by conducting Wikipedia searches for key terms and determining those that did not have pre-existing articles. Once Dr. Webb approved their topics, students drafted articles in their Sandboxes. Each class member critiqued three of their fellow classmates' articles at the draft and live stages, evaluating a total of six different topics. Dr. Webb also provided feedback on the students' article Talk pages, modeling constructive feedback for the class.

The goal of Dr. Webb's Wikipedia assignment was to provide students with a publication-like experience based on the scientific method that would foster deeper learning of topics and ideas covered in the course. Students had a high-stakes investment in the outcome of the project because they knew their work would be visible to a global audience. For the Fall 2014 Plate Tectonics class, 21 students created 25 new Wikipedia articles, which collectively received more than 454,000 views. Student additions included the article "Lwandle Plate" (2014), which received almost 4,000 views on the day it was featured in the "Did you know...?" section on the main page of Wikipedia.

The assignment engaged students in the scientific method: asking questions, evaluating, problem-solving, and providing evidence for ideas at multiple steps. The result was better science communication and understanding for both the class and the Wikipedia community.

Assignment 3: Collaborating, Critiquing, and Interacting

The advancement of science depends on the collaborative construction and development of scientific knowledge. Exploration and experimentation in the sciences are rarely individual efforts (Hara, Solomon, Kim, & Sonnenwald, 2003). Instead, many people collaborate in the process of communicating science, contributing ideas, discussing theories, challenging results, and shaping the presentation of findings. The peer-review process, which is at the foundation of science communication, is by its very nature collaborative, calling upon those with expert knowledge to assess the accuracy and reliability of information.

Collaborating on a group project is a key activity that helps students develop communication and problem-solving skills, teaches conflict resolution, increases academic achievement, and gives students a more positive outlook on learning (Colbeck, Campbell, & Bjorklund, 2000; Forte, 2015; Smith & MacGregor, 1992). Furthermore, collaborating on a project creates a knowledge community in which each contributor can advance the overall project. In exchange for collaborative participation, individuals build confidence in their understanding of a topic (Smith & MacGregor, 1992) and become part of a network of people with shared interests.

Wikipedia is a collaborative group project that fosters participation on a global scale. It draws upon a large-scale peer review process where a diverse community of contributors with a variety of expertise helps validate content (Kaplan & Haenlein, 2014). This collective experience improves the quality of scientific content available online by employing crowd-sourced knowledge construction.

Including a peer-review component in a Wikipedia assignment is an excellent way to facilitate collaboration. As student editors construct contributions in their Sandboxes, classmates can offer critiques on the associated Sandbox Talk page. Rubrics provided by the instructor or designed collaboratively by the class help students learn to critique content systematically and shape constructive assessment to improve contributions. After editing online, classmates can meet face-to-face to share their thoughts about content and organization, offering opportunities to address different reactions to feedback while assisting student editors. Through feedback and revision, students develop an appreciation for the peer-review process, learn to accept criticism, and modify their contributions to meet standards of scientific rigor. Students also learn to provide written constructive feedback, further increasing their ability to participate in knowledge construction. Giving and receiving peer feedback allows students to hone their critical skills in a supportive environment. This process continues to reinforce students' sense of authority by interacting with others to construct knowledge (Camihort, 2009). Being able to constructively exchange feedback while supporting and defending their stance is necessary for students to be successful in science (more about feedback training see Chapter 12).

This review component was interwoven throughout the Natural Disturbances and Society course. Each student was randomly assigned to review two of their classmates' articles and provide critiques based on criteria outlined in a rubric. Dr. Carmichael reviewed both student contributions and critiques, providing an additional layer of feedback.

Students in the class demonstrated excellent skills at providing feedback. One student, Rachel, said it was her "duty to make a difference, not only with her own contributions, but also to support her fellow classmates." Figure 13.3 illustrates the quality of Rachel's feedback, where she summarized the overall strengths of her classmate's contribution, highlighted specific areas to address during revision, and provided a detailed critique based on the assignment rubric. While students in the class were only required to provide feedback for two of their peers, Rachel and others enjoyed the exercise and joined other online contributors in improving contributions.

Feedback [edit source]

Overall, your contributions to this article are significant and well-written. You add a lot of pertinent information to the article. I do think you could edit the original Legal Response section to further improve the article and to make sentences flow better. Some wording could be changed to make it read more civilian e.g. change 0600 to 6 a.m. Also, I think adding in either a sentence or an entire section about the everlasting repercussions of the oil spill on the ecosystem is necessary. For example, you state that the oil-filled barge is still in the ocean. Does this mean the ocean is at risk for another spill if that ship were to be damaged in some way? Are species still affected today? Have the wildlife adapted to changes in the coral reef? Did endangered species die out, and were new endangered species created? All of these are important questions that need to be addressed, as they relate to the full consequences of the oil spill and the ecosystem's recovery (you could even put these answers/information in the conclusion section I mention adding later on).

Clear, logical organization and structure [edit source]

The article is presented in a good structure and is easy to read and to follow. I like the inclusion of the wildlife and environment sections as they pertain to the class. One suggestion I have for better sentence flow is to move the Legal Response section because the end of the Cleanup of Oil subsection reads well as an ending to the whole article as compared to the Legal Response section and ending.

Writing adequately developed and supported by evidence, examples (specifics) [edit source]

Though you do include a good amount of examples, I would like you to elaborate a bit more. Why were some of the Most Affected Species "least resilient" to the oil spill and what precisely was the effect on these animals? Explain further the effects on tourism caused by the oil spill. What parts of the Oil Pollution Act exactly did the cleanup adhere to? Elaborate on the problems experienced by the cleanup crews that you briefly mention in the Cleanup subsection. After listing multiple affected species or organisms, go back and focus on one as an example. This will help to provide a real case of how the oil spill impacted one specific organism. This is more effective than a long list of examples. Also, you should expand on the Affected Environment section. What exactly about the environment was affected? Was it temporary or long-lasting(if so, how long)? Something needs to be included to further elaborate on this important part of the anthropogenic disaster.

Well written: clear, concise sentences comprehendible at a high school level [edit source]

Your sentences were well-written and most everything was easy to understand at a general level. One thing for improvement is perhaps explaining in more detail why the remaining oil could not be retrieved from the ship safely and why some animals were not saved using the soap cleaning treatment. Doing this concisely and in layman's terms would help for a more well-rounded, clear understanding of the disaster.

Factual, not persuasive writing: neutral, unbiased [edit source]

The writing is completely unbiased and is presented in a nonpartisan, neutral manner. There was no persuasion used in the article.

Appropriate section headings [edit source]

All of the new section headings were titled well, although like I mentioned previously I would rearrange the Legal Response section (or you could add a short conclusion section). One important addition that I believe must be made is the detailed inclusion of the effects on mankind and human economy brought about by this oil spill. Though you briefly mention that tourism was affected, you do not include much else on the impact on humans (legal section aside). I think adding in a section for the impact of local communities as well as possible side effects (did any humans die or suffer from exposure to oil?) would be extremely beneficial to understanding the full impact of the disaster.

Appropriate references provided, including scientific peer-reviewed articles [edit source]

Your references are peer-reviewed journals from a multitude of sources. My one question is if you can make all of the sources linkable as this would help future research into this subject. If not, that is understandable. In using these references in citing, you leave out citations for the beginning of the Wildlife section, which is important as you mention a lot of numbers. Also, there is a large paragraph in the Cleanup subsection in which you do not cite. Citations in these instances would serve as useful tools to be sure the information is factual and supported by evidence.

In-line links appropriate [edit source]

You did a good job of placing in-line links to appropriate sources that help to provide further context for the page. You could include more in-line links in the beginning when you discuss specific sea life and specific historical places affected by the oil spill as this would create more useful, illustrative information for the reader.

No grammatical or spelling error [edit source]

As a note, make sure every time you mention the ship's name , you italicize it. Also, when you use the "J" in the name of the ship, a period should be placed after it. Other than that, I went ahead and corrected minor grammatical and sentence structure errors for you. Check the history page. Also, be sure to reread sentences for correct use of verbs and punctuation.

Rlambert1893 (talk) 06:52, 18 March 2014 (UTC)

Review from Dr. Becky [edit source]

Rlambert1893 has provided you with extensive feedback for improving your article and I agree with her evaluation. Overall, you have made an impressive contribution and I encourage to keep going. The lead section is impactful and grabs the reader's attention. Look at this style and apply it to the rest of the article. Improvements can be made by supporting statements with citations, including peer-reviewed scientific references. Carefully evaluate the organization of information for clarity and remove redundant and unnecessary phrases. There are several places that lend well to internal and external links. Including those offer the reader additional resources for a deeper understanding about this oil spill. Work on making more specific comments and less generalities. Some sentences read as run-ons. Finally, be sure that the connection between ecological effects and societal effects is clear- you do get into this, but it needs further development. I am looking forward to the next version of this article!! B.J.Carmichael (talk) 22:08, 19 March 2014 (UTC)

Headline [edit source]

For my headline I said that I would remove any unnecessary information. I did that by removing any words or phrases that did not contribute anything to the sentence that they were in. I did this the most in the Legal Response section. I also combined some sentences to make things less redundant. I combined three sentences together in the Effects on Wildlife section.
19:08, 1 April 2014 (UTC)

Figure 13.3 Rachel's feedback provided for the article "Morris J. Berman oil spill" during the Spring 2014 semester. An assignment rubric was used to organize critique ("User Sandbox Talk Page," 2014). CC BY-SA 3.0.

Shyrece, another student in the class, said, "The project helped me understand course material because I could not fake it. If what I said was half-baked bologna, I would get called out. I actually had to understand my topic and be able to effectively convey the information. The Wiki[pedia] project pushed me to understand the material better than any test."

Along with peers, students may receive comments and suggestions from individuals around the globe about ways to improve their Wikipedia contributions. Though the experience can be intimidating, it allows students to participate in the publication experience while learning course content. This exchange teaches students how to adopt critical feedback, leading to improved contributions. Students learn about the process of scientific writing and publication, personally interacting with a community of editors, which in turn builds confidence in topic knowledge and prepares them for interactions with Wikipedia, scientific, and social communities at large. While critique by the global audience is not automatic, when provided, it helps reinforce the importance of accuracy of contributions and increases the stakes of these assignments. For example, the Talk page of the "2012 Kamaishi earthquake" article (2014) documents an exchange between a student and an editor, who points out the lack of relevance of the source.

> I added information that researchers found could have led to the 2012 earthquakes in Kamaishi. This journal article can be found here [on the Talk page]...
>
> Student editor

> I'm not clear that there is any implication that they were in any way precursors to the 2012 event. The paper doesn't mention this at all. Without a clearer link I think that this section lacks relevance.
>
> Wikipedia editor

Students in Dr. Alex Webb's Plate Tectonics course also received feedback from a Wikipedia editor who is interested in geology.

> I am happy to report that this appears to be your own work made without copying others' writings. This would be much better with some diagrams. Examples of real structures would be great. It would be good to have more references. At least one is a review, and the other is highly relevant. 5 or 6 references could be the right number to match your peers! "surface of the crust" is a bit confusing, as salt lakes will have a crust, but do you mean crust of the Earth?
>
> User: Graeme Bartlett

Because Wikipedia is visible to anyone with an internet connection, students sometimes have concerns about how content they add will be received. Sierra, who created a Wikipedia article as an independent project for a bacterial ecology course, said, "I did not enjoy the feeling of impending doom that comes with submitting anything [to Wikipedia] and thinking it would get pulled/deleted, come under super harsh criticism from 'internet trolls.'" Breea, a Natural Disturbances and Society student, had the same concern: "I was hesitant to edit a Wikipedia article because it is something that can be accessed by millions of people worldwide, and I was worried about making a mistake or getting a citation from the Wikipedia administration." Despite these reservations, students' recognition that their work is open to the public encourages them to develop a better product, and the skills they gain learning to communicate effectively with other members of the Wikipedia community help them become better at receiving and responding to criticism. Receiving editorial critiques to students' contributions is real-world training for the rigorous scrutiny and reviews that science writers face professionally.

Benefits of Wikipedia Assignments

Students and professors alike recognize the benefits of using Wikipedia lessons in the classroom. Given a choice of technologies, including Google tools and TED ED Lessons, 29% of students in Natural Disturbances and Society courses over four semesters reported enjoying using Wikipedia and planned to use it again in the future. Additionally, Wiki Ed reports 97% of instructors would teach with Wikipedia again because it improved literacy skills in a collaborative setting (Salvaggio, 2016a). Dr. Thrash cites multiple course goals as being met through the Wikipedia project: "namely developing critical thinking ability, improving reading comprehension with primary literature, and exposing students to modern technological elements of information transfer."

As Breea, the student quoted above, states,

> ...having to be responsible for the information to the extent of creating your own content for the Wikipedia articles is a more challenging and more rewarding experience [than traditional classroom assessments such as exams]. It involves understanding the course material at some level, conducting research on the material, and eventually, writing on the topic for other users, which requires a higher level of understanding. To truly contribute to an article, one cannot simply regurgitate information from class or a source; to make the information accessible to other users, it requires synthesis, which requires a high level of understanding.

Students as Authorities

Students often struggle with feeling they are not authorities on a topic, and Wikipedia provides an opportunity to increase their sense of authority as scholars (Salvaggio, 2016c). Individuals who contribute to Wikipedia show increased self-confidence in their grasp of subject matter (Yang & Lai, 2010), a trait documented in LSU students. As contributors, students develop self-confidence in their understanding of science and ability to employ scientific material to communicate. Shyrece, a student quoted above, reported "becoming more familiar with scientific jargon and research-style writing" by contributing to Wikipedia. Students also gain an appreciation for their current breadth of knowledge and can recognize what remains to be understood. Another student, Eileen, said that editing Wikipedia "transformed" the way she consumed information, helping her become more adept at verifying the credibility of information. By contributing to Wikipedia, students increase self-confidence in understanding course content and are more willing to share their knowledge with a large audience.

Higher Level Learning Strategies

The Association of American Colleges and Universities (Kuh, 2008) recognizes that the integration of high-impact practices (HIPS) promotes deeper learning of content and strengthens development of information literacy. Wikipedia assignments provide an opportunity to promote deeper, interdisciplinary learning. Learning occurs when students combine ideas from multiple classes and publicly demonstrate synthesis and application of knowledge in a project that evolves and fosters discussion beyond the classroom setting (Kuh & O'Donnell, 2013; Prince & Felder, 2007). Applying knowledge and skills in the digital landscape capitalizes on students' critical thinking capabilities and increases the likelihood students are engaged in HIPS. Such assignments incorporate several levels of Bloom's Taxonomy, including understanding, analyzing, evaluating, and creating (Perkins, 2008) information. Further, Wikipedia-based assignments can be designed to address recall and reproduction, skills and concepts, strategic thinking, and extended thinking, as described in Webb's Depth of Knowledge (Aungst, 2014; Webb, 1997). Such assignments challenge students to think critically through content creation, provide opportunities for students to draw connections among ideas learned in class, justify contributions, and produce new work to expand understanding.

Tips for Incorporating Wikipedia Successfully in the Classroom

Faculty who elect to use Wikipedia assignments in their courses must plan carefully and be invested in all stages of the process to help students achieve desired learning outcomes. The following tips for designing effective Wikipedia assignments are based on Wiki Ed suggestions, Wikipedia protocols, and our experience using Wikipedia in the courses.

Creating a Course Page via the Wiki Education (Wiki Ed) Dashboard

The goal of Wiki Ed is to recognize the value of student research by making scholarship available beyond the classroom, enabling students to share knowledge with the global community. Connecting to Wiki Ed ensures access to online training tools, instructor orientation, editing resources, and personal assistance from knowledgeable staff. Students can access online resources and handouts such as "Editing Wikipedia," "Moving out of your Sandbox," and discipline-specific resources, like "Editing Wikipedia articles on Environmental Sciences," which are all available at wikiedu.org. Wiki Ed offers educators assistance with assignment design and evaluation, online and printed training resources, and metrics to track student involvement. An important resource for Wikipedia-based assignments is Wiki Ed Dashboard. The Dashboard is a landing spot for all members of a class, housing an overview of course assignments, resources and tutorials that guide students through each step of the project, and a platform for quick interaction with students, including direct access to their Sandboxes and articles to which they are contributing. Use of the Wiki Ed Dashboard also connects professors and students to Wiki Ed staff, facilitating contact when issues arise.

Ensuring Students have Individual Wikipedia User Accounts to Track Their Contributions

Individual student accounts help professors track student progress, provide professors with direct access to students' content additions and modifications, and validate students as Wikipedia editors.

Establishing Clear Expectations and Rubrics

Expectations for Wikipedia activities should be clearly outlined at the beginning of the course and accompanied by a rubric that will be used to evaluate contributions. Reminding students to focus on content development in their Sandboxes helps ensure appropriate information is added. Coding and formatting issues can be resolved later.

Participating in the Editing Process with Students and Testing the Projects

Instructors should be involved in all facets of the assignment, from editing articles to interacting with global editors, to model excitement and engagement for the students. Remember: incorporating Wikipedia-based assignments in the classroom is a process and, just like writing a paper, flexibility and critique are necessary.

Searching for Topics to Determine If They Exist in Wikipedia

Students should search for their intended topics via Google and Wikipedia before creating new article pages. Google searches provide a more in depth, thorough search compared to the Wikipedia internal search. Searching an intended topic also helps students discover possible knowledge gaps to determine if there is room to edit and improve the selected topic. Searching first saves energy later.

Checking in with Students Periodically to Monitor Their Progress

Short, in-class conversations encourage students to ask questions and voice concerns about their assignment. These check-ins can address editing concerns or serve to clarify scientific concepts.

Including Peer-Review on the Sandbox Talk Page

Students can be assigned to edit each other's articles before they go live. Participating in peer-review teaches students to give constructive feedback on content and construction, while building a supportive classroom community. Such activities prepare students for addressing comments from global Wikipedia editors and give them practice justifying their contributions.

Moving Student Contribution into the Live Space of Wikipedia

Student work should not remain in Sandboxes; rather it should be placed in the live article as soon as possible to receive feedback from editors and expose students to the editing process. To increase interaction, consider moving student contributions into the Wikipedia mainspace at least a month prior to the end of the semester.

Encouraging Interaction with Global Wikipedia Editors

Experts and enthusiasts edit Wikipedia on a range of topics. Notice which specific editors are contributing to your students' pages and reaching out to them via their User Talk pages. Some editors may be willing to provide feedback to your students or even suggest existing pages that require attention.

Providing Opportunities for Students to Reflect on the Process

Receiving student feedback can help professors streamline and modify the Wikipedia assignment for future courses. Short, low-stakes reflection essays are ideal for students to share thoughts about the process, examine how their skills have changed, and provide suggestions to improve the experience for future student editors.

Conclusions

Wikipedia is a unique resource and, when incorporated into the classroom, gives students ownership of their work, improves their understanding of scientific topics, strengthens communication skills, and builds their confidence to participate in science.

As Wikipedia contributors and editors, students work to highlight points of confusion in existing course content, crafting new ways to illustrate concepts. An integral part of the learning process is researching what is currently known. Students combing through the aggregation of references (Cox, 2014) within Wikipedia articles are challenged to decipher meaning and determine if the content agrees with external peer-reviewed scientific research (see Assignment 1). As students explore article content, they develop the credentials to evaluate existing information and contribute new information (see Assignment 2). Content creation requires that students locate and assess written material, hone their skills in presenting information, and adopt and provide critical feedback (see Assignment 3). Wikipedia provides an authentic opportunity for students to participate in the collaborative process of science communication, while concurrently increasing the accuracy and reliability of Wikipedia.

References

"2012 Kamaishi earthquake." (2014). *Wikipedia*. Retrieved April 26, 2017, from https://en.wikipedia.org/wiki/2012_Kamaishi_earthquake.

"2013 Colorado floods." (2013). *Wikipedia*. Retrieved April 26, 2017, from https://en.wikipedia.org/wiki/2013_Colorado_floods.

Anderson, M., Hitlin, P., & Atkinson, M. (2016, January 14). Wikipedia at 15: Millions of readers in scores of languages. *Pew Research Center*. Retrieved January 14, 2016, from www.pewresearch.org/fact-tank/2016/01/14/wikipedia-at-15/.

Aungst, G. (2014, September 4). Using Webb's Depth of Knowledge to increase rigor. *Edutopia*. Retrieved February 25, 2016, from www.edutopia.org/blog/webbs-depth-knowledge-increase-rigor-gerald-aungst.

Barkley, E. F., Cross, K. P., & Major, C. H. (2005). *Collaborative learning techniques: A handbook for college faculty*. San Francisco, CA: Jossey-Bass.

Bean, J. C. (2011). *Engaging ideas: The professor's guide to integrating writing, critical thinking, and active learning in the classroom* (2nd ed.). San Francisco, CA: Jossey-Bass.

Brailes, A., Koskinas, K., Dafermos, M., & Alexia, G. (2015). Wikipedia in education: Acculturation and learning in virtual communities. *Learning, Culture and Social Interaction, 7*, 59–70.

"Brown booby." (2005). *Wikimedia Commons.* Retrieved April 25, 2017, from https://commons.wikimedia.org/wiki/File:Brown_booby.jpg.

Burns, T. W., O'Connor, D. J., & Stocklmayer, S. M. (2003). Science communication: A contemporary definition. *Public Understanding of Science, 12*, 183–202.

Camihort, K. M. (2009). Students as creators of knowledge: When Wikipedia is the assignment. *Athletic Therapy Today, 14*(2), 30–34.

Campaign: Louisiana State University. (n.d.). *Wiki Edu.* Retrieved March 1, 2017, from https://dashboard.wikiedu.org/campaigns/Louisiana_State_University/programs.

Citing sources on Wikipedia. (n.d.). *Wikipedia.* Retrieved February 2, 2016, from https://commons.wikimedia.org/wiki/File:Citing_your_sources.pdf.

Colbeck, C. L., Campbell, S. E., & Bjorklund, S. A. (2000). Grouping in the dark: What college students learn from group projects. *The Journal of Higher Education, 71*(1), 60–83.

Colorado Oil and Gas Conservation Commission. (2013, November 26). COGCC flood response. *COGCC 203 Flood Information.* Retrieved March 5, 2016, from http://cogcc.state.co.us/Announcements/Hot_Topics/Flood2013/COGCC2013FloodResponse.pdf.

"Condado Beach, San Juan, Puerto Rico." (2012). *Wikipedia.* Retrieved April 25, 2017, from https://en.wikipedia.org/wiki/File:Condado_Beach,_San_Juan,_Puerto_Rico.jpg.

Core content policies. (n.d.). *Wikipedia.* Retrieved January 3, 2017, from https://en.wikipedia.org/wiki/Wikipedia:Core_content_policies.

Cox, J. (2014). Why people trust Wikipedia more than the news. *Motherboard.* Retrieved December 5, 2015, from http://motherboard.vice.com/read/in-defense-of-wikipedia.

Denver Post. (2013, October 8). E. coli found in Colorado flood zones, but no oil, gas contamination. *The Denver Post.* Retrieved March 5, 2016, from www.denverpost.com/2013/10/08/e-coli-found-in-colorado-flood-zones-but-no-oil-gas-contamination/.

Dewey, C. (2016, June 20). The surprising reason some college professors are telling students to use Wikipedia for class. *The Washington Post.* Retrieved June 20, 2016, from www.washingtonpost.com/news/the-intersect/wp/2016/06/20/the-surprising-reason-some-college-professors-are-telling-students-to-use-wikipedia-for-class/.

Forestry Commission. (2017). *Pine processionary moth – Tree pests and diseases.* Retrieved March 5, 2015, from www.forestry.gov.uk/pineprocessionarymoth#symptoms.

Forte, A. (2015). The new information literate: Open collaboration and information production in schools. *International Journal of Computer-Supported Collaborative Learning, 10*, 35–51. doi:10.1007/s11412-015-9210-6.

Generous, N., Fairchild, G., Deshpande, A., Del Valle, S. Y., & Priedhorsky, R. (2014). Global disease monitoring and forecasting with Wikipedia. *PLoS Computation Biololgy, 10*(11), 1–16, e1003892. doi:10.1371/journal.pcbi.1003892.

Haigh, C. A. (2011). Wikipedia as an evidence source for nursing and healthcare students. *Nursing Education Today, 31*, 135–139.

Hara, N., Soloman, P., Kim, S., & Sonnenwald, D. H. (2003). An emerging view of scientific collaboration: Scientists' perspective on collaboration and factors that impact collaboration. *Journal of the American Society for Information Science and Technology*, *54*(10), 952–965.

Heilmann, J. (2011). Why we should all edit Wikipedia. *University of British Columbia Medical Journal*, *3*(10), 32–33.

Horrigan, J. B. (2006). The internet as a resource for news and information about science. *Pew Research Center*. Retrieved December 3, 2015, from www.pewinternet.org/2006/11/20/the-internet-as-a-resource-for-news-and-information-about-science/.

How do you measure the difference that open knowledge makes? (2015). *Wiki Education Foundation*. Retrieved February 25, 2015, from https://wikiedu.org/changing/classrooms/.

Jenkins, H., Purushotma, R., Weigel, M., Clinton, K., & Robison, A. J. (2009). *Confronting the challenges of participatory culture: Media education for the 21st century*. Cambridge, MA: The MIT Press.

Kaplan, A., & Haenlein, M. (2014). Collaborative projects (social media application): About Wikipedia, the free encyclopedia. *Business Horizons*, *57*, 617–626.

Kuh, G. (2008). *High-impact educational practices: What they are, who has access to them and why they matter*. Washington, DC: Association of American Colleges and Universities.

Kuh, G., & O'Donnell, G. (2013). *Ensuring quality & taking high impact practices to scale*. Washington, DC: Association of American Colleges and Universities.

Li, S., Daudin, J.J., Piou, D., Robinet, C., & Jactel, H. (2015). Periodicity and synchrony of pine processionary moth outbreaks in France. *Forest Ecology and Management*, *354*, 309–317. doi:10.1016/j.foreco.2015.05.023.

Liggett, S., Jordan, K., & Price, S. (2011). Mapping knowledge-making in writing center research: A taxonomy of methodologies. *The Writing Center Journal*, *31*(2), 50–88.

"Lwandle Plate." (2014). *Wikipedia*. Retrieved April 26, 2017, from https://en.wikipedia.org/wiki/Lwandle_Plate.

"Morris J. Berman oil spill." (2014a). *Wikipedia*. Retrieved April 25, 2017, from https://en.wikipedia.org/w/index.php?title=Morris_J._Berman_oil_spill&oldid=595815917.

"Morris J. Berman oil spill." (2014b). *Wikipedia*. Retrieved April 25, 2017, from https://en.wikipedia.org/w/index.php?title=Morris_J._Berman_oil_spill&oldid-606267325.

Moy, C. L., Locke, J. R., Coppola, B. P., & McNeil, A. J. (2010). Improving science education and understanding through editing Wikipedia. *Journal of Chemical Education*, *87*(11), 1159–1162. doi:10.1021/ed100367v.

Perkins, D. (2008). Levels of thinking in Bloom's Taxonomy and Webb's Depth of Knowledge. *Ohio Department of Education*. Retrieved February 25, 2016, from https://education.ohio.gov/getattachment/Topics/Teaching/Educator-Evaluation-System/How-to-Design-and-Select-Quality-Assessments/DOK-Compared-to-Blooms-Taxonomy.pdf.aspx.

"Pine processionary." (2016). *Wikipedia*. Retrieved April 26, 2017, from https://en.wikipedia.org/wiki/Pine_processionary.

Prince, M., & Felder, R. (2007). The many faces of inductive teaching and learning. *Journal of College Science Teaching, 36*(5), 14–20.

Purdy, E., Thoma, B., Bednarczyk, J., Migneault, D., & Sherbino, J. (2015). The use of free online educational resources by Canadian emergency medicine residents and program directors. *Canadian Journal of Emergency Medicine, 17*(2), 101–106.

Qualley, D. (1997). *Turns of thought: Teaching composition as reflexive inquiry.* Portsmouth, NH: Boynton/Cook.

Report card. (2016, August). Retrieved August 3, 2016, from http://reportcard.wmflabs.org/#core-graphs-tab.

Salvaggio, E. (2016a, March 25). 97% of instructors would teach with Wikipedia again. *Wiki Education Foundation.* Retrieved March 29, 2016, from https://wikiedu.org/blog/2016/03/25/fall-2015-survey/.

Salvaggio, E. (2016b, March 28). Five reasons a Wikipedia assignment is better than a term paper. *Wiki Education Foundation.* Retrieved March 29, 2016, from https://wikiedu.org/blog/2016/03/28/five-reasons-a-wikipedia-assignment-is-better-than-a-term-paper/.

Salvaggio, E. (2016c, July 2016). 7 ways the Year of Science is already making a difference. *Wiki Education Foundation.* Retrieved July 20, 2016, from https://wikiedu.org/blog/2016/07/20/7-ways-wikipedia-year-science/.

Smith, B. L., & MacGregor, J. T. (1992). What is collaborative learning? In A. S. Goodsell, M. R. Maher, V. Tinto, B. L. Smith, & J. MacGregor (Eds.), *Collaborative learning: A source book for higher education* (pp. 10–30). University Park, PA: National Center on Postsecondary Teaching, Learning, and Assessment at Pennsylvania State University.

"User Sandbox Talk Page." (2014). Retrieved April 25, 2017, from https://en.wikipedia.org/wiki/User_talk:BradLandry/sandbox.

Webb, N. (1997). *Research monograph number 6: Criteria for alignment of expectations and assessments on mathematics and science education.* Washington, DC: Council of Chief State School Officers.

Wiki Education Foundation homepage. (n.d.). Retrieved February 28, 2017, from https://wikiedu.org

Yang, H., & Lai, C. (2010). Motivations of Wikipedia content contributors. *Computers in Human Behavior,* 26, 1377–1383. doi:10.1016/j.chb.2010.04.011.

List of Contributors

Jonathan Buehl is an Associate Professor and the Vice Chair of Rhetoric, Composition, and Literacy Studies in the Department of English at the Ohio State University. He is the author of *Assembling Arguments: Multimodal Rhetoric and Scientific Discourse,* and his essays have appeared in *College Composition and Communication, Technical Communication Quarterly,* and *Landmark Essays on Archival Research.* With Alan Gross, he edited *Science and the Internet: Communicating Knowledge in a Digital Age.*

Lauren E. Cagle is an Assistant Professor of Writing, Rhetoric, and Digital Studies and affiliate faculty in the Environmental Studies program at the University of Kentucky, where she teaches courses on environmental rhetoric, technical communication, and academic science and social science writing. Her research focuses on environmental rhetoric, technical communication, disability studies, and feminist theory, and she is especially interested in debates about climate change by non-technical stakeholders in the public sphere. Cagle's work has been published in *Technical Communication Quarterly*, the *Journal of Business and Technical Communication*, and *Kairos.*

Becky J. Carmichael is the Science Coordinator with Communication across the Curriculum and teaches with the Rodger Ogdon Honors College at Louisiana State University (LSU). She received her B.S. from Purdue University and Ph.D. from LSU, both in biological sciences. She focuses on science communication and pedagogy in higher education by helping faculty and students strengthen their science communication skills. Becky collaborates with Wiki Education and SciFund through teaching and outreach projects, and serves as speaker coach for TEDxLSU and has created a podcast to highlight science occurring at LSU and the people conducting the research.

Carleigh Davis is a doctoral candidate at East Carolina University. Her research interests focus primarily on the application of meme theory in technical and scientific communication. Specifically, she is interested in the ways in which memetic proliferation facilitates the development of cultural rhetorical spaces in digital forums. In the past she has used this theoretical framework to analyze the development and efficacy of hacktivist groups, "fake news," and modern folklore, demonstrating the ways in which digital affordances allow ideas and cultural elements to reproduce and build upon one another while also restricting any that fall outside the established norms of a given space. She regularly teaches ENGL 3820 (Scientific Writing) and ENGL 3880 (Writing for Business and Industry) by relying on a culturally inflected pedagogy and working to help students situate their technical communication skills in the broader contexts of their fields of interest.

William T. FitzGerald is an Associate Professor of English at Rutgers University-Camden, where he directs the Writing Program and the Teaching Matters and Assessment Center. He is the author of *Spiritual Modalities: Prayer as Rhetoric and Performance* (Penn State Press, 2012) and co-author of the most recent editions of *The Craft of Research, A Manual for Writers of Research Papers, Theses, and Dissertations,* and *The Students Guide to Writing College Papers,* all published by the University of Chicago Press.

Erin A. Frost is an Assistant Professor of technical and professional communication in the Department of English at East Carolina University. She regularly teaches Scientific Writing—the junior-level course described in her co-authored chapter in this volume. Her research interests include health and medical rhetorics, gender and feminism in technical communication, environmental rhetorics, and risk communication. She has an employment history as an investigative journalist; this experience taught her the importance of civic engagement and social justice, and she uses her journalism experience to inform both her teaching and research. Her current research projects focus on the communication of scientific and medical phenomena (for example, goals of medical imaging and effects of environmental toxins on health) to nonexpert audiences. Her work has appeared in *Communication Design Quarterly*, *Composition Studies*, *Computers and Composition: An International Journal*, *Journal of Business and Technical Communication*, *Peitho*, *Present Tense: A Journal of Rhetoric in Society*, *Programmatic Perspectives*, and *Technical Communication Quarterly*.

Maria E. Gigante is an Assistant Professor in the Rhetoric and Writing Studies program in the department of English at Western Michigan University. Her research concerns both current and historical cases of visual communication between scientific communities and non-expert public audiences. Specifically, she argues for more scholarly attention to the role of images in introducing science to uninitiated audiences. A monograph on cases of visual popularization is forthcoming with the University of South Carolina Press. Her work can also be found in *Rhetoric Review* and the *Journal of Technical Writing & Communication*. Gigante is interested in pursuing practical applications of her research as well as collaborative projects between STEM fields and the humanities. She has developed and taught courses in visual rhetorics and writing in the sciences.

Lindsey Harding is the Director of the Franklin College of Arts and Sciences Writing Intensive Program at the University of Georgia. She serves as the faculty advisor and editor for *The Classic*, the Writing Intensive Program's journal of undergraduate writing and research. Lindsey is also a co-coordinator for Write@UGA, a month-long spotlight on writing that involves a guest speaker event series, a program and publication exhibit, and a series of writing pedagogy workshops. In the Department of English, she teaches advanced critical, creative, and digital writing courses. Her research and writing interests include writing-in-the-disciplines pedagogy, composition and rhetoric, creative writing, and digital humanities. Currently, she is studying strategies for teaching writing in large enrollment courses, the Five Minute Teach as an instructional strategy and training tool for writing instructors in the disciplines, and the use of annotation and reflection with instructional materials to facilitate a collaborative learning environment and activate writing process pedagogy. She has published critical essays in *Photographies*, *Teaching English in the Two-Year College*, and *Harlot*. In May 2015, Lindsey graduated from the University of Georgia with her Ph.D. in English. In May 2011, she graduated from Sewanee University's School of Letters with her M.F.A. in creative writing. She earned her B.A. in English from Columbia University in 2004. She can be found online at www.lindseymharding.com.

Steven B. Katz is the R. Roy and Marnie Pearce Professor of Professional Communication, and a Fellow of the Rutland Institute of Ethics, at Clemson University. At Clemson, he is the Co-Founder and Director of the Writing in the Disciplines (WID) Initiative with Lesly Temesvari (Professor of Biology and Associate Dean of the College of Science), with whom he also co-teaches a continuously running multi-semester course in Popular Science Journalism. He is the co-author with Ann M. Penrose of *Writing in the Sciences: Exploring Conventions of Scientific Discourse*, as well as several other books. Dr. Katz has written articles on the rhetorical ethics of scientific, technical, medical, and environmental communication, and has been the recipient of the National Council of Teachers of English Award for Best Article on the Theory of Scientific and Technical Communication; some of this work has been reprinted, and it has been cited extensively, as well as critiqued. Katz also has presented many papers on biotechnology communication, and has delivered workshops and/or consulted nationally and internationally with biotechnology and pesticide scientists and industries on communicating ethically with the public.

Metha M. Klock is an instructor at the University of Louisville. She received her B.A. from Sarah Lawrence College, and her M.S. in forestry and Ph.D. in biological sciences from Louisiana State University (LSU). She has worked in book publishing, in land management for two non-profit, volunteer-based habitat restoration organizations, and in science communication at LSU's Communication Across the Curriculum Studio. Klock is the author of multiple articles on species invasions describing her research on the life history and management of a highly invasive plant species in the southeastern United States, Chinese privet (*Ligustrum sinense*), and the mechanisms influencing invasion of Australian *Acacia* species. Klock has won numerous awards and grants for her research, and has travelled extensively to locations such as Costa Rica, Australia, and Norway to conduct field work. Metha enjoys hiking, reading, biking, traveling to new places around the world, and spending time with her family and friends.

C. Claiborne Linvill is the president of Rock Creek Communications, LLC, where she offers marketing consulting, digital content development and writing and editing services. She has more than 13 years of experience working in marketing and public relations for corporations and nonprofits, and has also taught public relations classes in the Communication Department at Clemson University. Claiborne graduated *magna cum laude* with a B.A. in Communication from Wake Forest University. She earned an M.A. with honors in Professional Communication from Clemson University, where she completed her thesis under the direction of Steven B. Katz.

Kate Maddalena is an Assistant Professor of Writing at William Peace University, where she teaches courses in composition, communication, and professional writing. She is a scholar of media theory, rhetoric, and STS, and she studies how open-ended tools and units of measure work inductively to make knowledge in the environmental and life sciences.

Scott A. Mogull is an Associate Professor of Technical Communication in the Department of English at Texas State University in San Marcos, TX. His research interests include scientific and medical communication. He has published research in both microbiology and technical communication and has recently published a book on scientific writing entitled, *Scientific and Medical Communication: A Guide for Effective Practice* (also available from Routledge). For nearly a decade, Mogull has worked in the biotechnology and molecular diagnostics industry as a scientific communicator, product manager, marketing manager, and coordinator of global technical information. Since 2008, he has been on the editorial board of *Technical Communication Quarterly*, the journal of the Association of Teachers of Technical Writing (ATTW). He has a Ph.D. in Technical Communication and Rhetoric from Texas Tech University, a M.A. in Microbiology from The University of Texas at Austin, and a M.S. in Human Centered Design and Engineering from the University of Washington.

Kathryn Northcut is a Professor of technical communication in the Department of English and Technical Communication at the Missouri University of Science and Technology. She teaches courses in technical communication across all student levels, specializing in visual theory, visual rhetoric, and qualitative research methods. She is a co-director of the department's BS and MS degree programs in Technical Communication. She has published in various journals including *Technical Communication Quarterly*, *Programmatic Perspectives*, and the *Journal of Technical Writing and Communication*. With Eva Brumberger, she co-edited *Designing Texts: Teaching Visual Communication*, which won the 2015 CCCC Award for Best Original Collection of Essays in Technical or Scientific Communication.

Gwendolynne Reid is an Assistant Professor and the director of the writing program at Oxford College of Emory University. Her qualitative research examines digital and multimodal composing practices in the disciplines. Her research interests include writing across the curriculum, scientific writing, genre studies, and writing program administration. Her writing can be found in *Across the Disciplines*, *WPA-CompPile Research Bibliographies*, *Composition, Rhetoric, and Disciplinarity*, *Twenty Writing Assignments in Context*, and *Contingent Faculty Publishing in Community*.

Colleen A. Reilly is a Professor of English at the University of North Carolina Wilmington, where she teaches courses in professional and technical writing including technical editing, scientific writing, and document design. Her current research relates to curricular design, digital research methods, entrepreneurship, and genders, sexualities, and technologies.

Gregory Schneider-Bateman is an Associate Professor of English at the University of Wisconsin-Stout. His research explores the intersections of rhetorical theory and the public understanding of science, particularly in the context of science and natural history museums. He's visited informal science institutions throughout the US and Europe and presented his research at numerous national and international conferences.

Liz Studer graduated from the University of Georgia with a master's degree in Entomology and an Interdisciplinary Certificate in University Teaching. There, she worked on the conservation implications and use of a federally threatened tiger beetle species, *Habroscelimorpha dorsalis,* as an ecological indicator species of human impact in beach ecosystems. In addition to research, she was actively involved in pedagogical training as a member of the Future Faculty Program, UGA Women's STEM Mentors, and the Introductory Biology Program Curriculum Committee. She has designed and organized two teaching-in-the-discipline pedagogy symposiums for the Entomological Society of America entitled, "From Pupae to Pupils: Teaching Entomology in a Changing World" and "Teaching Teachers: A Primer on Designing Graduate Teaching Seminars in Entomology." As an undergraduate student, Liz graduated Magna Cum Laude from the University of Colorado in 2010 as a double major in Anthropology and Ecology & Evolutionary Biology. Liz is currently pursuing her Ph.D. in Ecology at Dartmouth College focusing on forest biodiversity, conservation, and ecology. Her dissertation research works in collaboration with the Hubbard Brook Long-Term Ecological Research Program to understand historical changes in arthropod diversity, abundance, and food web structure to predict how ecosystems will react under changing climate and anthropogenic pressures in the future.

Denise Tillery is a Professor of English and Associate Dean of Liberal Arts at the University of Nevada, Las Vegas. She teaches courses in professional writing, grammar, and the rhetoric of science. Her areas of research include environmental rhetoric, program and curriculum development in technical communication, and gender and scientific rhetoric. Tillery has published in *Technical Communication Quarterly*, *IEEE Transactions on Professional Communication*, the *Journal of Business and Technical Communication*, *Present Tense*, and *Rhetoric Review*, in addition to chapters in edited collections. She also co-edited the collected volume *The New Normal: Pressures on Technical Communication Programs in an Age of Austerity* (Baywood, 2015), and is currently at work on a book on scientific commonplaces in environmental discourses.

Candice A. Welhausen is an Assistant Professor of technical communication at Auburn University. Before becoming an academic she was a technical writer/editor at the University of New Mexico Health Sciences Center for the department formerly known as Epidemiology and Cancer Control. Her research focuses on data visualizations in the field of public health and epidemiology with an emphasis on how these visuals shape the ways that knowledge about disease, illness, and health is constructed. Her work has been published in the *Journal of Business and Technical Communication*, *Project on Rhetoric of Inquiry*, and *Technical Communication*.

Han Yu is a Professor of technical communication in the English Department, Kansas State University. Han teaches courses in engineering communication, scientific communication, and business/workplace communication. She also coordinates the Department's Graduate Certificate in Technical Writing and Professional Communication and its internship program. Han's research focuses on visual communication, intercultural technical communication, and science/health communication. She published in various journals in the field and edited (with Gerald Savage) *Negotiating Cultural Encounters: Narrating Intercultural Engineering and Technical Communication* (IEEE/Wiley). She is the author of *The Other Kind of Funnies: Comics in Technical Communication* (Baywood/Routledge) and *Communicating Genetics: Visualizations and Representations* (Palgrave Macmillan).

Index